REF

QB
981
.C86
1981

Cronin, Vincent.

The view from planet
Earth

DATE			

THE VIEW FROM PLANET EARTH

MAN LOOKS AT THE COSMOS

VINCENT CRONIN

THE VIEW FROM PLANET EARTH

MAN LOOKS AT THE COSMOS

WILLIAM MORROW AND COMPANY, INC.

New York   1981

Grateful acknowledgment is made for permission to quote from the following works:

Waiting for Godot by Samuel Beckett, reprinted by permission of Grove Press, Inc. Copyright © 1954 by Grove Press, Inc.

Endgame by Samuel Beckett, reprinted by permission of Grove Press, Inc. Copyright © 1958 by Grove Press, Inc.

The Bald Soprano by Eugene Ionesco, reprinted by permission of Grove Press, Inc. Copyright © 1956, 1965 by Grove Press, Inc.

Library of Congress Cataloging in Publication Data

Cronin, Vincent.
   The view from planet Earth.

   Includes index.
   1. Cosmology. I. Title.
QB981.C86   521'.5   81-4056
ISBN 0-688-00642-6   AACR2

Printed in the United States of America

First U. S. Edition

1   2   3   4   5   6   7   8   9   10

For
CHANTAL

Acknowledgments

The following have kindly given permission for quotation from works in copyright: Penguin Books Ltd for lines from Chaucer's *The Canterbury Tales*, trans. Nevill Coghill (Penguin Classics, Revised edition, 1977) pp. 292, 60, 85, 475, 473, © the Estate of Nevill Coghill, 1958, 1960, 1975, 1977, and lines from *Aquinas* by F. C. Copleston (Pelican Books, 1955), © F. C. Copleston, 1955; Heinemann Educational Books and Harcourt, Brace, Jovanovich, New York, for lines from *The Little Prince* by Antoine de Saint-Exupéry, © Editions Gallimard 1946; Jonathan Cape Ltd and the Estate of Edward MacCurdy for lines from *The Notebooks of Leonardo da Vinci*, edited and translated by Edward MacCurdy; Lawrence and Wishart Ltd for lines from Karl Marx and Frederick Engels: *Collected Works I*; Wildwood House for lines from H. E. Gruber's *Darwin on Man*; John Calder (Publishers) Ltd and Grove Press, New York, for lines from *The Bald Prima Donna* by Eugene Ionesco; Thames and Hudson Ltd and The New York Graphic Society for lines from *The Complete Letters of Van Gogh*; Harrap and Co. Ltd for lines from *The Wright Brothers* by F. C. Kelly; The University of Chicago Press for lines from *The Collected Poems of Elder Olson*; Weidenfeld and Nicolson Ltd for lines from Norman Mailer's *A Fire on the Moon*; Faber and Faber Ltd for lines from *T. S. Eliot's Collected Poems 1909–1962*; Faber and Faber Ltd and Grove Press, New York, for lines from *Endgame* by Samuel Beckett; the Author's Literary Estate, The Hogarth Press Ltd and Harcourt, Brace, Jovanovich, New York, for lines from *A Writer's Diary* by Virginia Woolf.

Contents

the first men on the moon – celebration of Holy Communion – impressions of life in another world – reactions to the moon landing – what the rock samples revealed – changed place of moon and Earth in our world-picture.

Illustrations

Helios and stars depicted as boys. Painting on an Apulian vase of the fifth century BC (*British Museum*)

Atlas supporting the zodiac. Roman statue in the Villa Albani, Rome. (*Mansell Collection*)

The Ascension. Top band of ivory panel made in Rome, late fourth or early fifth century. (*Bayerisches Nationalmuseum, Munich*)

Constellations and signs of the zodiac. Illustration to Cicero's version of Aratus's *Phaenomena*, copied in Salzburg in 818. (*Bayerische Staatsbibliothek, Munich*)

The Presentation in the Temple. Mid-eleventh-century pen and ink drawing from the Annals of St Germain des Prés. (*Bibliothèque Nationale*)

Englishmen pointing to the comet of 1066. Scene from the Bayeux Tapestry. (*Mansell Collection*)

The human body influenced by signs of the zodiac. From a fourteenth-century German calendar, Codex Germanicus 32. (*Bayerische Staatsbibliothek, Munich*)

Dante and Beatrice in the heaven of Mars, by Giovanni di Paolo. Yates Thompson MS 36, folio 160 r. (*British Library*)

Dante surveys the seven spheres, by Giovanni di Paolo. Yates Thompson MS 36, folio 169 r. (*British Library*)

Dante and Beatrice gaze at the souls of the blessed, by Botticelli. (*Staatliche Museen, East Berlin*)

An astronomer observing a meteor with a quadrant. From the *Nusrat-namah*. Sixteenth-century MS in the Topkapi Library, Istanbul.

Sphaera Civitatis. Diagram from John Case's book of that name, 1588. (*Huntington Library, San Marino, California*)

Gentlemen measuring the height of buildings and stars. Late sixteenth-century print.

A watercolour illustration by William Blake to Edward Young's *Night Thoughts*: Night the Fifth. (*British Museum*)

Our solar system among other similar systems. Engraving in Leonhard Euler, *Theoria motuum planetarum et cometarum*, 1744. (*ETH Bibliothek, Zurich*)

Cornfield by Moonlight, by Samuel Palmer. Private collection. (*Geoffrey Shackleton*)

NOTE

There is a Chronological Table on page 327. Billion is used to signify one thousand million. Readers not familiar with metric measure may find the following approximate equivalents useful:

1 centimetre = 0·39 inch
1 metre   =   39 inches
1 kilometre = 0·6214 mile.

A rough-and-ready way of turning kilometres into miles is to multiply by 5 and divide by 8. Other useful measurements:

1 light year = 9·46 million million kilometres
                    (5·88 million million miles)
Freezing point of water = 0° Centigrade
Boiling point of water = 100° Centigrade
Absolute zero = −273° Centigrade.

Introduction

Almost as far back as records go, men have looked upwards at the sky, marvelling at the sun, moon and stars, trying to situate their Earth and themselves in relation to those orbs and their changing paths. They have thought they perceived some surprising things, such as a tall, slim lady doing a back-bend over the sky, and they have asked some searching questions: Were the heavenly bodies created by a God, and what, in the context of space, did man amount to?

Since it seems to be the nature of man to try to impose order, which is a form of mastery over his surroundings, he has felt himself challenged to provide an explanation of the apparent rhythm or of a spectacular phenomenon, and he has constructed for himself a world-picture, whether it be that of the Alexandrians, with their gods inhabiting the planets, or our own, with the big bang and *Star Wars* on a silver screen.

These world-pictures we know as cosmologies. The word has two main meanings: (i) the science concerned with the universe, and (ii) a view of the universe, not necessarily scientific, though usually based on scientific findings. It is the second meaning that concerns us in this book. A cosmology in this sense is not just a sum of knowledge, true or false, about the cosmos; it is a selection of such knowledge, made partly at the conscious level, partly at the subconscious, and it is arranged most often visually to answer certain questions and satisfy certain needs judged vital. It is, almost literally, a world-picture, and it affects men's beliefs, values and even behaviour.

Not all those who have constructed cosmologies, let alone those who have been influenced by them, have been scientists. The present book is an attempt by a non-scientist to trace concisely the most influential of man's various pictures of the cosmos, and how those pictures have affected his view of himself and his origin.

It is an exciting subject, and a demanding one. New ground has had to be broken, the gaps that still exist between the two cultures and between science and religion have had to be bridged. Throughout I have been deeply conscious of my limitations. Because the subject is big I have had to be, in the first place, selective. I confine myself to the West, and within the West I have described what I believe to be important. I start with the Greeks, for they, drawing on Babylonian observations, were as far as we know the first people to work out rational, unifying views of the cosmos. Thereafter inevitably my choice reflects my personal interests and also the physical limitations of one volume.

With the beginning of the nineteenth century science drew apart from humane studies, with the result that several of the cosmologies then and since have done less than justice to man's spiritual side. Today in particular some find the predominant world-picture, so stirring at one level, unduly narrow at another, yet see no way of validly enlarging it. The concluding chapters look at how this dilemma arose, and how it might be resolved.

I have avoided technicalities and mathematics; because of that and of my need to compress, often scientific niceties have had to be simplified. With a subject such as this it would be easy to lapse into abstractions, so I have tried to keep close to each world-picture by approaching it chiefly in terms of knowable men and women, and whenever a world-picture has been a presupposition for the arts or literature I have felt free to draw on the works of architects, painters and writers.

In making an interdisciplinary book a writer is particularly beholden to friends and to specialists. It is a pleasure for me to thank Basil Rooke-Ley for keeping me abreast of new scientific discoveries, astronomers at the Royal Greenwich Observatory who allowed an outsider to watch them at work and talked to him about some of their projects, Heather Couper of the Greenwich Planetarium for help with astronomy, Nigel Henbest for help with astrophysics, Professor A. J. Meadows for advice in the planning stage, Charles Cuddon, who discussed the book with me as it was being written, Fr Joseph Crehan for help with the early Christian world-picture, Cecil Clough of Liverpool University for advice on Renaissance cosmography, Winifred Gérin, who gave me details about the Brontës, Bernard Hamilton of Nottingham University, who suggested to me the parallel between *The Book of Revelation* and some science fiction, Eileen Agar for telling me about a painter's approach to the moon, and Professor W. R. Bennett Jr of Yale University for information about a statistical approach to cosmic order.

CHAPTER ONE

The Greeks

A seafaring people who navigate by the stars and feel at ease with nature – their gods, fashioned in man's image, but immortal – stories about the constellations – attitude to comets and eclipses – Anaxagoras's world-picture and his arraignment for 'impiety' – Pythagoras's cosmos based on number – Plato models his ideal republic on the order he observes in the heavens – Aristotle separates lowly Earth from the aether-made heavenly bodies, circling the sky attracted by an unmoved mover.

In the spring of 440 BC, when the white marble columns of the Parthenon gleamed new, Sophocles scored a success at the open-air drama festival in Athens with *Antigone*, which tells how the daughter of Oedipus, in defiance of an order by the tyrannical King of Thebes, gives decent burial to her brother Polynices. Though it ends with three suicides, the play is a celebration of man and, early on, the chorus of Theban elders chant these lines:

> Wonders are many on Earth, and the greatest of these
> Is man, who rides the ocean and takes his way
> Through the deeps, through wind-swept valleys of perilous seas
> That surge and sway.

The chorus mention other skills, such as farming and hunting, but navigation gets first place, and this was as it should be. Greece is a collection of islands and peninsulas separated by water, and in ancient times communication and trade depended on sturdy ships manned by skilful sailors.

Not having the compass, the Greeks navigated by the sun, and at night, when the sea was usually calmer, by the stars. So dependent were they on the stars that if the sky clouded they would row into harbour. We know what their world-picture was like because Homer gives it in the *Iliad* in the form of an actual picture, or rather bas-relief, wrought in bronze, tin, silver and gold, by the lame god Hephaestus on a shield for Achilles. It showed 'Earth, sky and sea, the indefatigable sun, the moon at the full, and all the constellations with which the heavens are crowned, the Pleiades, the Hyades, the great Orion, and the Bear, nicknamed the Wain, the only constellation which never bathes in Ocean Stream, but always wheels round in the same place and looks across at Orion the Hunter with a wary eye.' On the shield Hephaestus made other scenes, of town and country life, but lastly, 'round the very rim of the wonderful shield he put the mighty Stream of Ocean.'

This is a seafarer's world-picture, in which sea and sky have first place. As they sailed the Mediterranean in small ships of perhaps a dozen tons, bringing tin, lead and amber from the west, gold and silver, timber, pitch and wax from the Black Sea, how would such men have envisaged the sky? They thought of it as solid, either brass or iron, supported by pillars. Between it and Earth lay a misty air, and a purer upper air called aether. They knew that the sun rises in a different place, follows a different course, and sets in a different place according to the time of the year. It follows an oblique path and in their experience was never directly overhead. Day and night are of equal length in spring and late summer, and they found that the best time for making a voyage was during the fifty days after the latter date.

The moon they believed shone with its own light, and since the Mediterranean is almost tideless they had no reason to connect it with tidal movement. They reckoned time from one crescent moon to the next; twelve lunar months made a year of 354 days, and to bring this into line with the solar year of $365\frac{1}{4}$ days, each city-state inserted into its calendar, as needed, an extra, intercalary month.

About four thousand stars are visible to the naked eye from Mediterranean lands. Of these, the Greeks grouped about one thousand in 48 constellations. This act of grouping they called astronomy, the word *astronomos* meaning 'one who arranges stars'. They knew that the sidereal day is about 4 minutes shorter than the solar day, hence that the constellations change position in the sky throughout the year, but reappear regularly in successive years, thus providing a visual calendar. In the eighth century BC the poet Hesiod used it to advise farmers:

> When first the Pleiades, children of Atlas, arise,
> begin your harvest: plough, when they quit the skies.

The times he indicates are in early May and early November.

In a land warm enough in summer to sleep outside the nightly movement across the sky of the Pleiades or other star groups was familiar enough to be useful for time-telling, as in these lines by the poet Sappho, born in 612 BC:

> The moon is gone
> And the Pleiades set,
>   Midnight is nigh;
> Time passes on,
> And passes, yet
> Alone I lie.

The Greeks knew, as Homer says, that certain stars are visible throughout the year, most notably the Bear (of which our Plough is part) and the North Star, which in 440 BC was 12 degrees from the North Pole, compared to 1 degree today (over the years a slow but continuous shift of the direction of the Earth's axis results in a displacement westward of the stars). But the Greeks did not attach much importance to the North Star, for they thought primarily in terms of east and west, not as we do of north and south. For this there were several reasons: the shape of the Mediterranean, the line of their main trade routes, and their dependence on the rising and setting of heavenly bodies.

Only in the fifth century did the Greeks become interested in certain stars that move irregularly. They called them wanderers – our word 'planets' – and found five of them, the ones we call Mercury, Venus, Mars, Jupiter and Saturn.

The Greeks did not venture far north and had no first-hand experience of summer days lengthening with distance from the equator. So until quite late they accepted the evidence of their senses, and believed the Earth to be a flat disc – surrounded by sea.

Men's bodies, the Greeks believed, were made of the same stuff as Earth, and their life-breath of air. Intelligence and a sense of honour and justice set them quite apart from animals: according to Hesiod, Zeus had laid down that beasts and fishes should devour one another, but that men should govern their relationships by law. Such faculties might or might not be named 'soul', but always they were conceived of as in harmony with the body.

About the next world the Greeks thought very little. A minority believed that the dead passed as shadows to Hades, under the Earth. But Pericles in his great funeral oration on the Athenian dead does not mention an afterlife, and an epitaph of Athenian soldiers killed in 432 BC says only this: 'Their souls were received into the air of heaven; their bodies into Earth. By the gates of Potidaea they were slain.' Decent burial such as Antigone gave Polynices was due to man's dignity as a person, not as a rite ensuring survival.

Seafarers are practical, resourceful men, who know their own worth. Superstitious they may be, but they are not inclined to view the forces of nature as oppressive or monstrous, least of all in the temperate climate of the Mediterranean, so we find their gods made in the image of man, with human passions though immortal, and dwelling on an actual snow-capped mountain, Olympus.

Lord of the whole sky was Zeus, thunderer, wielder of lightning, who once in a rage hurled Hephaestus down from heaven, laming him. But Zeus was not omnipotent. When he took as his mistress a mortal girl, Callisto, his wife Hera deprived Callisto of speech, changed her into an ungainly bear and drove her into the forest. Zeus vainly sought her, and it was only long afterwards that he found her, and her little bear son Arcas. In pity for all they had suffered he raised Callisto and Arcas to the sky, as the Great and Little Bear.

The sun the Greeks considered to be a god, though much inferior to Zeus. They called him Helios, and later Apollo – a vigorous youth crowned with sunbeams who drives a chariot drawn by winged horses. His sister Eos, Dawn, rosy-fingered and saffron-robed, opens the eastern gates of pearl and flashes across the sky to announce her master's coming, while Helios's other sister, chaste Selene, the moon, directs her chariot downward. As Helios traverses the early morning sky, boys leap or dive out of the way: they are stars that twinkle, then vanish.

The Greeks were marvellous story-tellers and about those stars they grouped in constellations they told stories that have become part of our legacy in the West. A typical one concerned the giant hunter Orion. Wearing a lion's skin and a belt, carrying a club and sword, Orion scoured the forest, his faithful dog Sirius at his heels. One day he met a group of nymphs, the seven Pleiades, daughters of Atlas, who stands in the West helping to support the roof of heaven. Orion sought to approach them, but the coy nymphs fled. Orion pursued them until, their strength failing, they called on Artemis – a later name for the moon goddess – who changed them into snow-white pigeons. As they winged skyward a second transformation overtook them: they were changed into a cluster of seven bright stars. For long they shone undimmed. But when Troy fell into the enemy's hands all grew pale with grief, and one, more timid and impressionable than the rest, withdrew from sight to hide her anguish from the curious eyes of men.

Orion was so powerful a hunter it began to look as though he would exterminate every species. One day when he was stalking game in Crete, Earth brought forth the scorpion, which stung him and for a time checked his activities.

Artemis fell in love with Orion. Her brother Apollo disapproved and summoned Artemis to him. To allay suspicion he began to talk of archery and, under the pretext of testing her skill as a markswoman, bade her shoot at a dark speck rising and falling out at sea. Artemis took her bow, feathered her arrow and sent it with such force and accuracy that she touched the point and saw it vanish beneath the waves, only to learn that it was the head of Orion, who had been swimming in those waters. She mourned his loss with tears, vowed never to forget him, and placed him as a constellation in the sky, where we can see his girdle and lion's skin, his club and sword. She placed his faithful dog Sirius beside him and, well away from both, as another constellation, she set the scorpion.

Such stories as these, and of Perseus rescuing Andromeda, are apposite, tender and human. Because they are familiar to us, we accept them as normal. But in fact they were the exception at that epoch. They are in striking contrast to the crude and lurid cosmic myths of the Greeks' neighbours. In the temples of Egypt, for instance, priests ceremonially cut to pieces the figure of Apep, the fearful dragon-snake who daily endangered cosmic order by attacking the boat of the sun at dawn and at sunset.

The Greeks' star stories show an easy intercourse between gods and men, heaven and Earth. Yet this intercourse could occur only with the consent of the gods. If a man on his own sought to soar, he became guilty of hubris. Icarus, provided with wings by his father Daedalus, escaped from the Cretan labyrinth but flew so near the sun that the wax holding his wings melted and he fell and was drowned; Bellerophon tried to reach Olympus on the white winged horse Pegasus, whereupon Zeus sent a gadfly which stung the horse, causing it to throw Bellerophon, who fell and was blinded, while Pegasus continued upward and became a constellation.

The Greeks felt at ease with the sky, as they did with the Earth and sea. And if they recognized that such splendid bodies as sun and moon must be gods, they did not prostrate themselves before them. In fact, they considered it the mark of a barbarian to sacrifice to sun, moon or stars.

Yet the Greeks had two fears. One was of comets. In the fourth book of the *Iliad* Homer tells how Pallas Athena darts down from Olympus to Troy, her descent like that of a comet showering innumerable sparks, which is 'a portent of disaster to sailors at sea or soldiers on the battle-field'. Their other fear was of eclipses. Pindar, writing of an eclipse of the sun, asks 'Is it a signal of war, or a portent of famine, does it mean a heavy fall of snow, or is the sea to overflow the land, or fields be icebound, or the south wind bring rain, or a deluge overwhelm the world and drown all men?'

Certainly to many an eclipse indicated that the gods were angry. In 413 BC an Athenian army commanded by Nicias was trying to extricate itself from an invasion of Sicily which had gone wrong. Nicias was on the point of sailing from Syracuse when an eclipse of the full moon took place. 'Most of the Athenians, deeply impressed by this occurrence, now urged the generals to wait; and Nicias, who was somewhat overaddicted to divination and practices of that kind, refused from that moment even to take the question of departure into consideration, until they had waited the thrice nine days prescribed by the soothsayers.' The delay gave the Syracusans time to attack, they drove most of the Athenian ships ashore and inflicted a severe defeat. But that was a rare case of superstition. As a rule the Greeks did not go much in awe of anyone or anything.

How did their cosmology help to shape the Greeks' way of life? The first thing to emphasize is that by comparison with the cosmologies of other contemporary civilizations, the Greeks' was unoppressive. They were not bound by the position of stars at their birth, like the Babylonians, nor did their city-states, like the empire of China, go in fear of the inauspicious conjunction of planets. They did not bow down to an overpowering Sun God, like the Egyptians, nor lose their selves, like the Indians, in the deeper quicksands of pantheism. A Greek felt at ease in a cosmos where by his skill he could turn winds and stars to good account. Able to navigate where he wished, he developed to a marked degree a sense first of physical, then of moral freedom. This led to a strong sense of individuality, which could sometimes harden into tyranny, but which also made possible the supreme Greek achievement: democracy.

The Greek too was able to bring his keen aesthetic sense to the night sky. He marvelled at its beauty and its regularity, saw these as marks of perfection. It may be that frequent upward gazing to something higher and, as he thought, superior, fostered in him an urge to excel: led him to emulate in heroic action such figures as Perseus and Heracles, who were visible overhead, having been placed after their death among the stars. In the stables of the Greeks a winged horse is forever pawing the ground, whinnying to be up and away.

That aspiring element we find in their lyric poetry and in their sculptors' idealized treatment of the body. The most beautiful parts of the Parthenon, the pediment and frieze, were placed high, for men accustomed to look upwards, and that typical Greek athlete, the discus-thrower, is bent on overcoming gravity. That such an attitude in the arts had close links with their world-picture the Greeks themselves recognized by making astronomy – Urania – one of the nine Muses.

According to Aristotle, the origins of man's idea of the Divine were twofold: the phenomena of the sky, and the phenomena of his own soul. The Greeks' intimate knowledge of the night sky certainly contributed to their intelligent speculation about order and its cause, and later to their belief that there exist laws above manmade law. Antigone buries her brother partly for his sake, partly because she recognizes that God's law obliges her to do so, but the binding force of that law and other divine laws was guaranteed, so to speak, by the regularity of events in the sky.

The Greeks approached natural phenomena at another level also. As Bowra has well said, 'Just because the Greeks believed that the world of gods and of men is one, they had no difficulty in believing that what they saw around them had a divine as well as a physical side and that ultimately the two are not distinct.' A larch tree was a source of planks for building a fast ship and also the abode of a slim dryad. The sun was fire, but also Helios; Orion a group of stars and also a hunter. At once clear-sighted and starry-eyed, the Greeks asked that any explanation of cosmic events should satisfy not just their intelligence, but their whole being, including their awareness of the divine in the familiar objects of nature. This is perhaps their most distinctive characteristic, and may be seen in the work of four great thinkers.

*

In Ionia, that part of Greece which is now the west coast of Turkey, a number of men in the sixth century BC began to reflect on the continuity and periodicity of the celestial bodies. Homer had used the word *cosmos* in two senses; to mean order of any kind, and an ornament, especially of women. One of the Ionians, Heraclitus, merged the two senses and applied the word to the beautiful world order of the celestial bodies. He and others then tried to explain the cosmos in terms of an *arche*, a beginning or first principle. Heraclitus thought the *arche* was fire, Thales thought it was water, Anaximander air.

The most influential of these Ionian thinkers, Anaxagoras, was born about 500 BC in the town of Clazomenae, and at twenty came to Athens, where he was the first to teach philosophy. This was forty years before *Antigone* was produced, and Athens' greatness lay in the future. Little is known about Anaxagoras the man, save that he gave away his property in order to study, and when in later life he retired to a small town and the rulers asked what privilege he would like, he replied that the schoolchildren should be given a holiday on the anniversary of his death – a request which was granted.

When Anaxagoras was about thirty-three, in full daylight a meteorite fell near the small town of Aegospotami. Brown in colour and big enough to fill a wagon, it caused a sensation and was still a tourist sight five centuries later. Almost as much of a sensation was caused by Anaxagoras saying that the meteorite had fallen from the sun. The sun, he explained, was not fire, but an incandescent stone 'bigger than the Peloponnese'; the stars too were fiery stones, but their heat was not perceptible because they are so far away. The revolution of the stars takes them under the flat Earth, which is supported by air. As for the moon, it is a body like the Earth and reflects the light of the sun.

Early in time a rotating movement or 'vortex' of the aether separated primeval 'seeds' into the varied particles of matter we see now, and continues to carry sun, moon and stars round in their circular courses. Anaxagoras was struck by the regularity of this vortex. He could not believe that it resulted from the random distribution of this or that element. So the *arche* of things, he declared, is not water or fire or air, it must be Mind – eternal and separate, omniscient and perfect.

This was a novel view, and it caused the Athenians to nickname Anaxagoras *Nous*, meaning Mind, probably in derision. On the other hand it attracted certain leading citizens, including the statesman Pericles and the playwright Euripides, both of whom became his pupils, and young Socrates. But to Aristophanes this talk savoured of hubris, and he decided to make fun of it in the person of Socrates. In *The Clouds* he depicted Socrates suspended in a basket gazing skywards and declaring his profound belief not, as the audience expects, in Zeus but 'in wide space and in the Clouds', while his pupil asserts, 'There is *no* Zeus. Young Vortex reigns, and he has turned out Zeus.'

Later Socrates rejected Anaxagoras's cosmology, because it did not go far enough. Whereas Anaxagoras's Mind merely set the vortex rotating, Socrates wanted to deduce from the notion of a providential Mind whether the Earth is flat or round, whether it is at the centre of the universe and why, and he was disappointed to find that Anaxagoras postulated mechanistic causes such as 'airs and aethers and waters and many other absurdities' to account for astronomical phenomena. Things like that, Socrates complained, cannot be regarded as the real cause, and people who think in that way are misusing the term 'cause'. Is mechanism an inadequate explanation, is purpose an unjustified projection? – variations on this early debate will recur throughout our History.

Unlike Socrates, most Athenians thought Anaxagoras went too far. His sun made of stone seemed to them blasphemous, and how could a single Mind be squared with their pantheon? This unease came to a head during the anxious days of war with Persia, and the Athenians arraigned Anaxagoras on a charge of impiety. It was only through the eloquence of Pericles that he was not put to death, but he was sentenced to pay a fine and to quit Athens. He retired to Lampsacus, on the Dardanelles, where he died in his early seventies.

Though it mars Athens' fine record of freedom of speech, this sentence should not blind us to the influence of Anaxagoras. As the first Greek to argue coherently from the orderly movements of the heavenly bodies to a supreme Intelligence, he prepared the way for monotheism. But before monotheism could mature in Greece, much heart-searching had to take place about the relationship of natural forces to the gods, of Zeus to Mind, and indeed about whether the Supreme Mind was not a mere projection of man's mind. We have a glimpse of this in Euripides's play *Trojan Women*, when the widow of the King of Troy, enslaved by the Greeks, in the confusion of her grief goes so far as to identify Zeus with the upper air:

> Thou, Earth's support, enthroned on the Earth,
> whoever thou mayst be, hard to discover,
> Zeus, be thou nature's law or the mind of man,
> I pray to thee: for by a noiseless course
> thou guidest human fate in righteousness.

Many such doubts and dilemmas there were, but under the eddies what may be termed the mainstream Greek world-picture flowed strong and clear: it declared the Earth to be the centre of an orderly cosmos, and that after the gods man is its central figure; his spirit at ease in his body, a man's aim should be to run the race of life and seek to win first prize.

*

This view did not go unchallenged. The man whose teaching initiated the challenge, Pythagoras, predates Anaxagoras, but his views became influential in Greece only after Anaxagoras's death. The son of a gem-engraver, Pythagoras was born in the Ionian island of Samos about 570 BC. As a young man he travelled, and in middle age emigrated from Samos, because it was ruled by a tyrant, to the Greek colony of Croton in southern Italy. Here he became the leading citizen and gave the town a constitution. He may also have designed and introduced the unique incuse coinage which was Croton's earliest money. He married, had two children, and founded an esoteric school of philosophy to which belonged the wrestler Milo, six times Olympic champion. After some twenty years in power Pythagoras was driven from Croton by a revolution, and eventually reached another south Italian town, Metapontium, where, forced to take refuge in a temple of the Muses, he starved himself to death.

In describing Pythagoras's beliefs it is convenient to begin with his family's trade. From boyhood Pythagoras would have handled gem-stones and noticed that they came in regular geometric shapes: a beryl as a hexagon, a garnet as a twelve-sided crystal, and so on. Now the Greeks pictured numbers like the geometrically arranged dots on dominoes or dice, so it would have been natural for Pythagoras to think of each kind of gem he handled as composed of a different number. He began looking for other examples of number underlying reality. Experimenting with the seven-stringed lyre, Pythagoras found that the pitch of a note depends on the length of the string that produces it, and that concordant intervals in the scale are achieved by simple numerical ratios, for example an octave is produced by the ratio 2 : 1, and a fifth by the ratio 3 : 2. Having found that not only crystals but sounds could be explained numerically, Pythagoras concluded that the *arche* of all things is number.

This was a brilliant insight, which would be strikingly confirmed by the modern discovery that the properties of a piece of matter are determined by the number of protons and electrons in its constituent atoms. Its effect on Greece was profound. It initiated quantitative science; it set sculptors depicting the human body proportionally: seven and a half heads equal the total height; it set architects building temples in which the width and height of columns and the space between them follow strict mathematical proportions.

Yet, very curiously, having declared that the *arche* of things is number, Pythagoras did not follow it up in a scientific direction. Quite the opposite. He travelled in Babylonia and Egypt, where he learned Egyptian and acquired secret religious lore from the priests. Here he seems to have imbibed number mysticism and cosmic ideas which flout sense experience.

Pythagoras's cosmology goes like this. The centre of things is not the Earth, but a central fire, or hearth – also called Zeus's watchtower – which we cannot see but which provides the sun with its light. Round it revolve the sun, the moon, the Earth and the five planets then known. With Zeus's watchtower that makes nine. Since the perfect number is ten – the sum of 1 plus 2 plus 3 plus 4 – the cosmos, which appears to be perfect, must have ten heavenly bodies. So Pythagoras added a tenth, anti-Earth, which we cannot see.

The celestial bodies and the stars, as they whirl round, produce a noise, the tone of which varies according to the distance of each from the centre. The sum of those tones is the music of the spheres. Why don't we hear this heavenly symphony? Because, said Pythagoras, from birth we are accustomed to hearing the music as a background, and since there is never absolute silence to contrast it with, our senses are unaware of it.

Having constructed so colossal an edifice, Pythagoras might have been expected to stop. Not at all. He went on to declare that both the stars and the human soul are composed of the very fine air called aether, and like the stars the soul is divine and immortal. This was contrary to Greek tradition, which held that only the gods are immortal, and man's soul is at best a residuum, compared by Homer to a shadow or a bat.

The body is unworthy of the soul. Having drawn on Babylonia for his cosmology, Pythagoras drew on India for his next doctrine: transmigration of souls. The soul passes from this body to that – Pythagoras himself claimed to remember four previous incarnations, including one as the warrior Euphorbus during the Trojan War. The soul may even inhabit an animal's body. This teaching angered many Greeks, and the poet Xenophanes told the following story against Pythagoras. The philosopher passes by when a puppy is being whipped. 'Stop,' he says. 'It is the soul of a friend. I recognize it by its yelp.'

A cosmology such as this, which downgraded the Earth, divinized the stars and claimed that man's soul was a star-like substance ill at ease here below, naturally produced an ethic quite different from the traditional Greek ethic. Life, said Pythagoras, is not a race, it is a spectator sport. Man must study the principle of order revealed in the universe, especially in the celestial bodies. He must attune his soul to that order, detach it from earthly things, and so his soul will eventually escape from the cycle of birth and attain immortality.

This was indeed a revolutionary doctrine. It opened the way to mystery cults, magic, extremes of asceticism and, most notably, to puritanism. Pythagoras himself seems to have been a puritan. His robe, we are told, was white and spotless; his quilts were of white wool. He was a vegetarian and abstemious about wine. Asked when a man should consort with a woman, he answered, 'When you want to lose what strength you have.'

He seems also to have cultivated the role of mystery man. He lectured at night and for their first five years his disciples were not allowed to set eyes on the master. He imposed on them apparently irrational taboos, such as 'Do not look in a mirror beside a lamp,' 'Do not eat beans,' 'When you get up in the morning roll the bedclothes together and smooth out the place where you lay.'

Pythagoras will always be esteemed for his insight into the importance of number in nature, and into the soul's immortality. And his cosmology has made an enduring appeal to those of like mind with Shakespeare's Lorenzo:

> There's not the smallest orb which thou behold'st
> But in his motion like an angel sings,
> Still quiring to the young-eyed cherubins;
> Such harmony is in immortal souls;
> But whilst this muddy vesture of decay
> Doth grossly close it in, we cannot hear it.

Nevertheless Pythagoras's views were probably on balance harmful. He was anti-rational in positing Zeus's watchtower, anti-Earth and stellar music for which he had no evidence; by divinizing the stars and thereby causing a population explosion of gods he scaled down man, and he followed that by initiating the lugubrious tradition according to which man's soul is merely a passing tenant of his body, and the good life consists in detaching oneself from earthly activities.

*

Some of Pythagoras's ideas were developed by the next important cosmological thinker, Plato, and used by him in modified form as part of a world-picture of unrivalled power and beauty. Plato was born in Athens in 428 BC. His family was distinguished and on his father's side claimed descent from the sea god Poseidon: thus the link between divine and human, such a feature of Plato's thinking, was built into his genealogy. As a young man he probably knew Pericles, then at the height of his influence, and he became a close disciple of Socrates who, we have seen, enlarged the scope of Anaxagoras's Mind. From Socrates Plato imbibed the conviction that to be a good ruler you must have a sound education. This half-truth was rendered plausible by the fact that a single Greek word, *episteme*, means both knowledge-that and knowledge-how. Plato travelled to Sicily, where by means of careful tuition he tried to turn the tyrant of Syracuse, Dionysius, into a philosopher-king. The results proved disappointing, but Plato persisted in his belief that a good education will produce good rulers and made it one of the central principles of the Academy, the great school of science he founded in Athens.

Plato grew up in an age which rejected the idea of a flat Earth. The sight of the top of a ship's mast appearing before the hull when the ship is approaching shore, and the round shadow cast by the Earth on the moon during a lunar eclipse had convinced men that the Earth is spherical. Plato not only accepted this but held that a sphere is the basic cosmic shape. The sky is not a pillared roof, as Homer had believed, but a hollow sphere, and sun, moon, planets and stars are also spheres. They revolve round the Earth at different speeds.

Plato was aware that three of the five planets then known – the ones we call Mercury, Venus and Mars – sometimes seem to stop and to move backwards. In effect they execute elongated loops in the sky relative to the fixed stars, and they do so because they and the Earth move in elliptical orbits with the sun at one focus. No one in ancient times suspected that truth, and in Plato's day two views were current. One held that the planets are indeed wanderers, and their movements as random as a firefly's flight. The other held that planetary divagations were in principle explicable by a complex system of circular movements, and this was the opinion Plato adopted. He very firmly took the line that all celestial bodies move in circular orbits. Indeed, he declared that such movements are the supreme manifestation of geometric principles governing the universe. For whereas Pythagoras held that the *arche* of things is arithmetic, Plato, if pressed, would have declared that it is geometry. Karl Popper convincingly suggests that the Pythagoreans lost confidence in the power of number when they discovered that certain numbers, such as the square root of three, are irrational, that is, they cannot be expressed as the ratio of two whole numbers, and Plato repaired the damage by devising a cosmology based on solid geometry. Certainly Plato inscribed over the portal of his Academy 'Let none but geometers enter here.'

Plato held that the motions of celestial bodies were too regular to occur by chance and were caused by divine beings. Just how he explains in the form of a vision in the final pages of the *Republic*. On the knees of Necessity is an adamant spindle, which revolves. The central part of the spindle consists of eight whorls, representing the orbits of the moon, sun, five planets and fixed stars. The seven inner whorls rotate with varying speeds in the opposite direction to the movement of the whole. On each of the whorls – a Pythagorean touch – stands a Siren utter-ing one note and all together constituting one harmony. The three daughters of Necessity, the Fates, sit round about and keep the spindle and its eight whorls turning with their hands.

In another dialogue, the *Statesman*, Plato suggests that the Fates, or God, occasionally lose control of matter, whereupon there is a reversal in the motion of the universe. Time starts to run backwards. The dead come out of their graves and start living again – growing each day younger. Old men find their white hair growing grey, then black; middle-aged men lose their beards and become downy-cheeked; the young become children, and eventually disappear as tiny infants! This passage has become topical today, when some scientists hold that if man succeeds in travelling faster than light he will indeed reverse Time and grow steadily younger.

Plato believed the movement of the heavenly bodies is supremely beautiful, therefore wholly satisfying to the soul. The soul, he says in the *Phaedrus*, is like a winged charioteer driving a team of winged horses. The horses belonging to the souls of gods are all good, but a human soul has one good horse and one evil. This makes it difficult for it to follow the procession of souls, headed by Zeus, to the rim of heaven. The souls of the gods pass outside heaven and stand upon its back, contemplating perfect Beauty and the rational principles of the cosmos.

Other souls share in this vision in different degrees, according to the success of their drivers in controlling their horses; but many get their wings broken, fall back to Earth and are incarnated in human bodies. Such souls, however, may regrow their wings through the perception of physical beauty and the consequent recollection of the perfect Beauty they glimpsed in their supra-celestial vision.

Plato suggests that if a man lives well he goes after death to dwell happily for ever in a star. He expresses this belief in verses which Shelley translated under the title 'To Stella':

> Thou wert the morning star among the living,
>   Ere thy fair light had fled;
> Now, having died, thou art as Hesperus, giving
>   New splendour to the dead.

Such a cosmology, expressed mainly in myths or beautiful images, led to an ethic not unlike Pythagoras's. The body is a hindrance in the acquisition of knowledge, and the wise man must rise above it. 'Platonic love' is better than physical passion, philosophy than the active life. But Plato took the consequences of his cosmology further than Pythagoras by applying it to politics.

A good ruler's first duty is to understand that the cosmos is rational and for this he must study astronomy, especially the mathematical principles underlying stellar movements. Only then will he be able to realize the cosmic order in political form. The best State will consist of three distinct classes: the ordinary people, the soldiers, the guardians, and for the last class marriage and private property will be abolished, since these foster private possessive emotions. The task of the guardians will be to see that the State continues on its rational course and that no innovations creep in. Plato implies that the stars in all their beauty, order and unchangingness are to be the model of a precisely interlocking political system, which will allow the highest class leisure to drive their winged horses, so to speak, to the rim of heaven, contemplate the cosmic order and ensure that the State remains on course. That Plato was in earnest about the importance of astronomy is clear from the fact that he made it a set subject in his Academy; this was an innovation in Greek education, the staple of Sophocles's schooling, for instance, having been gymnastics and music.

Our recent experience of totalitarianism has completely discredited Plato's political system. We see now, as our nineteenth-century forbears generally failed to do, that Athenian democracy is in almost every way preferable to the Utopia Plato derived from his cosmology. That having been said, it is none the less true that Plato's world-picture, perhaps because expressed with such visionary intensity, has been one of the most influential ever devised.

*

Plato's most brilliant pupil, Aristotle, was born in Stagira, where his father was doctor to the King of Macedonia. For twenty years he studied in the Academy, and Plato called him 'the intellect of his school'. Yet Aristotle was more down-to-earth than his master: he liked wearing showy rings and noticed such details as that theatre-goers eat more sweets when the acting is bad than when it is good. We do not hear of Plato having a wife, but Aristotle married, and after his wife's death took a mistress, who bore him a son. He was tutor for seven years to Alexander the Great, living to see his pupil's conquest of north-west India, and the consequent enlargement of man's view of the Earth.

This doctor's son was primarily a marine biologist, and a very good one. He dissected at least fifty animal species, including sea-urchins and starfish, and his detailed description of the octopus's reproductive system remained unsurpassed until the nineteenth century. He was chiefly interested in how an animal develops, decays and dies; and in its ability to move and to breed true to type, powers he attributed to its soul.

When this biologist looked at the sky what chiefly struck him was that the movement of the orbs is quite different from animal movement. It is regular, apparently circular, and repetitive. Aristotle concluded that the orbs are not composed, as animals are, of earth, water, air and fire, but are made of aether, the very pure upper air not found here below, to which he attributed the distinctive property of moving in circles.

Since Plato's day an astronomer named Eudoxus had come up with an explanation of the planets' anomalous movements. For the plane circles of earlier astronomy Eudoxus substituted concentric spheres, or rather spherical shells, each rotating on a different axis; the paths of moon, sun and planets as seen from the Earth are the results of the component motions of the various spherical shells. Adopting an expanded version of Eudoxus's theory, Aristotle held that moon, sun, planets and stars are attached to spheres, 55 in all, composed like them of aether, and in physical touch with each other. The circling of these spheres on different axes produces the periodically varying movements of the celestial bodies. But why do the spheres circle in the first place? Because, answers Aristotle, each is attracted by the perfection of a divine unmoved prime mover.

At first Aristotle spoke of 55 unmoved movers, one for each sphere, but later he said, 'It is better to think the first mover to be one rather than many.' This was a giant step from polytheism to monotheism, and Aristotle merely comments that hypotheses must be kept to a minimum. Why, he does not say; probably he took it as axiomatic. Indeed this preference for a single cause is distinctive of Greek thinking, and not found to the same degree elsewhere. The Persians, for example, preferred the dualism of a Good Principle and an Evil Principle, which later infected the West as Manicheism, and the Chinese have long shown a preference for the complementary duo Yin and Yang.

In the Greek predilection for a single cause I believe there are two components: the thrifty, even cheese-paring attitude of men who dwelt in a sunny but poor land and instinctively made the most of every olive, every grape; and, secondly, the rare unification of mind and body the Greeks achieved in their personal lives – that quality we recognize when we say that the Greeks first attained individuality – which would have made it natural for them to seek the one behind the many.

To return to Aristotle's single unmoved mover. He is pure mind, the *arche* of all things. He did not create the world, which has existed always, indeed he is indifferent to it, but his perfection is such as to keep the spheres turning, rather as the sight of a beautiful woman makes men turn their heads. So sun, moon and stars are not themselves gods, as Plato had said, but in a literal sense followers of God.

Because the stationary Earth, its atmosphere and man are of a different composition from space and spatial bodies, there is a gulf between the two realms, and it is not bridgeable as Plato had claimed. Any changing rectilinear movement in the sky must therefore belong to Earth's atmosphere, not to aether, hence Aristotle's insistence that meteors are hot, dry, inflammable exhalations from Earth which under certain conditions burst into flame. He believed, incidentally, that the Milky Way is of the same nature as meteors.

The unbridgeable gap between the changing sublunary world and unchanging outer space where God is limits man's activities. Whereas Plato had claimed that by contemplating the stars man's winged soul could go into orbit and so share in their perfection, Aristotle denied this. We are too different from the stars, he said, in composition and movement. The personal soul is dependent on the body; it is foolish to think it can soar away from the body or indeed survive it. So man is cut off from space and has to fulfil himself here on Earth.

He does this by living virtuously and exercising his highest faculty, intelligence, in what Aristotle calls *theoria*. The word is often translated 'contemplation', but I think a more exact rendering would be 'the practice of natural philosophy', in other words, the making of a world-picture, but one that takes as much account of starfish as of stars.

After the soaring, quasi-mystical teaching of Pythagoras and Plato, Aristotle brought Greek cosmology back to Earth. Yet he too made astronomy a set subject in the school he founded, the Lyceum – not, like Plato, as a guide to politics, but in order to contribute to the intelligent man's world-picture. The stimulus given to astronomy by these two philosophers was to lead to important discoveries, as we shall see in the next chapter.

I think the richness of Greek speculation is evident from the thinkers I have discussed. There were others, less influential but also original, such as Democritus, who taught that the cosmos is made of an infinite number of hard, indivisible atoms in eternal but purposeless motion, and Epicurus, who derived from that a scale of values and way of life.

The views of Aristotle, as those of Anaxagoras and Pythagoras, Plato and Democritus, are like model dresses by the Paris *grands couturiers*. A few women wear them, quite a lot copy them, but the majority merely adjust their waist-line and their hem-line, and go on dressing according to tradition. So before turning to later antiquity it may be well to remind ourselves of the classical tradition. The Greeks held the Earth to be the centre of the cosmos, which by and large was a friendly place, even its sky at night peopled with familiar figures, who had once roamed Greek hills and sailed Greek seas. Man, after the gods, was the most important being, and through insights poetical and religious, biological and astronomical, he sought to understand what happened overhead. He had no doubt at all that it could be understood. Yet the cosmos was touched also with something divine, so that in learning more about it man might also hope to learn more about the gods or God.

CHAPTER TWO

From Alexandria
to Bethlehem

Rise of Alexandria – Archimedes's planetarium – Zeno teaches that Mind in the shape of pneuma pervades the cosmos – improved astronomical measurements – Posidonius claims that tides are caused by the moon – the physics and ethics of Roman Stoicism – Seneca's career and writings – growth of astrology – worship of sun, moon and stars – the Jewish world-picture – the early Christian world-picture and calendar.

On an unspecified day between 247 and 222 BC the world's biggest ship sailed from Sicily on her maiden voyage. A three-master of some 4000 tons, the *Woman of Alexandria* had thirty cabins, tessellated floors depicting scenes from the *Iliad*, a gymnasium, flower-beds and grapevines on her main deck, a bathroom with three bronze tubs, and a boxwood-panelled library with a concave ceiling depicting the vault of the sky. She carried grain, Sicilian salted fish and 600 tons of wool – gifts from King Hieron of Syracuse to King Ptolemy III of Egypt.

In due course the *Woman of Alexandria* sailed into harbour at the world's most advanced city. Alexandria had taken over from Athens, now in decline, its tradition of philosophy and astronomy. It was a bustling city of wharves and corn magazines, royal palaces, Temples of Poseidon and of Serapis, a wrestling arena, a museum and library, two great obelisks, later to be known as Cleopatra's Needles and now in New York and London, and the Mausoleum of the city's founder, Alexander. Especially remarkable was Alexandria's lighthouse, 80 times the height of a man, built of marble and lighted by a wood fire, one of the seven wonders of the world. The Pharos, as it was called, epitomized man's growing technological skill and his ability to navigate confidently in an expanding world; visible at night from afar like a man-made star, it symbolizes the second phase in man's knowledge of his cosmos.

The person who had designed the sky-ceiling of the ship's library and succeeded, where others had failed, in launching the great ship, was Archimedes of Syracuse, a kinsman of King Hieron. By turning a single ingeniously contrived windlass he had eased the ship down the slipway, and on this occasion uttered a famous boast: 'Give me a place to stand on and I will move the Earth.'

Archimedes was born about 287 BC, the son of an astronomer. He studied in Alexandria, where he invented a screw pump for raising water which continued to be used in Egypt until the twentieth century. On his return to Syracuse he engaged in astronomy, mathematics and engineering. About his concentration many stories are told. He would draw mathematical diagrams on the ashes of his fire while cooking and on the oil on his body while at the bath.

According to a story by Leonardo da Vinci Archimedes travelled to Spain, where he aided Ecliderides, King of the Cilodastri, in a maritime war against Britain by inventing a device for spouting burning pitch on to enemy ships. Ecliderides and the Cilodastri do not appear elsewhere in history, but Leonardo's story may be true, for we know that during antiquity Archimedes's screw pump was in use in Spain's silver mines; Archimedes may have installed it there himself and on that occasion helped in the war. When the Romans besieged Syracuse Archimedes invented catapults and grappling machines which held off the legions for two years. He was eventually killed by a Roman soldier.

As a mathematician Archimedes demystified the circle and the sphere. He showed that these bodies, which because they have no beginning or end had seemed to Plato expressions of God, could be measured. He declared a circle's circumference to be πd, where d = diameter of the circle, and calculated upper and lower limits for π. He circumscribed a sphere within a cylinder, and calculated the ratio of the latter to the former as $\frac{3}{2}$ – a discovery which, at his request, was depicted on his tomb.

Archimedes constructed the first known planetarium. Made of brass and powered by water, it reproduced the movement of the heavenly bodies round the Earth, and the phases and eclipses of the moon. Marcellus, the general who captured Syracuse, shipped it to Rome, where Cicero admired it two centuries later. The planetarium marks an important step in man's approach to the night sky. He now had before his eyes a model of the cosmos such as led him to picture the movement of the sun, moon and planets as powered by a single material source.

Archimedes is best known for his principle that a body plunged in a fluid loses as much of its weight as is equal to the weight of the displaced fluid. He also conducted experiments with levers, fulcra and balances, and so demonstrated how weight can be the cause of action at a distance.

Archimedes's most ambitious work was to make reasonable guesses about the dimensions of the cosmos, and then to calculate the huge number of grains of sand required to fill space. He estimated that 10^{63} grains of sand would be required. The net effect of his labours was greatly to demystify the cosmos, and to bring it within the span of man's calculating powers.

Now during Archimedes's lifetime a new philosophy was attracting attention. Stoicism was so named because first taught about 312 BC in Athens' painted porch or Stoa by a Cyprus-born Phoenician named Zeno. In opposition to Epicurus's view that the universe is a product of chance, Zeno declared that the universe is held together by Mind. Mind and matter, however, are not distinct, and Mind manifests itself visibly in a material compound of air and fire called pneuma, or warm air, which impregnates everything. Whereas Epicurus taught that in a chance universe man can do no better than pursue happiness, Zeno urged his followers to conform to the purposes of Mind. Zeno was the first to crystallize the notion of duty by using a word hitherto meaning 'bearing down on' to signify 'morally incumbent'. Just as the pneuma in physical things preserves a certain right tension, so man must keep himself in right moral tone – 'up to the mark' – in order to perform his duty.

Zeno's physics were largely conjectural and very sketchy. But discoveries such as Archimedes's and, as we shall see, of his successors seemed to provide hard evidence for that early hunch that all things are somehow inter-related in a reasonable way. Under their impetus Stoicism in the third and second centuries BC was to take on muscle and gradually become the most plausible philosophy of later antiquity.

The next important cosmologist, Eratosthenes (276–196 BC) was a friend of Archimedes and curator of Alexandria's great library. His versatility won him the nickname *pentathlos*, a term from athletics meaning all-rounder. In classical Greece such an epithet would have been superfluous, but specialization was now creeping in, and versatility was sufficiently rare to attract notice.

[36]

Eratosthenes was the first man to measure the circumference of the Earth. He found that at noon on Midsummer Day in Syene (now Aswan) the sun shone directly down into a deep well, without casting any shadow. At the same time an upright rod at Alexandria, 5000 stadia due north, cast a shadow which showed the sun's rays inclined to the vertical at an angle equal to a fiftieth of a complete circle. From this Eratosthenes deduced that the circumference of the Earth measured along the meridian through the two cities is 50 times 5000 = 250,000 stadia. The exact length of a stadium in unknown, but is usually assumed to be 157·7 metres, which yields the astonishingly precise value of 39,425 kilometres. The best modern estimate is 40,076 kilometres.

Having provided man with an accurate idea of the size of his habitat, Eratosthenes also devised an improved geography. In place of the old circular land mass, with Delphi at its centre, surrounded by ocean, he envisaged an inhabited land mass more than twice as long as wide, comprising Europe, Africa and Asia. He believed that near the Equator, and beyond it, intense heat makes life impossible, therefore he placed the land mass in the northern hemisphere, and held that it occupied only one-sixth of the whole globe.

By making observations of the moon from different points Alexandrian astronomers calculated the distance of the moon from the Earth, coming very close to the correct figure of 384,000 kilometres. They also tried to calculate the distance of the sun, by waiting for the moment when the moon is exactly half full, and measuring the angle at the observer's eye between sun and moon. An error of less than three degrees in the angle led to a large error in the distance, hence they believed the sun to be 18 times farther than the moon, whereas the correct figure is almost 390 times.

The best astronomer of antiquity, Hipparchus (190–120 BC), further increased man's confidence in face of the cosmos by compiling a star catalogue, which gave accurate positions for one thousand stars. He also observed that the moon's speed varies according to its position in its orbit, and that the orbit is inclined at about 5 degrees to the plane of the ecliptic (the sun's apparent path), so that it is sometimes north of the ecliptic, sometimes south. The correct explanation of the first effect – that the moon's orbit is elliptical – was unknown in antiquity, but Hipparchus explained it so accurately in terms of combinations of circular motions that his successors were able to predict the time of a lunar eclipse to within an hour or two.

From being able to predict lunar motion it was only a short step to thinking of the moon as a body like any other. Posidonius, the man who contributed most to this step, was born in Syria about 135 BC. He studied at Athens under the Stoic philosopher Panaetius, travelled in the Western Mediterranean and settled in Rhodes, where he held high political office and was ambassador to Rome. His many pupils included Cicero. That he was a man of action as well as intellect is suggested by his vigorous style, interest in the oddities of human behaviour and rejection of the Stoic belief that virtue is enough to make a man happy: Posidonius said health, money and strength are also necessary. Presumably he thought of himself as happy, for he enjoyed all three until old age, when he suffered from gout.

In the Spanish town of Cadiz Atlantic breakers pound a white sand beach and there is a difference here between high and low tide of about one metre. Posidonius visited Cadiz and studied the difference. He also observed that the tidal River Guadalquivir, near Cadiz, overflowed when the moon was new. He concluded that tides are caused by the circling of the moon. 'When the moon rises the extent of a zodiacal sign [30 degrees] above the horizon, the sea begins to rise and encroaches noticeably upon the dry land, until the moon reaches the middle of the sky. Conversely, when the moon descends the sea slowly withdraws ... The monthly cycle means that the high tide is strongest at the new moon.'

Posidonius declared that the movement of the moon increases or decreases the tension of the pneuma between it and the ocean, so causing the water to ebb or flow. Though it ignored the sun's role, this was quite a good explanation for an age unaware of gravity.

To conclude later antiquity's increasingly reasonable view of the moon, we may notice that Plutarch, a century after Posidonius, was to pronounce the moon to be similar in composition to the Earth and despite its thin air perhaps inhabited; it remains in orbit without falling because it is like a stone whirled in a sling.

Posidonius declared the sun's diameter to be $39\frac{1}{4}$ Earth-diameters, evidently on the basis of a too low estimate of the sun's distance, for the true value is 109. But the point is, this was the largest figure so far proposed. It tended to make men see their world as a relatively small body, readily affected by its big neighbour.

On his travels Posidonius noticed how nations varied in character and behaviour. Italians lived frugally, content to dine off pears, nuts and water, whereas in Gaul he saw men gnawing joints of meat 'like lions', and human heads nailed as trophies to their houses. Again applying the Stoic principle of material interconnection, Posidonius attributed such differences to the sun. According to the angle at which it shines, the sun imparts different degrees of 'vital heat', and these in turn impart variety of values and behaviour. Posidonius also took an interest in volcanoes, attributing them to pneuma trapped in the Earth, and he described how a new island rose volcanically out of the Aegean.

Posidonius claimed to be a Stoic but he drew on some non-Stoic ideas, notably one propounded by Plato in the *Timaeus*, that the cosmos is itself a vast living organism. This is a retrograde notion, unworthy of Posidonius at his scientific best, but as it was to reappear – in Plutarch, for instance, and during the Middle Ages as the origin of 'humours' – it deserves mention. The heart of the cosmo-organism is the sun; it sends out heat and light. The Earth is its stomach, the sea its bladder. They send up exhalations which become fuel for the fiery sun and the stars. The moon is the liver, returning exhalations to Earth in purified form after use. The whole organism is penetrated, controlled and unified by a World-Soul.

*

What was the effect of these scientific discoveries of later antiquity on man's view of the cosmos? As he came to see space in terms of Earth-distances, the moon as a body like the Earth with a motion that can be predicted, and the stars' circular paths as subject, in principle, to measurement by man, he considered the cosmos to be not unlike his own habitat. Rejecting Plato's view of the celestial bodies as gods, and Aristotle's view that they are different in kind from our world, he thought of sun, moon, stars and Earth as somehow homogeneous. They were perhaps parts of a whole, as suggested by Archimedes's planetarium and, if Posidonius's theory of the tides were true, perhaps capable of acting on one another at a great distance.

The later Stoics formulated such views into a coherent system. The cosmos, they said, is a continuum of pneuma, all parts of which are in dynamic interplay. Everything is physically interdependent, from the farthest star to the nearest grain of sand. Everything is a link in a vast chain of causality. The causality is reasonable, not random, because it is permeated by Mind in the form of Fate. Eventually our cosmos will be consumed by fire, to be replaced by a new cosmos. This too will go up in flames, and the cycle will continue eternally.

[39]

How does man fit in? According to Alexandrian physiologists, he inhales a little of the cosmic pneuma into his lungs, whence it passes to the heart, enters his blood and becomes a special kind of pneuma, the 'vital spirit', which keeps his body alive. Arriving in the brain, it is again changed, this time into 'animal spirit' which, distributed through the nerves – thought of as hollow – ensures coordinated action. So every one of us is connected physiologically with the cosmic pneuma, itself permeated by Fate.

Such a world-picture weighs very heavily on the individual. He may be a free agent, as Stoics taught, but freedom to do his duty is plainly limited by the vast interconnectedness around him, and by the fact that he is not wholly separate from the causal chain. Roman Stoics therefore settled for a solution along these lines: Recognize the interconnectedness of all events, including those of your own life; usually they will be adverse, and often the most you can do is adopt an attitude of courageous endurance. Prepare yourself by learning self-control.

We can trace the link between world-picture and way of life in the career of Lucius Annaeus Seneca. The second son of rich parents of Spanish origin, he was born in Cordoba in AD 1. He studied rhetoric in Rome, and philosophy under a Stoic with austere Pythagorean leanings. He became a vegetarian. But abstinence from meat was misconstrued as abstinence from pork, hence as conversion to Judaism, a proscribed religion, and after twelve months Seneca resumed eating meat. He continued, however, to abstain from eating oysters and using a soft mattress.

Seneca had poor health. He cured a youthful attack of tuberculosis by a visit to Egypt, but he was always to suffer from asthma, which suggests a nervous disposition. He entered public life and made his name as an orator, adopting a very terse style, justly criticized by the Emperor Caligula as 'stones without lime'. He made a happy marriage to a woman younger than himself, Paulina, and had one son, who died in boyhood. Then he began a foolish affair with Caligula's sister, for which he was exiled to Corsica. During eight years there he studied winds, thunder, lightning. On his return he became tutor, then chief adviser to Nero. At first he tried to curb Nero's cruelty, saying to him once, 'However many you put to death, you will never kill your successor.' Later he flattered Nero in the hope of appealing to his better feelings. When it came to the test he seems not to have shown the moral bravery and indifference to wealth he preached to others. He acquiesced in Nero's murder of his mother and used his influence at Court to amass in four years the huge fortune of 300 million sesterces, one of the outward signs of which were 500 banqueting tables of citrus wood with legs of ivory. He is said to have suddenly called in 40 million sesterces from Britain, and thus to have caused Boudicca's insurrection.

The Stoics believed earthquakes, like volcanoes, are caused by pneuma trapped in the Earth forcing a way out, and took a special interest in them as dramatic examples of the interconnectedness of nature. Seneca wrote his first book on earthquakes, and aged sixty-two returned to the subject in *Questions about Nature* by describing the earthquake of 5 February 63, which destroyed part of Herculaneum. A friend of Seneca's who happened to be taking a bath when tremors began 'asserted that he saw the tiles paving the bathroom floor separate, then rejoin. At one moment, when the pavement opened, the water was absorbed through the joints; the next, when the pavement closed, it was forced out, bubbling.' Earthquakes, says Seneca, are not caused by angry gods, they result from inherent imperfections, which may occur anywhere. So those who have fled Campania for other regions are ill-advised. Science for the Stoic has to drive home a moral, and Seneca concludes this passage by saying that death may occur at any time, and we should prepare for it now.

Whereas the classical Greeks had admired the beauty of the heavens and the world, Seneca typified later antiquity by concentrating on the littleness of the world as scientists like Posidonius had shown it to be – 'like a threshing-floor, where ants congregate' – and on its imperfections as revealed by Roman travellers, who ventured farther afield than the Greeks: 'covered mostly by sea and, even where it rises above the sea, an ugly waste either parched or frozen.' This was not a habitat where man could feel at ease; he must expect insecurity, adversity and suffering.

These Seneca depicted in eight tragedies, modelled on the more terrible Greek tragedies, but with added horrors. What interested Seneca was the struggles of human beings in desperate straits holding out until they can bear no more, as in the line of *Oedipus: 'fatimus agimur; cedite fatis'* – 'we are driven by fate; yield to fate.'

In *Letters on Morality* Seneca tackles the problem of evil. 'Much in life is sad, horrible, hard to bear,' someone complains. 'Yes,' God might reply, [by God Seneca means Reason permeating the cosmos] 'and since I could not exempt you from such things, I have armed your minds against them all. Bear them bravely. That is how you can surpass God;

He is exempt from enduring evil, while you show yourself superior to it ... Above all, I have seen to it that nothing should detain you against your will. The way out is open. If you do not wish to fight, you can flee. I have made nothing easier than to die.' And after death? Seneca wavered between belief and disbelief in the soul's survival.

One pleasing aspect of Stoicism followed from the view that man is part of a cosmic whole, and more specially of the human race. Seneca advocates humane treatment of slaves and in an essay *On Mercy* written for, and dedicated to, Nero, he praises mercy as something which is 'in the truest sense self-control and an all-embracing love of the human race even as of oneself.'

After murdering his mother Nero became deaf to such pleas and at the age of sixty-one Seneca thought it prudent to retire. But his known opposition to tyranny made him suspect and three years later, on the discovery of a conspiracy against the Emperor, Nero sent a messenger to Seneca with the order of death.

Seneca's philosophical thinking had in a sense been a preparation for this moment and the scene that followed might have come from one of his own tragedies. He accepted the order with composure and was deeply moved when his wife Paulina elected to share his fate. Both stretched out their hands to the centurion, who slit the veins in their wrists. When his wife lost consciousness Seneca ordered her wrists to be bandaged, and she survived: for the rest of her life her cheeks were startlingly white. He meanwhile slowly bled to death.

Seneca's life illustrates a general pattern. The average Roman, a landsman, had no call to use and esteem the stars' orderly movement, and he approached the cosmos through terrestrial phenomena. If a Stoic, he applied the dubious concept of pneuma – but with a heavy hand, thus blurring soul and body, person and cosmos. In effect, he became a pantheist. This curtailed his sense of freedom and zest for knowledge. The Stoic Emperor Marcus Aurelius was to express satisfaction that when he took to philosophy he did not waste time on history, syllogism or astronomy, while Seneca declared that recent inventions such as transparent windows, central heating and shorthand, were of no interest, since they could not help to strengthen or calm the soul.

Starting from a bold, panoramic view, Stoicism ended up narrow, centred on the individual struggling to hold out against Fate, with no help from God, who is within the cosmic flow. As Cicero said, Stoicism was a cold creed: it might convince the intellect, yet it failed to satisfy the whole man, his *joie de vivre* and his creative side. Cicero might have added that it burdened man too heavily, left each of us, like Atlas, with the whole weight of the universe on his shoulders.

*

Stoicism was a Graeco-Roman philosophy. While it thrived the conquests of Alexander, then of the legions, brought Eastern beliefs to Europe and with them a different world-picture. The chief source was Babylonia. The Babylonians were unusual in that they actually worshipped the sun, moon, planets and stars. It was they who gave the planets the names of gods and goddesses which we use today in Roman translation. But whereas Greeks and Romans usually only associated the deity with the orb, the Babylonians identified the two. They held that each important heavenly body or group of stars, as a deity, possesses not only certain quantitative or geometric features – the Greek view – but certain *qualities*. Sometimes these derived from observation: Saturn, because it moved most slowly of the planets, was apathetic and irresolute. Sometimes they derived from theology. Astarte, which we call Venus, was the goddess of love and possessed the attributes associated with sexual passion.

The Babylonians noticed that the sun, moon and planets move in a circular band of the sky – the zodiac – spreading some eight degrees either side of the ecliptic. They divided the zodiac into twelve equal parts or signs, and named each after one of twelve constellations through which the sun moves in the course of twelve months. They began with the constellation of Aries, into which about 500 BC the sun rose at the spring equinox, and ended with the constellation of Pisces. Because the Earth wobbles slightly as it spins on its axis, each year the constellations appear a little later, and today on 21 March the sun rises not into Aries but into the middle of the constellation Pisces.

The Babylonians held that every man's life is determined by the sign rising above the horizon at his place and time of birth, and by the qualities of the planets in or near the sign; these planets, we recall, being gods and goddesses with various degrees of power.

It is one of the paradoxes of ancient history that just when the Alexandrians were working out a rational view of space, they fell for the Babylonian world-picture. Astrology, zodiac and horoscope – these are late Greek words and testify to the popularity of the star-cult. It would have horrified a classical Greek. Then man had stood proudly erect, master of his life; now he was a pawn at the mercy of the star gods and goddesses who happened to have presided over his birth. Then Sophocles had written a strong chorus in praise of man's powers; now, ashamed to face up to himself, the late Greek penned feeble verses about swallows, locusts and mosquitoes. Then the confident sculpture of the *Charioteer of Delphi*; now *Laocoön and his Sons*, pathetic victims of forces greater than themselves.

Fatalism reigned. People would no longer take a bath, go to the barber, change their clothes or manicure their fingernails, without waiting for the propitious moment. Epitaphs stated the exact length of life to the very hour, for the moment of birth determined that of death.

Many accepted that they were prisoners of the zodiac. Yet some sought to escape. For that they had to force the hands of the star gods with magic. All sorts of things believed to have 'sympathy' with this or that star were used: the onion, the baboon, the sacred ibis, and formulae sometimes esoteric, sometimes touchingly naïve. One formula to compel the moon says: 'You have to do it, whether you like it or not.'

The Alexandrians revived Plato's suggestions that good souls after death go to the sky. So in the opposite direction to the traffic of astral influences went a soaring of virtuous souls. Posidonius, like Plato, believed such souls go to a star, whence they can exert a good influence on us below, while a first-century gem from Chalcedon in Asia Minor depicts souls dwelling in the moon.

Many of the best astronomers of later antiquity were firm believers in astrology, among them Hipparchus and Ptolemy of Alexandria (flourished AD 140), who accurately measured the planets' movements. Through them it became part of the cosmology of educated men, and went to reinforce the interconnection of all things, as taught by Stoic physics.

*

From Alexandria the Babylonian star-cult and astrology spread to Rome. Since early times the more superstitious Romans had divined from the sky, first from thunder, then from birds. If birds flew from the left, that was inauspicious: 'left' in Latin is *sinister*. Towards the end of the Republic Roman diviners began to look more at the astral bodies, and the Augustan poet Manilius wrote a long poem called *Astronomica*, four-fifths of which is in fact astrology. Manilius had a fine time squeezing trigons, quadrants, dodecatemories and paranatellonta into his hexameter verse, showing a taste for jargon which ever since has characterized astrology.

The direct link of heavenly bodies with the Earth, and their astrological influence here below received public endorsement during the principate of Augustus, when the Romans abandoned their eight-day week, replacing it with the seven-day week established by the Babylonians, wherein each day was named for the planet that ruled its first hour. The week began with the day of Saturn, followed by the days of the sun, the moon, of Mars and so on. We have preserved the first three as Saturday, Sunday and Monday, and three others too by way of the Norse: Tyr, who is Mars; Thor, Jove; and Freya, Venus.

The Roman gods, as described for instance by Ovid, were pretty rather than powerful. They lacked the massiveness to withstand the cosmic gods that came in with the Eastern world-picture. The Egyptian goddess, Isis, became specially popular with the Romans. Originally the Moon-goddess, her sway grew with time. In the last chapter of Apuleius's *Golden Ass* the hero, Lucius, who has been changed into an ass, bathes seven times in the moonlit sea, and as Isis, 'queen of the stars and greatest of deities', rises from the water, he implores her help. During a festival in her honour, when the faithful in white linen garments carry torches and shake bronze, silver and golden rattles, Isis restores Lucius to human shape. She has told him that she has power to prolong his life beyond the span ordained by fate if he will enter her service and remain chaste; the novel ends with Lucius accepting these terms and being initiated into the Egyptian mysteries of her who rules heaven, Earth and the underworld.

The Sun-god too had many votaries. In 218 Heliogabalus, a fourteen-year-old boy-priest of the Syro-Phoenician Sun-god at Emesa, became Roman emperor. When he arrived in Rome the image of this Sun-god, a black meteorite set in precious gems, brought from Emesa, was set on a chariot and drawn by six milk-white horses through streets sprinkled with gold dust, and Heliogabalus declared it to be the supreme divinity. Half a century later Aurelian built a new sanctuary for the Invincible Sun and transferred there the images of Bel and Helios, trophies from captured Palmyra.

The most popular of the Eastern religions was Mithraism. Mithra, originally a Persian god, was associated with the sun and believed to cross the sky in a chariot drawn by white horses. His standing epithet, 'with broad pastures', suggests that he originally gave protection and grazing rights over large territories. His cult was celebrated in cave-shaped, usually underground buildings imitating the vault of the sky. Many reliefs and statues of Mithra slaying a bull testify to the popularity of the cult, especially among legionaries, and suggest that an actual bull was sacrificed, perhaps as a ritual rejuvenation of the year. Though Mithra was the most important god in the ritual, superior to him was a god of the whole cosmos, about whom little is known.

A neophyte of Mithraism took an oath – *sacramentum* – similar to the one required of army recruits, and a mark was branded on his body. Each of seven degrees of initiation was connected with a planet: Raven, Bridegroom, Soldier, Lion, Persian, Heliodromos – personifying the sun's daily course – and Father, who was the chief of a community. The birthday of Mithra was celebrated on 25 December, just after the solstice, when the sun is lowest and weakest.

Mithraism apparently worshipped cosmic forces and heavenly bodies. It held that the stars constantly act upon man, and one Mithraic dignitary calls himself on his epitaph *studiosus astrologiae*. But it also sought purification. After death the soul was believed to pass through the seven planetary spheres, divesting itself of sensuality and lust, until it became as pure as the gods whose company it entered.

Fundamentally different from the creeds so far considered was Judaism, which spread to the West, especially after 63 BC when Pompey captured Jerusalem. The core of Judaism was its belief in a Creator God. It is God who made all things, including man, 'By His word the heavens were made,' says the Psalmist, 'by the breath of His mouth all the stars.' God is quite above the heavenly bodies. So the Jews abominated the worship of sun, moon and stars, considering such cults idolatry.

The Jews believed that the Earth is shaped like a flat disc and rests on pillars: stretched above it is the sky or heaven, solid and also resting on pillars. This was not unlike the early Greek view. In or just beneath the heaven are the celestial bodies, which move across it: the sun is ordered by Joshua to stop moving. Above the heaven are waters. Beneath the Earth is Sheol, the abode of the dead, and 'the waters under the Earth'. The Jews were much less interested than the Greeks in physical phenomena as the expression of God or as a clue to His nature. Perhaps originally this was the attitude of a desert people to a landscape where life was rare; later the word of God, as spoken by His prophets, established a direct link between God and His people, so that the evidence of creation became superfluous. In other words, the Jews, possessing a single God, did not need the unitary view of the cosmos which the Greeks sought and found.

This God transcended His creation. The Jews were never in danger of pantheism, nor of considering man as merely a part of an interconnected Stoic-style whole.

God made man in His own image – the very opposite of the Greek view – and looks after his welfare. This was the highest view so far of divine concern for man. It surpasses even Pallas Athena's protection of Achilles or Zeus's translation of Callisto to the stars.

Yet the Jews were not puffed up by such a conception, for they believed in a multitude of beings superior to man – the angels. These serve and praise God, and enter into His counsels. Normally they are invisible, unapproachable and unaffected by human needs. They are often sent by God to intervene in men's lives, and then they appear to them in human form.

Belief in the existence of angels is found in the oldest as well as the latest book of the Bible, and is unlikely to have been a mythologizing of natural forces. Indeed the angels have personal characteristics, and their mission is often to individual men or women. Incidentally, there is no mention in the Old Testament of fallen angels or battles in heaven.

Angels were to remain, off and on, a feature of the Western world-picture. Their importance is considerable. Henceforth, for those who believed in them, what we should call space was never quite empty: in it lurked these superior invisible beings. And man therefore was never quite alone: when he praised God, he did so in concert with the angels.

More important, if angels were admitted to exist, it followed that there is an order of reality beyond what we can experience with our senses, and scientific knowledge is only one part of a greater whole.

*

The Jewish prophets foretold a Saviour, and it is the belief of Christians that he was born in Bethlehem during the reign of Augustus. At the time of his birth a star moved across the sky to a position above

Bethlehem: this may have been a nova, or new star, which Chinese astronomers observed for ten weeks in the second year of Ai-Ti; if so, the date of the Saviour's birth would have been 5 BC. At a time when star-worship and astrology were rife, the star or comet would have greatly interested three magi, perhaps Zoroastrians from Persia, who, Matthew tells us, followed it to Bethlehem. Shepherds on the nearby hills needed the visit of an angel to inform them of the birth, and this angel, after imparting his message, was joined by many others, who gave praise to God.

The Gospels are the historical record of a flesh-and-blood Nazarene who provided evidence that he was also the Son of God. They are concerned with a man's doings, not with the natural world. Yet, like the birth, key movements in that life were counterpointed at the cosmic level, while some of his miracles, such as the calming of the storm on Lake Tiberias, have cosmic overtones. They demonstrate that Christ is above nature. They also guarantee the truth of his words and prepare his apostles for events of an entirely new dimension, unsuspected by any natural philosopher: Christ's crucifixion and death, which redeems fallen man, and his resurrection. The crucifixion was counterpointed by three hours' darkness and an earthquake, and it was these cosmic disturbances which caused the centurion and his guard to say, 'No doubt but this was the Son of God.'

Yet the New Testament is little concerned with the cosmos as such. So Christians took over the Old Testament account of creation, modified it in the light of the Gospels, especially the first chapter of John, and propounded it in terms people of the day could understand.

Although we have no intimate portrait comparable to that of Seneca, we can piece together the beliefs of a second-century Christian living in Alexandria. Heir to the traditions of classical Athens, he looked for an *arche* or beginning. He found it in the act of love whereby out of nothing God brought the cosmos into being. That act of love in no way depended on its object, which is utterly contingent. But since the cosmos, and creatures in it, owe their being to God, they partake of His goodness and hence are to be accounted, as *Genesis* says, 'very good'.

Whereas his fellow-Christians of Antioch interpreted *Genesis* literally, our Alexandrian interpreted it allegorically. The six days, for instance, are not a measure of time but express the differing dignity of the things created. He rejected the view held by Aristotle that the cosmos has existed eternally, and the Stoic cycle of cosmoses.

Our Alexandrian differed from the classical Greeks in holding that the scientific world-picture is not the whole picture. Man cannot understand the cosmos solely by listing what it is made of or the immediate causes which make it as it is. The cosmos being the work of a Creator, man must investigate the Creator's purpose in making it. Even with the words of Revelation before his eyes 'natural man' cannot read the cosmos aright. To do that he must be touched by supernatural grace. But grace is given, not acquired. 'The ballplayer,' says Clement of Alexandria, 'cannot catch the ball unless it is thrown to him.' Once touched by grace, man becomes assimilated to the mind of Christ. Only then does he understand that the cosmos is moving towards God and will find its fulfilment in Him.

Living in an age when sun, moon and stars were worshipped as gods, our Alexandrian had to be cautious about cosmic speculation. Clement exemplifies this caution when he writes, 'What is the use of knowing the causes of the sun's movement?' But Jews were equally cautious. Rabbi Akiba (*c.* 50–132) taught: 'Whoever reflects on four things, it were better for him if he had not come into the world: what is above, what is beneath, what is before [in space as well as in time] and what is after.'

Our Christian, in his debates with pagans, sought to preserve the other-worldliness of his infant creed. In doing so he played down this world and this life. Origen (185–253) said we should execrate, not celebrate, our birthdays. For long it was concluded that early Christians, though from different motives, shared Marcus Aurelius's sentiment of being 'a stranger far from his true home'. But a more careful study of the sources now shows this to be only part of the story. Sometimes, on the contrary, Christians expressed a joyful appreciation of this world, as in a recently published, little-known homily 'On the Newly Baptized, for Easter Day', by the fourth-century Amphilocius of Iconium in Galatia:

When the season of spring replaces the scowls of winter, birds of all colours fly through the air and with sweet voices proclaim to men the delights of the time. Swallows with sweet chatter cleave the air at speed: they circle round men's heads like flowers and besiege their ears with their own clamour. Then can one see the sky bright and windless and men's faces light up to match the clear weather. Birdsong rejoices the ear and the cloudless sky brightens the eye. The glow of many-coloured flowers is a pleasure to see and the mingled scent of the plants revives our sense of smell. Yes, and all this delight is provided, Beloved, by an early perishable spring, while Christ, our divine and imperishable spring, covers the meadow of the Church with violets, roses and lilies of the Spirit; He brightens our eyes with faith and fills our hearts with scents that are divine.

That note of joy would be hard to match in pagan writing of the period; it flowed from a belief that this world is good, though less good than Christ, for whose second coming the Christian must prepare.

The test of any attitude to the natural world is its attitude to the human body. While pagans were denigrating the body as never before, describing it as a mere envelope, a prison, a cesspool, many Christians resolutely defended the wonder and sanctity of flesh moulded after the image of Christ. Against the Gnostics, who claimed that only soul and spirit constituted man, Clement is quite emphatic: God is the creator of man's body, 'flesh, marrow, bones, nerves, veins, blood, skin, eyes, pneuma, righteousness, immortality.'

Although his salvation had been accomplished at a certain moment in the past, our Christian re-enacted it in the Eucharist, and by commemorating it annually. Gradually to the solar, lunar and sidereal years was added a liturgical year. Here the Church recognized the extent to which man is a cosmic being, his mood and thinking influenced by the seasons. Early in the fourth century the Church decided to put Christmas immediately after the winter solstice, on 25 December, a date already chosen, as we have seen, by the Mithraites as the birthday of Mithra, and fixed Easter as the first Saturday night–Sunday morning after the full moon on, or immediately following, the spring equinox, which is 21 March. Though the dating of Easter was for a while disputed, the birth and death of Christ had been linked to appropriate cosmic events in the northern hemisphere.

The Church had already accepted the seven-day week of the civil calendar, while rejecting its astrological implications, because it tallied with *Genesis*'s six days of Creation, followed by the day of rest. It was the Emperor Constantine (ruled 306–337) who changed the day of rest from Saturday to Sunday, probably so that Christians in the legions could attend church. However, he continued to date years from the legendary founding of Rome according to a calendar devised by Alexandrians and adopted by Julius Caesar. The world would have to wait until 532 before years began to be dated from January after the birth of Christ; then officially as well as in practice the star of Bethlehem began to outshine the lighthouse and star-gods of Alexandria.

CHAPTER THREE

The Propinquity
of Heaven

Pliny's Natural History *presents marvels in a context of disorder, including stars that hop and twitter – influence on Gregory of Tours and on the Dark Ages generally – miracles as a means of combating disorder – Isidore's stars that have no light of their own – Bede's world-picture in which space is peopled by evil spirits – Louis the Pious and the comet of 838 – the comet of 1066 and the Norman invasion – Isidore's OT map of the world – Romanesque church architecture – Guibert and the saints and devils who haunt the air around us – Abbot Suger, Macrobius and the beginning of Gothic architecture – St Francis, Brother Sun and Sister Moon.*

In the year 585, on his way home from a diplomatic mission in Koblenz, a bishop dismounted at a hilltop monastery near Yvois, on what is now the Franco-Belgian border. His name was Gregory, his diocese Tours, the most important in Gaul: he was aged forty-seven and by the standards of the day well-educated. Not only could he read and write, he was later to compose a sensible, and valuable, *History of the Franks.* Gregory was not strong physically and a bit of a hypochondriac, but he had plenty of courage and during his twelve years as bishop had risked his life to defend the Church's rights against the bandit kings and queens of Gaul.

Gregory spent several days at Yvois as the guest of a holy Lombard monk named Walfroy. In his youth Walfroy had lived on a pillar like Simon Stylites. In obedience to his superiors he had forsaken his pillar and, as a preliminary to building a church, demolished a famous shrine of Diana. He told Gregory that his whole body had at once become covered with pustules, but he had anointed it with oil from the church of St Martin and the pustules vanished overnight. Gregory was pleased to learn this, because that church was in Tours, and he himself was devoted to St Martin.

One evening after sunset, happening to be outside the monastery, Gregory and Walfroy saw rays appearing from the four quarters of the Earth until they covered the whole sky. 'In the middle was a gleaming cloud,' Gregory later recalled, 'to which these rays gathered themselves as it were into a pavilion, the stripes of which, beginning broad at the bottom, narrow as they rise, and meet as it were in a hood at the top. In the midst of the rays were other clouds, flashing vividly as lightning.'

Though he had heard accounts of it before, this seems to have been the first time Gregory saw the *aurora borealis* or Northern Lights. One might have expected him to speculate about what caused it, and why it was not seen as far south as Tours, but no. 'It was a great sign,' he wrote, 'and filled us with fear, for we expected some disaster to be sent upon us from heaven.'

This is a most revealing comment. In Gregory's world murder, looting, torture, gouging out of eyes, were everyday occurrences, and it took a lot to frighten the bishop. Yet here he is, confessing that the beautiful roseate sky filled him with fear. Why?

Only three centuries have passed since the heyday of Alexandria, but there has been a great falling-off. The Huns have sacked Rome and deposed the last of the Western Emperors; Slavs have invaded Greece; impoverished Western Europe is cut off from the richer but now declining Greek-speaking East. In the flurry of siege and retreat first-hand knowledge of Greek learning has disappeared from Western Europe almost as totally as though a lobotomy had been performed. Only essential books have been copied on to vellum – itself a scarce luxury: Scripture, the Latin Fathers and a very few pagan Latin authors.

One of the pagan authors – the most influential during the Dark Ages and the one Gregory of Tours knew best – was Pliny, a soldier-barrister with a jackdaw mind and little of that common sense usually credited to Romans, who died while investigating the eruption of Vesuvius in AD 79. His *Natural History* is mostly about plants and animals, but the second of its ten books touches on cosmology.

Pliny opens with platitudes: the Earth is round and stationary, the sky spherical. Then he begins to go astray. The planets sometimes appear to halt in their courses: the reason is that they are deflected and lifted straight upward by the rays of the sun! Weird too is Pliny's view of the stars: 'On a voyage I have seen stars alight on the yards and other parts of the ship, with a sound resembling a voice, hopping from perch to perch in the manner of birds ... If there are two of them, they denote safety and portend a successful voyage.'

Pliny offers no information about what the heavenly bodies are made of, or how they remain in space. He quotes Pythagoras to the effect that the moon is 15,000 miles away, but himself adopts a defeatist attitude to distances: 'the world is immeasurable.' Pliny is pleased with this conclusion, for in general his sky is a place without order, either mathematical or physical, a forum for wonders such as the comet which appeared after Julius Caesar's death, a place where almost anything can happen, for God, should He exist, 'takes no interest in our world.'

The haphazardness seen in the heavens Pliny looks for, and finds, on Earth. He notices, and records, such things as a plant that cures 53 ailments, apes that play chess, elephants that understand their master's language, and a man 9 feet 9 inches tall. Hearing of a married woman in Argos named Arescusa who changed sex, grew a beard and took a wife, Pliny promptly included her (or him) in his book.

From Pliny's largely erroneous compendium of marvels and disorder Gregory of Tours devised a world-picture that would square with the Bible and his own needs as a prelate. The first thing to be said about it is that sun and moon, planets and stars, were of no interest in themselves. The rising of certain constellations was a convenient 'clock' for knowing when to get up and sing the night office known as Vigil, but otherwise the stars were ignored. Gregory was not aware of immense space between them and the Earth. He was not really aware of space at all, save as a stage for phenomena such as comets, lightning 'coiled like snakes' and the *aurora borealis*, all inexplicable, therefore portents. Gregory therefore did not think of God as being above and beyond space. He thought of God – and his saints – as very close, within touching distance of Earth. Representations of the Ascension from this period show Christ on a hill, and God's hand emerging from a low cloud at the same level, about to lift him up.

On this basis Gregory divided earthly experience into two categories: events deriving from nature and human nature, mainly of a disorderly and unpleasant kind, such as storms, floods, fires, contagious diseases, murders and wars; secondly, supernatural – but always visible and tangible – places or objects, such as churches, shrines, saints' tombs and relics which kept at bay or corrected the first category by means of what Pliny had called marvels and Gregory called miracles.

Gregory saw life as a battle between nature and miracle. It is the miracle, not nature, which is the principle of order, and this explains why three-quarters of Gregory's writings are records of miracles. Here are some of them.

When Gregory was a boy fire broke out on the family estate in Auvergne. It was harvest time, flames swept the sheaves and stacks of straw, the harvesters panicked. Gregory's widowed mother Armentaria was at table and as soon as she was told hurried to the fields. She was in the habit of wearing round her neck a gold reliquary shaped like a pea-pod containing the ashes of martyrs, and this she lifted in the direction of the flames. 'Immediately the fire stopped,' says Gregory, 'so completely that there was hardly a spark to be seen among the mass of burned straw.'

Gregory inherited the reliquary. One day he was riding from Burgundy to Auvergne when he was surprised by a severe storm. Thunder growled from a dark cloud and the sky was torn by lightning. 'Taking the reliquary from my neck,' says Gregory, 'I lifted it towards the cloud, which at once divided into two, one passing to the right, the other to the left, without doing us any harm.'

Gregory adds a footnote against himself. He was a young man at the time and, he explains, puffed up with pride; he boasted to his travelling companions of his virtue that had brought this grace upon him. At once Gregory's horse shied and he was badly thrown. 'I perceived that this had come of vanity and it was enough to put me on guard henceforth against being moved by the spur of vainglory.'

But the miracle *par excellence* was the cure. Gregory tells how, a young man in his twenties, on his way to Tours to become Archdeacon, he fell seriously ill of a purulent fever. He could eat nothing and grew very weak, but he insisted on being carried to St Martin's tomb, where he spent three nights in prayer. After this he slept for a few hours and woke wholly recovered. For the first time in his life he could drink wine and enjoy it. Thereafter, if he had a swelling on his tongue, he licked the rails round St Martin's tomb and the swelling was relieved. Once, when a fishbone stuck in his throat, he prayed in St Martin's little church just outside the walls, and the bone disappeared.

These and many similar goings-on, which seem so naïve today, followed quite logically from Gregory's world-picture, based on Pliny's 'disorderly' cosmos as modified by a reading of the New Testament. They illustrate values in the early Dark Ages, and their cosmological aspect is well summed up in a famous miracle which Gregory evidently liked since he commissioned a painting of it for his cathedral: St Martin, arriving in Gaul, encountered a great number of pagan idols; he knelt and prayed for their destruction; whereupon a huge column was suddenly launched from the sky and shattered the idols.

This survey of the cosmology of the Dark Ages, where sources are few, necessitates travel – first across the Pyrenees and south to Seville. The bishop of that town for almost 40 years was Isidore (*c.* 570–636); the most learned scholar of his day, Isidore wrote two books which passed at that time for science, *Etymologies* and *A Treatise on Nature*.

Isidore drew heavily on Pliny the disorderly, but also on compendia and digests containing out-of-context snippets from other Latin authors. Isidore's books are a jumble of such snippets, for the most part erroneous, and these he usually interprets as he interprets Scripture – allegorically. Sometimes however he applies number symbolism, at other times fatuous etymology. Here are some examples, the first from *Etymologies*:

> Night comes about either because the sun is wearied by its long journey and when it reaches the edge of the heavens its fire dies out, or because it is driven beneath the Earth by the same force that elevates its light above the Earth and the resulting shadow of the Earth causes night.

In his *Treatise on Nature* Isidore adds:

> Night, we believe, is for rest, not for work of any kind. In Scripture night has two meanings: first, ordeal and persecution; second, the darkness of a blind heart. Night gets its name, *nox*, from *nocere*, to harm, because it can harm the eyes.

The moon consoles man at night, continues Isidore, and the moon takes eight years to go round the Earth! Its seven phases correspond to the Church's seven meritorious graces. Of the Pleiades Isidore writes, 'These stars, seven in number and shining brightly, symbolize the saints who shine with the Holy Spirit's seven-fold gift; being close to one another but not touching, they signify those who preach God, close in charity, but distant in time.'

Isidore parts company from Pliny and Gregory of Tours on one highly important matter. Drawing on a snippet by the Augustan writer

Hyginus, Isidore declares: 'The stars have no light of their own; they are said to reflect the sun, as the moon does.' Stars that in themselves are dark – it is an astonishing belief that symbolizes well both the ignorance of the Dark Ages and the darkness that came over their sky. Indeed, for much of this period the heavenly bodies fade out.

It is typical of Isidore's muddle-headedness that the key to his world-picture is tucked away near the end of a chapter 'On the Course of the Sun'. Here Isidore declares, 'The cosmos has been made in the image of the Church.' One realizes that Isidore, totally at a loss to understand space and the heavenly bodies, is falling back on the one thing he can understand: Scripture, interpreted allegorically, that is, in accordance with Patristic snippets, however much these may be contradictory. For instance, in a single paragraph Isidore states that Venus is an emblem of Christ because in the *Apocalypse* Our Lord declares, 'I am the morning star'; but equally, since its Latin name is Lucifer, Venus is an emblem of Satan.

On Isidore's view of the world – and it was to be dominant for almost six centuries – space and the heavenly bodies lose whatever shreds of importance they may still have possessed, and become mere emblems. An astronomer comes to mean not someone who observes the stars but a cleric who calculates the Church's feasts from lunar tables. When St Columban (d. 615), an Irishman who founded monasteries in Gaul, wrote to the Pope: 'The Irish are better astronomers than you,' he was referring to a dispute between the Celtic and Roman Churches about the date of Easter.

The next man with a recognizable world-picture is Bede. Born in 672 or 673 in Northumbria, at the age of seven Bede was given into the care of Benedict Biscop, an ex-soldier who had become a monk and founded the new monastery at Wearmouth. Bede was soon transferred to the sister-house of Jarrow: we have to imagine not stone buildings but wattle and clay huts. Bede was ordained deacon at 19, six years before the canonical age, and priest at 30. He spent his whole working life teaching, and writing, chiefly history, arithmetic and chronology. In about his sixty-fourth year he fell ill of a respiratory disease. On the eve of Ascension Day 735 he called the monks to his cell and distributed to them his few possessions – a little incense, a little pepper, and some linen napkins. The monks wept, but Bede rejoiced, saying that he was about to see 'the King in his Beauty'. At his own request he was seated by the monks propped up on the floor of his cell, and with his face towards the sanctuary he sang the *Gloria Patri*, then breathed his last breath.

Bede drew heavily on Pliny and Isidore and his world-picture resembles Isidore's. Living near the North Sea, Bede worked out, first-hand, a more correct view of the tides than Isidore. On the other hand, the Englishman believed that thunder was a portent, and that one could divine from it according to its direction and the day of the week. Thunder in the east portended much bloodshed in the ensuing year.

Bede was unaware of the stars, as was the author of *Beowulf*, for though much of that epic happens at night it does not contain the word 'star'. The Vikings, like the Greeks, were sailors, but they navigated during the summer months when, in northern latitudes, the stars are not easily visible.

On to what may be called the physical blankness of space Bede and his contemporaries, if Christian, tended to project phenomena of theological import. Here, from his *Ecclesiastical History of England*, is Bede's account of how the holy bishop Fursa had a vision in which he was carried by angels to a great height and told to look back at the world. 'He saw some kind of dark valley immediately beneath him and four fires in the air, not very far from one another. When he asked the angels what these fires were, he was told that they were the fires which were to kindle and consume the world.' Their names were falsehood, covetousness, discord and injustice. Bede heard about the vision from an aged Jarrow monk who had had it from Fursa's own lips in East Anglia. The old monk added that 'although it was severe winter weather and a hard frost and though Fursa sat wearing only a thin garment, yet as he told his story, he sweated as though it were midsummer, either because of the terror or else the joy which his recollections aroused.'

In the best known passage of Bede's *History* one of King Edwin's thanes compares man's life to a sparrow flying swiftly through a hall in the middle of which a fire burns on a hearth, while outside wintry storms of rain and snow are raging. The thane has in mind the transience of life, but his imagery shows also how he thought of the world. Man's knowledge of it is confined to a layer of space no higher than the rafters of the hall, through which the sparrow-soul passes horizontally. 'It enters in at one door and quickly flies out through the other.'

That this is a correct view of the thane's image is supported by Bede's statement elsewhere that the air is inhabited by evil spirits who there await the worse torments of the Day of Judgement. The northern dark, in short, is something frightening, better not thought about.

Another century passes. England's centre of learning has shifted from Northumbria to York, and from York the monk Alcuin goes to Aix-la-Chapelle to teach Charlemagne to read, and his son Pepin to understand the world. But things have not improved since Isidore's day: as appears in the following question and answer from one of Alcuin's writings, explanations now are in terms not of causes but of effects.

> PEPIN: What is the sun?
> ALCUIN: The splendour of the universe, the beauty of the sky, the glory of the day, the divider of the hours.

A little later, in 838 Charlemagne's grandson, King Louis the Pious, was spending Easter at Aix-la-Chapelle. Louis had made Lear's mistake of dividing his kingdom among his children, who had then waged war upon their father. Over Easter a comet appeared in the zodiacal sign of Virgo and moved across the sky to the sign of Taurus, where, after having been visible for 25 days, it finally disappeared.

Even more than the *aurora borealis* a comet was a phenomenon that caused universal fear. The anonymous chronicler who describes the comet of 838 is known as Astronomus, because he was an expert in celestial portents. He goes on to describe how King Louis the Pious asked him to study the strange star in order to make a report. Astronomus did so, was recalled by the King, stammered a few words, then fell silent. 'Your silence can mean only one thing,' said Louis. 'You consider the comet signifies a change in the kingdom and the ruler's death.' This was Pliny's view, repeated by Bede and widely held. To which Astronomus replied by quoting Jeremiah: ' "Be not dismayed, as the heathen are, at the signs of heaven." '

The King reflected on this. More than most rulers, he had reason to fear revolution and as a man he tended to indecision. But he was not called Pious for nothing. 'We can never sufficiently wonder at God's mercy or praise it,' he finally said. 'For he deigns to warn us, lazy impenitent sinners, with such signs as this.' Louis ordered wine to be served, then he and his Court spent the night in prayer. In the morning he distributed alms to the poor and the clergy. So, having responded to the celestial warning, the King rode south to the Forest of Arden to hunt. Astronomus concludes his account of the comet: 'More game than usual are said to have been killed, and all the projects the King undertook at that time turned out successfully.' Louis was to live another two years and to die of natural causes.

Yet the sight of a comet continued to inspire terror, and two centuries later the reappearance of the comet King Louis had seen was to play its part in a decisive historical event, and even to be depicted by the ladies who embroidered the Bayeux Tapestry. At Bonneville sur Touques, or at Bayeux as the tapestry maintains, the enthroned Duke William of Normandy called his barons together to watch Harold, Earl of Wessex swear on two shrines containing bones of saints to aid William in securing the English throne. Harold then returned to his native land and upon the death of King Edward the Confessor accepted the crown of England.

'Then over all England,' says the *Anglo-Saxon Chronicle*, 'there was seen a sign in the skies such as had never been seen before. Some said it was the star "comet" which some call the star with hair.' It was indeed a comet, one of exceptional brightness, later identified by Edmund Halley and known since as Halley's comet. The tapestry depicts it like a modern space-ship, and six fearful men pointing it out. A courtier is shown hastening to tell Harold of the omen: in the tapestry border sinister, ghostly ships hint at coming disaster.

Whether the morale of Harold and his troops was deeply affected by the comet we can only surmise. But the tapestry, admittedly representing the Norman view, suggests that it was, and certainly points an implicit moral. Duke William landed in England, and at the battle of Hastings Harold was hewn down by one mighty blow from a broadsword. The comet had indeed portended a change of rule and the death of a king – one who had been guilty of sacrilege in defying the order imposed on reality by the bones of saints.

*

Apart from comets and eclipses – portents of pestilence – chroniclers of the Dark Ages have virtually nothing to tell us about celestial phenomena. They failed to record the very bright supernova, or exploding star, which appeared in the Bull in 1054 (leaving as its remains the Crab Nebula) and so impressed Court astrologers in Peking. Their silence is to be explained not, as is sometimes said, because they held Aristotle's view of the heavens as an unchanging realm where nothing new was possible (the return of Aristotle's view is described in the next chapter), but because they lacked any understanding of the stars' position in space, or their composition, or their movements, and so found them as unworthy of record as, say, the flashing of fireflies.

How did man of the Dark Ages picture Earth itself, the world as a whole, as distinct from portents? Isidore provided the model, again leaning on Pliny, but departing from the Roman in two important respects: he placed the centre of his habitable world in Jerusalem, and he gave the East pride of place, at the top of his map, because Paradise was believed to lie in the East. According to Isidore, the habitable world is surrounded by an O-shaped Ocean, so immense as to deter exploration and divided into three continents by a T-shaped body of water comprising Mediterranean, Red Sea and Black Sea. Hence the name, OT map, applied to early medieval representations of the globe. Even after mariners had begun to use the compass and hence to align landmarks on magnetic north, the eastern orientation prevailed. The Hereford Cathedral OT map of about 1280 has East at the top: it is advanced enough to depict men on skis in Norway, yet at the top, under the figure of Christ in Paradise, is a circular island, site of the Garden of Eden.

This eastern orientation of the world was very much in men's minds, and Hugh of Saint-Victor (d. 1141) was voicing a tradition when he wrote in his *Treatise on Noah's Ark* that civilization flows from east to west, even as the sun, and that when it reaches its westernmost limits – the Atlantic – the Day of Doom will occur. Oswald Spengler's *The Decline of the West* has a less original theme than its author supposed.

That 'flowing from east to west' found expression in the Romanesque church of around 1100. The church faced east – an old tradition – but east was identified with Paradise and man's beginning, while the western door, facing the setting sun, began to be decorated with the Last Judgment. The western doorway of the church of Conques, in southern France, begun in 1113, has the first great representation in sculpture of the elect and the damned. Christ is seated in front of his cross, to show he is a Saviour as well as Judge. On his right the elect approach, led by the Blessed Virgin, St Peter, and the pilgrim St James. Above their heads angels hold scrolls bearing the names of the virtues. On the other side are angels carrying book, censer, sword and lance, and beyond these are carved the tortures of the damned.

Since the cosmos, according to Isidore, was made in the image of the

Church, and since it was desirable for a church building to be a microcosm, Romanesque sculptors carved capitals with scenes from Scripture while architects extended the transepts to make them the arms of a cross, the nave and apse being the cross's upright. The great church of Santiago de Compostela, begun in 1070, was one of the first to have this cruciform ground-plan, and its cross shape became a model for the many new churches constructed in the twelfth century on the pilgrim routes leading from France to Santiago.

When men envisaged the world in the eleventh century, they were still uninterested in the sun, moon and stars save as emblems, and unaware of that third dimension we call astronomical space. The air was, so to speak, a void, and spiritual phenomena, real or imaginary, expanded to fill it. We speak today of Space-Time; at least some men of the Dark Ages, especially in northern latitudes, dwelled in what may be termed Spirit-Space, projecting on to space the saints and devils who preoccupied their imaginations.

One such man was Guibert, born in a château in the Beauvais region of France on 10 April 1053. Guibert was a sickly infant, not expected to live, and his mother promised, if he survived, to consecrate him to God. When the boy was eight months old, his father Evrard, a soldier, died, and there followed a predictably intense relationship between widowed mother and son. In time, honouring the vow made at his birth, Guibert entered the nearby monastery of Saint-Germer, and later was appointed Abbot of Nogent-sur-Coucy, in a pleasant well-watered setting where vines grew, though there are none today.

For his period Guibert was a sensible man. He considered that the Bible should be studied not allegorically but in order to furnish practical lessons, and he criticized the monks of Saint-Médard in Soissons for describing a relic as 'one of Christ's teeth' – Guibert held that it could not be, since Christ had risen and was now living in the flesh. In later life Guibert wrote a history of the First Crusade, but his most valuable work is his Autobiography, for what it reveals about men's presuppositions.

After the death of Evrard, Guibert tells us, his mother lived chastely at home but in time she declared a wish to become a nun. Her son and friends tried to dissuade her, quoting the tag, 'Let no prelate attempt to veil widows.' Then Guibert's mother, in a vision, met a beautiful lady of great authority, who offered her a costly dress, as though entrusting it to her for safekeeping, like a deposit to be repaid at the proper time. Guibert did not see the beautiful lady, but when he heard that his mother had seen her, he at once withdrew his opposition and allowed his mother to become a nun.

Here is another incident told by Guibert. A certain monk appointed to repair a road kept part of the money for himself and on his death-bed entrusted it to an infirmary servant. This servant hid the ill-gotten silver in his child's cradle. 'But at night when the child was put to bed, behold devils like little dogs leapt upon the child from the side and behind, beating on it here, there and all round, sometimes nipping it and making it cry out and weep. Being asked by both parents why it wept, it said it was being eaten up by little dogs.'

Guibert's mother heard about this. She questioned the child's father and obliged him to give up the money, whereupon, his conscience eased, he talked freely to her of other ways in which the dog-like devils had persecuted his child.

Another monk kept back alms for his own use. He fell ill of dysentery, declined to confess his sin, stretched out on his bed and there, says Guibert, 'was strangled by the Devil, as he lay on his back. You could see his chin and throat horribly flattened as though pressed violently down.'

We have to remember that Guibert was not a crank, and less credulous than most of his day, when we turn to his account of a certain Burgundian. This man kept a mistress but after some years he repented and decided to go on pilgrimage to the shrine of St James: Santiago de Compostela, where the great new Romanesque church was rising. But the man still harboured tender memories of his mistress and he took her girdle with him.

On the way the Burgundian met a stranger. 'Where are you going?' the man asked. 'To Santiago.' 'It is no good your going there,' the other said. 'I am that St James to whom you are hastening, but you have on your person something which is an insult to my majesty ... You dare to present yourself before me as though offering me the fruit of a good beginning, whereas in fact you are still wearing the girdle of that foul woman of yours.'

The pilgrim admitted it, whereupon the stranger claiming to be St James said, 'Cut off that member with which you have sinned, and then cut your throat.' He then disappeared.

The Burgundian had no doubt that the stranger, either real or more likely imagined, was St James, and we can only conjecture why. After five centuries of small-scale work the French were again carving recognizable figures, and presumably the Burgundian had seen a statue of St James, like the one above the west door at Conques, with his distinctive emblems: scallop shell, pilgrim hat and staff. Since he knew that space was occupied by saints and devils, it was not illogical for him to project his image of St James in such a way as to convince himself that he had actually met the apostle, and heard a command that was absolutely binding.

The Burgundian went to his lodging and did as he was bidden. 'Hearing the dying man's shriek and the splash of blood flowing, his friends awoke and bringing a light saw what had happened. They had Mass celebrated for his soul. After this it pleased God to heal the wound in his throat and restore the dead man to life.'

'I was brought before His throne,' the new Lazarus explained, 'in the presence of Our Lady the Mother of God, St James the Apostle also being there. When it was debated before God what was to be done with me, the Blessed Apostle, mindful of my intentions, sinner that I was and corrupt hitherto, prayed to that Blessed One on my behalf. She decreed that I, poor wretch, should be pardoned, inasmuch as the devil had contrived my ruin by masquerading as a saint. In order that I should amend my ways, and as a warning to others by God's command I came back to life.' His scarred and mutilated body guaranteed the truth of his account and served as a grim admonition. For men of the Dark Ages space was indeed, as Bede had declared, full of evil spirits.

*

Sunday, 14 July 1140 was a day of rejoicing in the Paris region, as King Louis VII laid the foundation stone of a new chevet, or series of apses, for the Benedictine abbey church of Saint-Denis. The man responsible for the plans was Abbot Suger, aged sixty, of whom a friend wrote: 'Small of body and family, constrained by twofold smallness,/ he refused, in his smallness, to be a small man.' Chief councillor to Louis VI and Louis VII, Regent of France during the Second Crusade, Suger was a hard-working self-made prelate with a head for business, an eye for beauty, a passion for gemstones, and a foible for perpetuating his name in verses affixed to work he had commissioned.

Suger had decided to build the apses because his Romanesque church had become too small and on important feasts women had been hurt in the crush round the exits. Suger's motive, in short, was utilitarian, and since except where jewels were concerned the abbot was careful with money, his flock assumed that costs would be kept to a minimum.

Yet four years later when the last stone of the chevet was cemented in place the people of Saint-Denis saw that the new building, in contrast to its Romanesque predecessors, had high piers, pointed arches and ribbed vaulting, so increasing considerably the height of the roof, and hence of the space within the apses. All this extra height and airy space called for much costly worked stone, yet it was of no practical use; why then had so hard-headed a businessman as Suger indulged in it? Furthermore why did this new style, which came to be called Gothic, spread so fast? Only twenty years later the cathedral of Notre Dame in Paris was to be begun in wholly Gothic style, with very high piers, and the whole centre of gravity raised from the choir stalls to shafts of sunlight beneath the pointed roof. Soon England and Germany, Italy and Spain, would be building cathedrals that no longer merely sheltered a congregation but were soaring enclosures of space.

Attempts have been made to explain Gothic in terms of money, technology, numerology, scholastic thought, but none satisfactorily answers that basic question why the roof was raised without necessity, and the amount of vertical space doubled, tripled, quadrupled.

The answer may well lie in an exciting book with a boring title which began to be influential in Suger's day: *A Commentary on Scipio's Dream*, by one Macrobius, a compiler, perhaps of African birth, who flourished about AD 400 and wrote in Latin. 'Scipio's Dream' is the closing section of a book by Cicero on political theory, and is written in imitation of the vision in the final pages of Plato's *Republic* described in chapter 1. The illustrious statesman-general Africanus appears to his grandson Scipio in a dream, wafts him to the stars and explains that it is there that he and all the heroically virtuous dwell after death.

Macrobius reproduced Cicero's text and added a commentary, parts of which bear on this quest. Using Greek sources, Macrobius stated that the stars shine with their own light, and each is larger than the Earth; he also attempted to assess the huge distances of space, saying for instance that the circumference of the sun's orbit is three and a half million miles.

Annotations on the manuscripts show that it was this part of Macrobius's commentary that most interested twelfth-century thinkers. One of them was Hugh, an Augustinian canon who taught in the influential school of Saint-Victor in Paris. In a book entitled *Practica Geometriae*, written before 1125, Hugh of Saint-Victor defined and discussed a new discipline, intermediate between geometry and astronomy, which he called *cosmimetria*. It is concerned, says Hugh, with the measurement of the dimensions of the terrestrial sphere and of the celestial sphere. He then gives actual figures for both, taken from the new data in Macrobius.

That was one way in which Macrobius's book proved influential – by suggesting that cosmic space could be measured. The second way was through Macrobius's Neo-Platonist interpretation of Scipio's dream. The key passage is in Book 1, section 9:

> Souls originate in the sky; moreover, this is the perfect wisdom of the soul, while it occupies a body, that it recognizes from what source it came ... The rulers of commonwealths and other wise men, by keeping in mind their origin, really live in the sky though they still cling to mortal bodies, and consequently have no difficulty, after leaving their bodies, in laying claim to the celestial seats which, one might say, they never left.

Suger's leading intellectual contemporaries, such as the poet Bernard Silvester and the philosopher William of Conches, are known to have fallen, like Hugh of Saint-Victor, under the influence of Macrobius's Neo-Platonism. Moreover, Suger kept abreast of the latest thinking, for he was praised as a man of letters at home in all subjects. Finally, since a great church was considered to be a replica in stone of the world created by God, a 'cosmic house', it was necessary that it should embody correct views of the cosmos.

These considerations may well have induced Suger to conceive a soaring high-roofed chevet, which would bring into his church a great measured chunk of luminous space, symbol of the higher, larger yet measurable cosmic space where the virtuous soul aspired one day to return. Though there is no statement by Suger himself on his architectural intentions, the following passage, where Suger describes his emotion when gazing at two jewelled ornaments in Saint-Denis, indicates the way in which the abbot conceived space:

> It seems to me that I see myself dwelling, as it were, in some strange region of the universe which neither exists entirely in the slime of the earth nor entirely in the purity of Heaven.

There is no need to labour the similarity of mood between this passage and Macrobius's call for wise men to 'live in the sky'.

To the further question why precisely Suger and his contemporaries responded to Macrobius, copies of whose book had lain about for a long time, the answer can only be speculative. One reason may be that the cathedral school of Chartres was making Platonism fashionable, another is that the long-held view of space peopled by devils had begun to weigh heavily on men's shoulders, and subconsciously they longed to throw it off.

Almost an illustration of this new approach is the statuary in the north porch of Chartres Cathedral, begun probably in 1194: the squat dumpy weighed-down men of Romanesque have been superseded by tall slim figures; even the shape of their heads has changed from round to long, and some of them are looking upward, just as worshippers within the cathedral now raised their eyes to the airy space beneath the nave's ribbed vault.

*

St Francis of Assisi (1182–1226) presents the last, and very appealing, world-picture of the Dark Ages. Its roots go far back in time, to a Roman Stoic, Claudius Aelianus (c. 170–235), who wrote a book to show the devotion, courage, self-sacrifice and gratitude of animals. A typical story concerns an elephant in India. Having discovered the wife of its keeper in the act of adultery, the elephant drove one tusk through the woman and the other through her lover, then left them lying dead amid the dishonoured coverings on the desecrated bed, 'so that when the keeper came he might note their sin and recognize his avenger.'

Aelianus's clever cooperative animals found their way into hagiography. According to Bede, St Cuthbert once spent the night in prayer up to his neck in the sea, at Coldingham. Then seals came and dried him with their fur and warmed his feet with their breath. Cuthbert blessed them and they returned to the sea.

Probably from such Lives and from the bestiaries, which also owed much to Aelianus, Francis of Assisi would have formed his view that animals are intelligent and, in a sense, man's brothers. We hear of him preaching to the birds and making a pact with the wolf of Gubbio, taking the wolf's paw and making him swear not to molest people of the town if they provided him with his keep.

Among the prayers of the Breviary in Francis's day was the so-called *Canticle of the Three Children*, dating from the first century BC.

The three children who have been ordered to be burned in a furnace by Nebuchadnezzar call on the diverse orders of creation to bless the Lord: the waters above the heaven, the sun and moon, the stars, lightning and so on, down to animals. This prayer, which was recited at Matins during Lent, would have been familiar to Francis. During a period of physical pain when he was suffering from an eye affliction the saint of Assisi amalgamated the thought underlying that prayer with his belief that all God's creatures were his brothers. The result was the *Canticle of the Sun*, of which these lines form the first half:

> Be praised, my Lord, with all thy works whate'er they be,
>   Our noble Brother Sun especially,
>   Whose brightness makes the light by which we see,
>   And he is fair and radiant, splendid and free,
>   A likeness and a type, Most High, of Thee.
>
> Be praised, my Lord, for Sister Moon and every Star
>   That thou hast formed to shine so clear from heaven afar.
>
> Be praised, my Lord, for Brother Wind and Air,
>   Breezes and clouds and weather foul or fair –
>   To every one that breathes Thou givest a share . . .

Unlike Suger, St Francis was not a scholar. He had no new scientific or philosophic insights into space. Yet, through his feeling for animals, he found his way to a fresh view of all created things. The sun was still partly an emblem of Christ, but it was also a brother; the moon still partly an emblem of the Church, but also a sister. They came into his ken in a new manner, and they shared in his affection. With the early Franciscans there is a marked decline in that fear of the air and of things in space which had begun with Gregory of Tours. A new mood of tenderness for nature, animate and inanimate, can be scented, and it has been preserved for us in the beautiful frescoes depicting the life of St Francis painted in the basilica of Assisi by Giotto of Florence seventy years after the saint's death.

CHAPTER FOUR

The Circles of Heaven and Hell

Gerard of Cremona finds and translates from the Arabic Ptolemy's Almagest
*– Ptolemy's mathematical cosmos – how it changed men's world-picture –
Dante's obsession with the stars –* The Divine Comedy *– Hell the centre
of the cosmos – Ptolemy the astrologer – influence of his* Tetrabiblos *–
Chaucer, civil servant, astronomer and astrologer – astrology in* The
Canterbury Tales *– Aquinas's cosmological argument for the existence of
God.*

In the tenth century fire swept Northern Italy, as town after town
was burned by Arab raiders based in Sicily, and by Magyar raiders
based in Hungary. Genoa in the west and Venice in the east built
navies to check pirates, but it was the arrival in 951 of the Saxon
Emperor Otto I – successor to Charlemagne's crown – and the presence
in Italy of Otto's son and grandson that effectively pacified Lombardy
and revived trade.

One of the towns to profit from that revival was Cremona. Standing
amid rice- and flax-growing fields on the banks of the Po, it was the
region's commercial crossroads. Cremona grew rich enough to assert
itself, in 1098, as a free, self-governing commune, owing merely nominal
allegiance to the Emperor. It had to defend itself against jealous and
stronger neighbours such as Milan, but emerged victorious and with
increased self-confidence.

In Cremona in the year 1124 there was born a remarkable man best
described as a discoverer, though what he discovered were forgotten
books and a forgotten cosmology. His name was Gerard – perhaps it
was given to him at the font of Cremona's new cathedral, begun seventeen years before his birth. While still in his teens Gerard became interested in books, not as most Cremonans understood that word –

volumes of civil and canon law that could justify a commune's claim to self-government over against Emperor and Pope – but in books bearing on science. From the few available in Cremona Gerard would have gathered that the Romans had been surpassed in science, especially in astronomy, by the Greeks, that the most esteemed of the Greeks was Ptolemy, who had lived in Alexandria in the second century AD, and that the most esteemed of Ptolemy's books was his *Mathematical Composition*, known familiarly and awesomely as *The Greatest*. It had been translated into Latin around AD 520 by the Roman statesman and author Boethius, but when Boethius was put to death by Theodoric the Ostrogoth, that translation had been immediately lost; had it survived, the Dark Ages' world-picture would have been very different.

Doubtless many Italians before Gerard had thought it would be a wonderful thing to find a copy of Ptolemy's book, but now, in rising Cremona, a new mood of adventure and self-confidence permeated the air, and Gerard, instead of merely sighing, decided to do something about it.

Where, in the early twelfth century, did one look for a Greek book? Alexandria was occupied by the Arabs; Athens a desolate shell: Byzantium was cut off by the schism between its Church and Rome. Gerard decided to travel to Paris, which had a reputation for learning. But once there he discovered that Parisians, like men in Cremona, were preoccupied with law, nor did he find his Ptolemy.

It was then that Gerard had a brilliant idea. Like Columbus after him, he decided that he would discover the east by travelling southwest. For in Spain, as the sea of Arab conquest ebbed, a tide-line of Arabic books remained, and some of these books might be translations of Greek authors.

Around 1144, now aged twenty, Gerard arrived in the sword- and silk-manufacturing town of Toledo. It had been reconquered by the Christians sixty years before, but was still Arabic-speaking. Once installed in Toledo Gerard would have met rivalry and suspicion. We know that in Arab-occupied Andalusia the supervisor of markets was asked to forbid 'the selling of Arabic books of science to the Christians' because, allegedly, Christian translators attributed these works 'to their bishops'. Nevertheless Gerard succeeded in obtaining copies of a number of books.

A word about Islamic science. With the riches of Greek knowledge at its disposal, it was well ahead of the West, notably in algebra, trigonometry and medicine, including herbal cures. Islamic cosmology set the Creator God of the Koran over a mainly Aristotelian universe; Islamic astronomy transmitted the work of the Alexandrians. This has left us many Arabic star names, such as Betelgeuse and Fomalhaut, the larger number of which are Greek names in an Oriental version. The last considerable Islamic astronomer had been a Persian, Omar Khayyam, engaged in 1074 by the Shah to correct the calendar. Omar Khayyam was also an algebraist, who did pioneer work on cubic equations, a freethinker and a poet. His *Rubáiyát*, in which he denounced with passionate pessimism a malevolent and inexorable fate controlling men's lives through the stars, was to be turned into English in the nineteenth century and widely read.

Having obtained books in Arabic, Gerard came to the most arduous part of his task. With no grammar or dictionary, he had to render into Latin sentences with a different structure and words for which at the time there were no exact equivalents. But with the help of a Christian Arab named Galippus, Gerard translated more than seventy books from Arabic into Latin which, if inelegant, was always intelligible. These included such key mathematical and scientific works as Euclid's *Elements*, Aristotle's *De caelo*, and the first three books of his *Meteorologica*, Al Farabi's *Commentary on Aristotle's Physics*, and a compendium of Ptolemy's cosmology by Al Farghani, better known as Alfraganus.

Most important of all, Gerard realized his ambition by finding a copy of Ptolemy's *Mathematical Composition*, which the Arabs called *Almagest*. A work containing many measurements in degrees, minutes and seconds, it would have been particularly hard to translate, because the West had not yet adopted Arabic numerals, yet Gerard managed to turn it, with his habitual scrupulous accuracy, into Latin, probably in 1175. He did not sign it, though, or any of his translations.

For forty years Gerard toiled at his self-imposed task of giving back Greek thought to the West. He died in Toledo at sixty-three and, modest in death as in life, was buried there in an unmarked grave.

About the Greek mathematical astronomer whom Gerard had restored to life we know just a little. Ptolemy, according to the Arabs, was abstemious, rode much on horseback, suffered from bad breath and dressed sprucely – the latter one can well believe since his book is a model of elegant reasoning. He had a head start over earlier astronomers in that he possessed the armillary astrolabe, an instrument for accurately measuring the position of the moon and planets. Ptolemy made many such measurements, noting them in degrees of latitude and longitude, and he culled many more from records.

Ptolemy accepted Aristotle's view that the celestial orbs were made of aether, the natural property of which was to move round the Earth in circles of unvarying speed. Ptolemy's dilemma began when his measurements seemed to suggest that sun, moon and planets moved with speeds that varied according to their positions in their orbits.

Ptolemy wrote the *Almagest* in order to reconcile his measurements with Aristotle's world-picture. He did this by devising a system of cunningly positioned wheels within wheels, whereby a subordinate wheel moving with its own speed plus the speed of the wheel on which it depended could account for a seeming increase in velocity. Thus, according to Ptolemy, the moon is held to move on an epicycle which is carried round on a deferent circle; the centre of this deferent circle, however, is displaced from the Earth itself. The centre of the epicycle moves around the deferent with a motion that is uniform not with respect to its centre, but to a third point, called equant.

By means of 40 circles and epicycles Ptolemy was able to account for all known astronomical measurements in terms of uniform circular movement, and to predict accurately the future positions of sun, moon and planets. He made everything intermesh harmoniously. It was a mathematical *tour de force*.

Ptolemy mainly concerned himself with positions and angles. But he did give distances for the sun and for the moon. When the moon lies in a straight line with the sun and the Earth, 'if the Earth's radius is taken as 1, we have concluded that the moon's mean distance is 59, and the sun's 1210.' Since the Earth's radius was held to be 5,200 kilometres, Ptolemy was saying that the Earth is slightly more than 6 million kilometres from the sun. This is only about one-twentieth of the correct value, but much nearer the truth than Pliny's figure of a mere 24,000 kilometres.

Ptolemy's model of the cosmos has been likened to a Ferris wheel. That comparison helps us to picture the motions but a mechanistic simile is not really appropriate, first because Ptolemy limited himself to mathematics and eschewed physics, and secondly because he once confessed, in an aside, to an unmechanistic view of the heavens: 'Each planet possesses for itself a vital force and moves itself.'

In more general terms Ptolemy, true to Greek tradition, claimed that mathematics alone yields unshakeable knowledge, and that mathematical astronomy helps men not merely to obtain such knowledge, but to improve their characters: 'where the sameness, good order, proportion and freedom from arrogance of heavenly things are being contemplated, this study makes those who follow it lovers of this divine beauty and instils – as it were makes natural – the same condition in their soul.'

It is not difficult to imagine the effect of such a book when it became known to the West in Gerard of Cremona's translation in the second half of the twelfth century. A whole new cosmos swept into being, with a new dimension of space. From being merely a shallow stage for portents, the sky took on immensity. The heavenly bodies, blacked out or blurred for so many centuries, were now seen to be the most orderly and harmonious objects in nature. The position of Mars next April, of Venus next May could be calculated according to Ptolemy's tables to within a few minutes. And to gaze at these mathematical marvels was not only interesting and useful but ennobling, almost a religious experience.

Here is Ristoro of Arezzo, an obscure cosmographer with a taste for antique Etruscan vases, brought up on Ptolemy's world-picture and writing in 1282: 'It would be a shame to live in a house and not know how it is built . . . not to consider the use of the wooden beams used in its construction. Man, with his upright posture and head held high, was designed by his Creator to look and listen, to know and understand this marvellous universe, and especially that noblest part of it above him: the heavens and their wonderful movements.'

The note of excitement and curiosity is unmistakable, but only with Dante do we encounter these in terms of a knowable human being. Dante Alighieri was born in 1265 in Florence, a medieval town of narrow streets enclosed by a high brick wall that was both the model and reflection of the pre-Ptolemaic world-picture. Here men slept in beds enclosed by curtains and artists painted in two dimensions; perspective had not been invented because it was not required.

Dante came of a family of money-lenders. In one of the local wars whereby Florence, like Cremona, asserted its independence, he served as a cavalryman. He learned the craft of writing poetry in the complex genres of the day, studied philosophy and read, perhaps in Gerard's translation, Alfraganus's compendium for the general reader of Ptolemy's world-picture.

Its effect on Dante was immediate. Of his early short poems nearly a quarter contain some reference to sun, moon or stars, planets or spheres, while the very first words of the *Vita Nuova*, the story of his love for Beatrice, allude to the circling spheres of the sun and of the stars.

Dante took an active part in local politics as a member of the party favouring close ties with the Emperor. In 1302, while he was absent on an embassy, the opposing party, which favoured ties with the Pope, came to power, arraigned Dante on what seems to have been a trumped-up charge of embezzling public funds, and condemned him to be burned alive should he ever return to Florence.

For the remaining nineteen years of his life Dante lived separated from the city he loved, his friends, his wife and home. Exile obliged him to take stock of himself in a larger context than the streets of Florence, and it was now that the Ptolemaic world-picture proved specially influential. The harmony and mathematical precision he knew regulated the sky Dante tended to look for also on Earth. He had become Greek in his thinking, by which I mean that it was not the unusual or startling that interested him but order, and in the misery of exile, like Einstein later, his first reaction was to look for a wider order to compensate for the disorder of which he had been the victim, and to express that order in the measured cadences of vernacular verse.

The upshot of such reflections was Dante's decision to depict justice on a cosmic scale in the form of a space journey which would also be a journey to the next life. In that journey the stars would symbolize the cosmic order: indeed Dante chose to end each of its three Books with the word *stelle*. The oddities which had fascinated an earlier epoch Dante would relegate to the margin: comets he mentions only twice, shooting stars twice, eclipses seven times: but there would be scores of allusions to the rhythmical progression of sun and stars, moon and planets, visible signs and guarantees of universal order.

Concurrently with Ptolemy, Dante had studied Aristotle, newly translated into Latin direct from the Greek, and adopted his view that there is a fundamental difference of composition between the spheres, stars and planets, which are composed of aether, and the Earth – small, dark and inert – made of the basest material and sunk, like dregs, to the bottom of the universe. Earth had re-entered the solar system, and that was thrilling, but compared to its superior neighbours it was a paltry thing.

According to Aristotle, high was noble, and the further down you went so imperfection increased. As you could easily prove by throwing a stone down a well, an object continued to fall even within the Earth, so it was logical for men of the Middle Ages to conclude that the place towards which everything tends to fall is Hell. A diagram at the end of *Image du Monde*, a popular cosmology in French verse written in 1246, shows a series of concentric spheres converging towards the Earth and, right in the middle of the diagram, at the centre of the Earth, is Hell. The Middle Ages, usually termed geocentric, would more accurately be described as 'Haidocentric'.

Dante not only subscribed to this view, but made the cosmological distinction between Heaven, Earth and Hell the basis of *The Divine Comedy*. Near a dark wood, on Good Friday 1300, Dante meets Virgil, and is led by him to the underworld. Hell is a conical cavity reaching to the centre of the Earth, around the sloping sides of which places of punishment are arranged in circles, so that the worst sinners are placed nearest to the apex of the cone. At the apex, in a frozen waste, dwells Satan, a bat-winged giant with three heads – a hideous travesty of the Trinity.

The travellers continue upwards on the other side. Dante turns and takes a last look at Satan, now upside down, while Virgil pants and strains in order to overcome gravity, believed to be much stronger at the Earth's centre than on the surface.

They arrive in Purgatory, a large conical hill rising out of the Ocean at a point opposite Jerusalem, centre of the habitable world. Souls dwell in seven terraces according to the seriousness of the sins they are expiating. At the top of the hill, inaccessible to fallen man, is the earthly paradise, where Dante is surprised to find that his brows are caressed by a breeze from the east. He is told that this is no wind, but the revolution of the atmosphere as it uninterruptedly follows the movement of the celestial spheres.

Beatrice had died in 1290 and, Dante felt sure, had gone to join the Blessed. At the beginning of the *Paradiso* she appears to Dante to guide him through the celestial spheres. All the redeemed dwell with God in the tenth Heaven, the Empyrean, but they enjoy different levels of blessedness, and in the course of the poem will manifest themselves to Dante in the sphere corresponding to their level.

The nature of a spirit being to rise, just as fire rises, Dante travels upwards into space as a disembodied spirit, swift as an arrow that strikes its mark before the cord has ceased to quiver. First he visits the sphere of the moon, then those of Mercury, Venus, the sun, Mars, Jupiter and Saturn. Here he mounts a mysterious golden ladder and finds himself in the heaven of the fixed stars, in the constellation of Gemini, his birth-sign. He looks down on the planetary spheres and sees in the distance the poor little disc of the Earth, which he compares, in Seneca's simile, to 'a threshing floor'.

Dante next ascends to the Primum Mobile, a ninth sphere added by the Arabs to Aristotle's eight in order to explain the diurnal motion of all the lower spheres. He sees the nine hierarchies of angels who direct the nine moving spheres, before soaring to the Empyrean, the tenth heaven which had been added in the previous century by Catholic theologians. Here Beatrice returns to her place among the blessed spirits who form the Celestial Rose, and in the last line of the poem Dante is encouraged by St Bernard to lift his eyes to *'l'amor che move il sole e l'altre stelle'*.

Leaving aside its literary genius, we have in Dante's poem a fresco of how an educated man of the early fourteenth century saw the world and life. Paramount is the fact, evident in that last Aristotelian line, that science and religion have joined hands. The cosmology of the scientists is the apt and harmonious setting for the truths of Revelation. There is no jarring note nor does the author have to avert his eyes from potentially embarrassing questions. Everything is answered, and in place; there is a satisfying unity more complete than any hitherto attained.

Yet far from being self-satisfied, this medieval world-picture is shot through with pathos, a quality absent from literature since Euripides. The reader pities these men who wish, too late, that they had made something better of their lives, and he winces too at the irony – the reverse of the irony in Homer but no less poignant – between God's high hopes for man and some men's failure to realize that homely image from the *Purgatorio*: 'We are born larvae and we must become angelic butterflies.'

A world-picture of graded spheres inevitably left its mark on the structure of society. The cosmos was seen as a divinely made guarantee of social hierarchies. Dante more than once compares the moon to the Emperor and the sun to the Pope. He means that both are divinely appointed, but independent, and in administering parallel forms of justice, civil law and canon law, they are upholding terrestrial versions of the order observable in the cosmos. Again, the completeness of the scheme is remarkable, though with hindsight we see not only the astronomical mistake on which the simile is based but the scheme's limitations and its scope for abuse.

On the purely debit side, even when allowance has been made for Dante's nagging sense of having been wronged, the world-picture tends to make the cosmos too exclusively a court of justice, and life a matter of future reward or punishment. The whole human race has, so to speak, been rounded up, marked with an appropriate brand and corralled. For medieval man, though he has found space again and the heavenly bodies, has found them in the form of spheres that are graded – a relic of Babylonian belief – and these cast an unwelcome shadow of status even in Paradise.

There is a related aspect. As a dweller in the lowest sphere, medieval man felt a sense of inferiority, not just spatial but psychological and moral, for he lived subject to a downdrag, the gravitational pull of Hell. This may well have induced in Dante and many of his readers a moral outlook to say the least dispiriting.

*

Quite a different but no less massive change took place in people's world-picture at the beginning of the Middle Ages, and this too resulted from the translation of a book by Ptolemy. Living where and when he did, and not being a Christian, Ptolemy believed firmly in astrology, and he left a detailed exposition of his views in the *Tetrabiblos, Four Books on the Influence of the Stars*, which was translated into Latin in 1138.

Ptolemy associates each planet with one or more of the elemental qualities: hot, cold, dry and moist. Heat and moisture are considered vivifying – fair enough, for one living near the Nile; cold and drought detrimental. Because its large surface catches many vapours from the neighbouring Earth, he considers Venus to be moist. He also associates each planet with one or more signs of the zodiac, and with a part of the human body: Venus, for instance, with Libra and Taurus, and with the abdomen.

Ptolemy characterizes each sign of the zodiac as masculine or feminine, mobile or stable, commanding or obeying, and so on. He then explains that every person's disposition is influenced by the qualities of the sign just above the eastern horizon at the hour of his birth, and of the planet within this sign.

There is more. Ptolemy was an expert geographer – the first to place the globe of the world in the 'string-bag' of lines of longitude and latitude – and he applied astrology to whole regions. He divided the

Earth into quadrants, each under the special influence of a third part of the zodiac. Britain, France and Germany are under the influence of Aries, Leo and Sagittarius, hence:

> They are independent, liberty-loving, fond of arms, industrious, very war-like, with qualities of leadership, cleanly and magnanimous. However, because of the occidental aspect of Jupiter and Mars, and furthermore because the first parts of the aforesaid three signs are masculine and the latter parts feminine, they are without passion for women and look down upon the pleasures of love, but are better satisfied with and more desirous of association with men. And they do not regard the act as a disgrace to the paramour, nor indeed do they actually become effeminate and soft thereby, because their disposition is not perverted, but they retain in their souls manliness, helpfulness, good faith, love of kinsmen, and benevolence.

Much of Western Europe homosexual – a somewhat sweeping generalization! But in fairness to Ptolemy it should be remembered that Hellenistic and Roman travellers had long characterized the Celts as homosexuals. So the Alexandrian was here doing little more than extending and adding substance to their view by linking it to stellar influence.

Ptolemy ends the *Tetrabiblos* by explaining how to predict from the stars a person's career, marriage, progeny, friends and enemies, even the manner of his death.

This detailed exposition of astrology by so eminent a mathematical astronomer had the most powerful effect on men of the thirteenth and fourteenth centuries. Some accepted it in its entirety, among them the Franciscan Roger Bacon, the founder of experimental science. Others were more cautious, particularly theologians of the Catholic Church who, like their counterparts in Islam, saw that it at least gravely curtailed free will. What may be called the orthodox Catholic view was stated by St Thomas Aquinas:

> The majority of men are governed by their passions, which are dependent upon bodily appetites; in these the influence of the stars is clearly felt. Few indeed are the wise who are capable of resisting their animal instincts. Astrologers therefore are able to foretell the truth in the majority of cases, especially when they undertake general predictions. In particular predictions they do not attain certainty, for nothing prevents a man from resisting the dictates of his lower faculties. Wherefore the astrologers themselves say that 'the wise man rules the stars' inasmuch as he rules his own passions.

That was approximately the view of Dante. He accepted that the stars have a general influence on men. He compared the stars to hammers, and the Earth to metal; and, again, to seals and the Earth to wax. Venus was 'the star of love', 'the fair planet which incited to love', and which exercised a real power over his own heart. But Dante had no time for technical astrology, not because he disbelieved in it but because he thought it impious to peer too deeply into the future, and he placed in the fourth circle of Hell the astrologer Aruns, who from his solitary cell had watched the stars and prophesied the victories of Caesar over Pompey.

As the fourteenth century advanced, if literature is a guide, people adopted a less cautious attitude than Dante. Many knew and practised astrological techniques, while a majority probably believed that every turning-point in their lives depended, wholly or in part, on the stars. Hence a very important change in the world-picture. The very shallow space of the Dark Ages had already been vastly increased in height by Aristotelian–Ptolemaic cosmology; now, to fill it, largely replacing saints and angels, came rays from distant planets and stars.

When an educated man of the fourteenth century looked at the night sky, he would have been aware of the orbs circling on 40 large and small spheres; equally he would have been aware of the position of each visible planet within the zodiac, and its favourable or unfavourable effects on him and his family. Even during the day he would have had such thoughts in mind, for the decorations in books showed signs of the zodiac and the planets, as did the ceilings of noble houses, and even, in Florence, the pavement round the baptismal font. If a soldier and an individualist, such a man might have seen himself, as in some continuous cosmic tournament, astride his horse, lance couched, tilting at the stars, fighting to make his life his own.

Such a conception of space naturally led to a new conception of time. Men of the Dark Ages, ignoring the stars and hence their influences, had lived in the present; now men began to live partly in the future, with all that that entails in anxiety and hope.

We have a good example of such attitudes in Geoffrey Chaucer, who was born in London about 1340 of a well-to-do family in the wine trade. After training in the law Chaucer entered the King's service, becoming controller of customs, and clerk of the King's works. He married Philippa, daughter of a Flemish knight, and had three children. As a young married man he lived on the second floor of Aldgate, one of

London's four gates, where his library contained sixty books, a large number before the invention of printing. In a portrait by his friend Hoccleve he is a portly man with a heavy nose and dark, hooded eyes, wearing his grey hair rather short, and with a moustache and small forked beard. He is dressed in a conservatively cut black hood and gown.

Like Dante, Chaucer thrilled to 'the new space', and in *The House of Fame*, which he wrote in 1383, he imagines himself seized in the claws of a golden eagle and carried through space to the region of the stars, whence he looks down on a 'pin-prick' Earth, and sees the Milky Way, which was now becoming so familiar to Englishmen that they referred to it, after their most bustling highway, as Watling Street.

When his son Lewis was ten Chaucer wrote for him a *Treatise on the Astrolabe*. The astrolabe is a circular star map which can be rotated about its north pole, resting on a plan of the sky as seen by an observer at some given latitude; though it could serve for observation, it was used mainly to calculate the positions of the brightest stars. Chaucer's little book, based on the Latin version of an Arabic treatise, is a lucid explanation of the instrument and has been called the earliest genuinely scientific work in English.

Chaucer is also probably the author of *The Equatorie of the Planets*, which describes the construction and use of another medieval instrument, the equatorium. Larger and more complicated than the astrolabe – Chaucer's calls for a main metal or wooden plate six feet in diameter – the equatorium was used for calculating the positions of the planets at a given moment; the unique manuscript, in Peterhouse, Cambridge, is thought to be in Chaucer's handwriting.

If the second booklet is indeed by him, the two together show that Chaucer possessed a sound knowledge of Ptolemaic cosmology and of the two instruments most in use by astronomers and astrologers. That he also mastered Ptolemaic astrology is clear from *The Canterbury Tales*. Chaucer once stated that he did not believe in 'judicial astrology' – the governance of State affairs by the heavenly orbs; whether or not he believed in other branches of astrology is a matter of dispute, but he certainly assumed such a belief in readers of *The Canterbury Tales*.

Memorable among Chaucer's Canterbury pilgrims is Dame Alice, the red-faced wife of Bath, in her broad hat, scarlet stockings and outer riding skirt wrapped around her ample hips. Lusty and life-loving, dishonest and a scold, she considers marriage a fine institution provided the wife has the upper hand. She has gone through five husbands, not counting 'other company' in her youth, and still has an eye for a shapely male leg.

The Wife of Bath explains such characteristics by the fact that she was born under Taurus, with Venus in the ascendant, in conjunction with Mars: 'For Venus sent me feeling from the stars/And my heart's boldness came to me from Mars.'

Red spots and moles on the skin were believed to be 'marks' of one's ascendant sign and dominant star. Chaucer tells us that Dame Alice bore the marks both of Venus and of Mars, the latter on her face, 'And also in another private place.'

If the planets at birth affect our disposition, throughout our lives they also control events. In another of Chaucer's works, *Troilus and Criseyde*, the heroine first sees Troilus when Venus is in her own 'house' (a particular zone on the sky), with no planet in bad aspect, hence Criseyde's love for Troilus is to dominate most of her life. When at the end of the poem she surprisingly betrays Troilus and takes Diomede for her lover, Chaucer explains that this too was written in her stars, though without particularizing.

Planetary influence extends even to days of the week; Chaucer gives an example in 'The Knight's Tale', which tells how two Theban knights, Palamon and Arcite, are imprisoned in a tower by their conqueror, Duke Theseus; they see Theseus's sister Emily walking in a garden below, are captivated by her beauty and both fall in love with her; that is, they come under the influence of Venus. As the knights' hopes rise and fall, Arcite recalls Venus's characteristics:

> Now up, now down, like buckets in a well,
> Just like a Friday morning, truth to tell,
> Shining one moment and then raining fast.
> So changey Venus loves to overcast
> The hearts of all her folk; she, like her day,
> Friday, is changeable. And so are they.
> Seldom is Friday like the rest of the week.

Arcite's horoscope is such that he is pursued by the enmity of Saturn, most fearsome of the planets, who thus describes his power over men:

> Mine is the prisoner in the darkling pit,
> Mine are both neck and noose that strangles it,
> Mine the rebellion of the serfs astir,
> The murmurings, the privy poisoner;

And I do vengeance, I send punishment,
And when I am in Leo it is sent.
Mine is the ruin of the lofty hall,
The falling down of tower and of wall
On carpenter and mason, I their killer.
'Twas I slew Samson when he shook the pillar ...

Palamon and Arcite fight for Emily's hand in a tournament at sunset on Tuesday, and this alas is Saturn's hour. A flash of fire from the Earth causes Arcite's horse to start and throw its rider, who injures his lungs. Saturn's special power over the body is to block up the humours normally expelled by breathing or perspiration, and he now causes the accumulation of hot and dry humours in Arcite's body. The unfortunate knight dies, and it is Palamon who wins Emily.

If the rays of one of the heavenly bodies could cause a man's death, it followed that these rays might be foiled, just as a poison is by the appropriate antidote. In 1388 Sir Robert Tresilian, Chief Justice of England, was convicted of treason, went into hiding and, having been caught, was dragged to the gallows. But Tresilian appeared unperturbed. ' "So long as I wear certain things about me," he asserted, "I cannot die." Immediately they stripped him and found certain charms – *experimenta* – and certain signs painted upon them, after the fashion of signs of the Zodiac, and one demon's head painted, and many names of demons were written. They were taken away, and he was hanged naked ...'

Chaucer's contemporaries believed that planetary influence extended to plants, and on this subject consulted the *Book of Secrets*, written in Germany about 1300. Venus's special plant is Peristerion, called Verbena. Its root heals swine-pox, cuts and haemorrhoids. It makes the breath sweet; 'it is also of great strength in venereal pastimes, that is, the act of generation.' 'Infants bearing it shall be very apt to learn, and loving learning, they shall be glad and joyous. It is also profitable, being put in purgations, and it putteth aback devils.' But it has to be gathered from the 23rd to the 30th of the month, 'and in gathering make mention of the passion or grief, and the name of the thing for the which thou dost gather it.'

From these few examples one realizes how complicated life would have been for a believer in astrology. He had to attend to the planetary associations of almost every living thing he encountered, and if a family man his concern would have extended also to celestial influences on his wife and his children.

For some the complication was even greater. Metals at this time were believed to be living things, and it was held that they too had been fashioned under the influence of heavenly bodies. Each was designated by its planetary sign: silver by a waning moon, copper by a circle with a cross underneath. A list of metals is given by Chaucer:

> Gold for the sun and silver for the moon,
> Iron for Mars and quicksilver in tune
> With Mercury, lead which prefigures Saturn
> And tin for Jupiter, Copper takes the pattern
> Of Venus if you please!

Now in Alexandria the Gnostics had made much of Aristotle's dictum that underlying the four elements there exists a basic matter, and in due course certain Arab thinkers interpreted this to mean that there exists a pure matter – 'the philosopher's stone' – of which the base metals are degenerate forms. If this philosopher's stone could be made in the laboratory, it would have power to transmute degenerate metals into gold or into silver, depending on whether its colour was red or white. The lore of making the stone was called al-kimia, the lore of Egypt, which became our world alchemy.

In 'The Canon Yeoman's Tale' Chaucer tells how a good priest was tricked by a wicked canon into paying £40 for a worthless alchemical formula. In the course of it he ridicules the hard-worked, frustrated yeoman, or servant, with his stocking cap, who for seven years has toiled in the laboratory of the canon he hates, conducting experiments which sometimes caused the metals to burst from the retort and 'leap into the roof'.

> As for proportions, why should I rattle on
> About the substances we worked upon,
> The six or seven ounces it may be
> Of silver, or some other quantity,
> Or bother to name the things that we were piling
> Like orpiment, burnt bones and iron filing
> Ground into finest powder, all the lot,
> Or how we poured them in an earthen pot?

Though Chaucer makes fun of alchemy, the principle of transmutation of metals, with or without reference to their planetary lords, seemed quite feasible to medieval men on the analogy of another well-known process involving a furnace, which Chaucer describes in his translation of the French poem *Romance of the Rose*:

> ... Behold we not
> What different form the fern hath got
> When 'tis by fire to ash reduced,
> And straightway thence clear glass produced
> By depuration, as we learn?
> And yet we know glass is not fern,
> And none would say that fern is glass.

Alchemy survived Chaucer's laughter in *The Canterbury Tales*, and alchemical practices were flourishing so much in England in the fifteenth century that they had to be forbidden by law. Even so, they continued clandestinely, and there was doubtless more to it than the desire to get rich. The changes he could effect in his furnace were among the very few of which medieval man was capable and doubtless gave him a sense of power, while the eventual transmutation of base metals into gold would be one means of outwitting those Big Brothers and Sisters, the planets. It is just as well he did continue with his earthen pots and glass vases, his sophic salt and sophic mercury, for the alchemist in due time was to become the chemist, Chaucer's canon Robert Boyle. As for the designation of metals by their planetary signs, this was to continue down until 1803, when John Dalton would tabulate metals and other known elements by their atomic weight.

One other aspect of medieval cosmology remains to be discussed. In the new universities theologians began to think not of what the stars tell us about men, but of what they tell us about God.

This line of thought goes back to the brass planetarium made by Archimedes, which Cicero saw in Rome and described in *De Natura Deorum*. If that planetarium, asks Cicero, were taken to barbarous regions like Scythia (north Asia) or Britain, who would doubt that it had been fashioned by an intelligent being? All the more reason, therefore, to believe that the heavens of which the planetarium is a model were fashioned by an intelligent Designer.

[83]

The rediscovery of Ptolemy's mathematical model of the cosmos, in-spiring even more awe than Archimedes's planetarium had done, and the need to supplement faith with reasonable arguments for the exist-ence of God led thirteenth-century theologians to perfect the cosmo-logical argument, as it is called. Aquinas stated it in both his *Summa Theologica* and his *Summa Contra Gentiles*. Here it is in a modern ex-ponent's summary:

> We observe material things of very different types co-operating in such a way as to produce and maintain a relatively stable world-order or system. They achieve an 'end', the production and maintenance of a cosmic order. But non-intelligent material things certainly do not co-operate consciously in view of a purpose. If it is said that they co-operate in the realization of an end or purpose, this does not mean that they intend the realization of this order in a manner analogous to that in which a man can act con-sciously with a view to the achievement of a purpose. Nor, when Aquinas talks about operating 'for an end' in this connection, is he thinking of the utility of certain things to the human race. He is not saying, for example, that grass grows to feed the sheep and that sheep exist in order that human beings should have food and clothing. It is of the unconscious co-operation of different kinds of material things in the production and main-tenance of a relatively stable cosmic system that he is thinking, not of the benefits accruing to us from our use of certain objects. And his argument is that this co-operation on the part of heterogeneous material things clearly points to the existence of an extrinsic intelligent author of this co-operation, who operates with an end in view.

The cosmological argument may be envisaged as one of the flying buttresses supporting the Gothic cathedral of thirteenth- and fourteenth-century faith. It was to continue to be influential, and disputed, right down to modern times, when part of it came to be paralleled by the ecological argument.

Though there are naturally differences of emphasis between Aquinas, Dante and Chaucer, they and their contemporaries shared a world-picture which is probably the most orderly to have been devised before or since. Certainly it was one of the most satisfying to the various aspects of man's personality. It satisfied his scientific knowledge, his religious views, his aesthetic sense, his kinship with other living things, his need to feel at home in nature. Perhaps it made him feel a shade too de-pendent on the stars, and made the slopes of Hell a shade too steep. But by and large it convinced and it was modified only in detail – by

Nicholas of Cusa, for instance, who claimed that there are other in-habited planets. Its very satisfyingness is probably why no one thought fit to question its basis – the fundamental assumption that seemed beyond challenge: that the sun goes round the Earth.

CHAPTER FIVE

Wayward Explorers

New interests in Florence – Leonardo, his character, gifts and ambitions – first flying machines – Columbus, his character and sense of religious mission – navigation by stars and sun – believed to have come from the sky – tricks Jamaicans with eclipse – the Magellan-Elcano circumnavigation – the Earth's configuration now known in essentials – Cortes discovers in Mexico a wholly different cosmology – Cardano, a typical Renaissance polymath – his view of the cosmos – weight of astrology heavier than ever – moon journeys – discussion about comets – Copernicus, his character as a key factor in his theory – Revolutions of the Heavenly Spheres *– nature of the 'Copernican revolution'.*

In the centre of the floor of the north transept in Florence's Cathedral is a gnomon, the largest and most exact sundial of its period. It was constructed, probably in or about 1468, by the Florentine astronomer and mathematician Paolo Toscanelli, and each year on 21 June beams from the sun at noon cast an exactly designated shadow on the brass dial, these beams coming from a conical opening in the lantern of the very high cupola, which Brunelleschi, using rediscovered Roman techniques, had built in 1434. This giant astronomical instrument within a place of worship typifies a new interest in the cosmos which arose in fifteenth-century Florence.

Why did it happen there and then? In a very complex answer three factors call for mention. Banking and production of excellent scarlet cloth had put money in the pockets of non-military laymen, whose intellectual interests naturally differed from those of priests; a republican government fostered civic humanism and acted as an incentive to individual effort; the rediscovery of classical authors revealed undreamed-of knowledge and theories, stimulating in particular a mathematical approach to reality in place of Aristotle's qualitative approach ('A body falls to the ground because it has a longing for rest').

When Toscanelli made his gnomon probably the most gifted of these enterprising Florentines was a sixteen-year-old apprentice in Verrocchio's painting studio, by name Leonardo da Vinci. Here he learned the technique of perspective, a recent Florentine discovery, whereby the way the size of objects recedes in space can be determined by geometry. Leonardo happened to be endowed with a very strong physique, and perhaps for that reason became more interested in making things than in depicting them. He got hold of the projects of the hoisting machines used by Brunelleschi in the construction of the cupola and used them as the basis of his own juvenile drawings of machines. At thirty he took service with the Duke of Milan as a military engineer and began to work out a cosmology.

To his cosmological thinking Leonardo brought certain peculiarities of temperament. One was a subconscious abhorrence of the natural processes of generation and reproduction. In a remarkable fable he describes a nut falling into the crevice of a great belfry and burgeoning there; its thickening roots tear the walls asunder until a great part of the belfry collapses. In another note he shows fear of sexual activity: 'The spider, thinking to find repose within the keyhole, finds death.' He disbelieved in the stars' influence on man but expressed it thus: 'All the astrologers will be castrated.'

With an abhorrence of normal sexual activity went a fascination with water. Leonardo never tired of making drawings of foaming waves and swirling eddies – as two others of homosexual temperament, John Ruskin and Gerard Manley Hopkins, were also to do. He clung to the old-fashioned view that the whole surface of the moon is covered with a sea. He was particularly interested by reflections of the sun and moon in water, and by mirror-images generally. He used mirror-writing for his Notebooks.

Leonardo was no less interested by the movement of light in space from distant objects to the retina. In a passage of rare lyricism he wrote that 'the image of the moon in the east and the image of the sun in the west at this natural point become united and blended together with our hemisphere ... Who would believe that so small a space could contain the images of all the universe? O mighty process! What talent can avail to penetrate a nature such as these? What tongue will it be that can unfold so great a wonder? Truly, none!' Medieval thinkers had sometimes marvelled at God's stars, but to Leonardo even more marvellous was man's eye, that could contain them.

From space Leonardo moved to a consideration of the air and wind, taking particular notice of all that ran counter to the usual direction. 'Slant' is a word that recurs, and he was fascinated by the fact that birds fly not *with* but *against* the wind. 'When the hand of the swimmer strikes and presses upon the water it makes his body glide away with a contrary movement; so it is also with the wing of the bird in the air.' 'A bird,' he concluded, 'is an instrument working according to mathematical law,' and having studied that law Leonardo went on to ask a question which perhaps only the inverted cast of his genius allowed him to ask.

Man is an Earth-bound creature: so taught the early and medieval Church, and made the point in a familiar sermon variously ascribed to Augustine, Ambrose and Maximus. The sermon tells how Simon the Magus, who figures in the *Acts of the Apostles*, on one occasion claimed to be Christ and asserted that he could ascend to the Father by flying. Using a magic that came from Satan, he suddenly raised himself into the air and began to fly. Then Peter began to pray and 'brought him down like a captive from high in the air, so that Simon fell precipitately upon a rock and broke his legs.' The fall was depicted in many churches, and its message was plain: Earth-bound man could fly only with Satan's help, and any such flight would end in punishment.

It was precisely this 'unnatural' character of human flight, and the metaphysical paradox implicit in it which acted as a challenge to Leonardo. Twice in his Notebooks he described feather beds thus: 'Flying creatures will support men with their feathers.'

'Dissect the bat, study it carefully, and on this model construct the machine.' Leonardo in fact envisaged two machines, one with birdlike fabric wings, the other a helicopter, to be powered by a man sitting upright and pedalling. 'I conclude that the upright position is more useful than face downwards, because the instrument cannot get overturned . . .' Its airscrew would be helical: 'turned swiftly, the said screw will make its spiral in the air and it will rise high.'

What may be called the inverted strain in Leonardo's thinking and his proclivity for paradox could lead to foolishness as well as to brilliance. For instance, Leonardo believed that the reason sun and moon sometimes appear larger on the horizon than overhead is that on the horizon they are 3,500 miles *further away*, for 'every luminous body grows larger as it becomes more remote.' The foolishness underlying his concept of flight is that man's weight invalidates any close comparison with a bird.

Leonardo, however, had the courage of his convictions. On a page where he sketched flying machines he made notes about suitable nearby places, evidently a kind of verbal flight map. Elsewhere he noted: 'The machine should be tried over a lake ... carrying a long wineskin ... so ... you will not be drowned.'

Did Leonardo try out a flying machine? There is one piece of evidence that he did. In a book written some forty years after Leonardo's death Gerolamo Cardano of Pavia says that Leonardo tried to fly but his experiment failed. If Cardano is right, probably Leonardo made his experiment in 1505 on a mountain overlooking Fiesole.

Though not immediately productive, Leonardo's inventions mark a turning-point in cosmological thinking, for they expressed a conviction that man can overcome his Earth-bound status and rise into space. But the brilliance of Leonardo's cosmology did not end there. It was he who first stated that the configuration of the Earth's crust has been caused by processes, principally fluvial, operating over immense periods of time, and that the highest peaks of the Alps and Apennines had once been islands in an ancient sea. For these statements he offered the evidence of marine fossils found there. As regards Earth's relation to the sun and moon he generally held to the traditional belief, but he did make – without offering evidence – two separate speculative statements of great interest. One says: 'The Earth is a star almost like the moon'; the second: 'The sun does not move.' The possible source of these will be considered later.

*

The second big challenge to medieval cosmology came by way of the sea in the person of another Italian living away from his home city. Christopher Columbus was born in Genoa probably a year before Leonardo, the eldest son of a wool-weaver. He early went to sea and when he was twenty-four his ship was sunk by a French task force off Portugal. Wounded and clinging to an oar, Columbus swam ten kilometres and landed near Lagos. From there he went to Lisbon and joined his brother in the business of drawing charts and maps.

As a result of Portuguese voyages down the west coast of Africa cartography was changing fast. Whereas medieval maps had shown the East at the top, Jerusalem in the middle, and the three continents clustered like a big island surrounded by a big ocean, the kind of map

Columbus and his brother drew had North at the top, separated the continents much more, and sometimes marked longitude and latitude in degrees. Following Ptolemy, they placed zero meridian in the 'Fortunate Islands', identified with the Canaries.

Besides drawing maps Columbus continued to go to sea – on one voyage he sailed in command as far as the Gold Coast – and won a name as a well-informed, resourceful sea-captain. In appearance he was tall, with red hair, blue eyes and a ruddy freckled complexion; a good leader of men, and agreeable unless crossed, when he would flare into a rage. He married a Portuguese noble lady, had a son, and after her death formed a relationship with another woman, by whom he had a second son.

Despite these differences, Columbus shared two important characteristics with Leonardo. His diaries and letters are full of light, air and space, and he too, driven by an idiosyncrasy, set himself to do something which on the available evidence was quite impossible.

Columbus was a deeply religious man, and his idiosyncrasy was a conviction that he, Christopher – Christ-bearer – was destined to play an important part in regaining the Holy Sepulchre from the infidel. He believed the medieval legend that somewhere in the East lived a powerful Christian king called Prester John, and that if only Christian Europe could enlist Prester John's help they could recapture the holy centre of medieval maps: Jerusalem. That was the motive force behind Columbus's ambition to cross the Western Ocean to Asia.

The authorities and experts said it couldn't be done. Ptolemy held that the length of Europe and Asia together was 180 degrees, which meant that the western Ocean too was 180 degrees wide. Since Ptolemy took a degree to measure 50 nautical miles, that meant a crossing of 9000 nautical miles – far too long, Columbus knew, for him to win any backing. Paolo Toscanelli of Florence, who had talked to Russians, believed Eurasia was much more extensive than Ptolemy said, and the western Ocean correspondingly smaller: about 7000 nautical miles. Columbus wrote to Toscanelli and copied his answer into his favourite bedside book, Cardinal d'Ailly's *Image of the World*. By combining Toscanelli with various other little-known writers, including the apocryphal second *Book of Esdras*, who stated that six-sevenths of the Earth is land, Columbus brought himself to believe that the world was considerably smaller than contemporary experts said. He estimated that the distance from the Canaries to Japan was a mere 2400 nautical miles – in fact the true distance by air is 10,600 – and only because he made this preposterously low estimate did he win backing for his first voyage west.

On 6 September 1492 Columbus sailed out of San Sebastian, in the Canaries, with some 90 men in three ships, the largest of which was of about 100 tons. Navigation was by means of a compass mounted on the poop deck, the needle of which pointed slightly east of the North Star. The height of the North Star above the horizon and its relation to the two outer stars of the Little Bear, when observed with a crude quadrant or an astrolabe, gave approximate latitude, while the height of the noon sun, when applied to tables of its declination, also yielded latitude. Longitude Columbus calculated by compass bearing, time measured with a half-hour glass and the ship's speed, on average about 8 knots.

Columbus held at least two erroneous views regarding the heavenly bodies. He believed that the planets' positions affected weather: for instance, that the opposition of the sun and Jupiter or Saturn produced violent winds. And he believed that a compass needle was attracted by some power in the North Star – he knew nothing about the Earth's magnetic field. When Columbus left the Canaries, his compass pointed slightly east of the North Star, but as he sailed across the Atlantic, it swung westward. (Today, because the Earth's magnetic field has changed position, the opposite happens: in Europe our compass points 12 degrees *west* of the North Star, and as the Atlantic is crossed swings eastward, until in the Caribbean it is about 5 degrees east of the North Star.)

On 13 September, as Columbus noted in his Journal, 'the needles pointed north-west quite a fourth of the wind, and the sailors were frightened and downcast and they would not say why.' In the then state of knowledge Columbus had no means of knowing what was happening, but he considered it imperative to restore his men's faith in their compasses. So he gave out, quite erroneously, that the North Star, like most other stars, was moving westward round the sky.

For thirty days the three small ships sailed west – longer out of sight of shore than any earlier voyagers. They exceeded the 2400 nautical miles called for by Columbus's estimate and still had no glimpse of land. The crews became alarmed and showed signs of mutiny. According to one account, Columbus persuaded his officers to hold course only by promising that if land were not found within three days he would turn home.

At 2 a.m. on the thirty-third day, having sailed 2800 nautical miles – about an eighth part of the Earth's surface – a crewman at long last sighted the island of Guanahani in the Bahamas. Columbus took possession of it for the Spanish crown, naming it San Salvador.

Columbus sailed on to Hispaniola – that part which is now Haiti – where he found a friendly, unclothed people who believed the source of all power and goodness to be in the sky. Since Columbus and his men wore clothes, possessed powerful ships and treated them in friendly fashion, they believed that the newcomers had arrived from the sky! But Columbus's misconception of the Hispaniolans was almost as great. He believed they were Asians, inhabiting islands near Japan. Despite three further voyages, Columbus was to remain firm in his conviction that he had arrived in Asia. He could not believe otherwise, being the prisoner of those too optimistic figures which had enabled him to win royal backing in the first place, for if Eurasia was as wide as he believed it to be there was simply no room at that latitude for an extra continent.

On his third voyage, one day in August 1498 an enormous tidal wave roared through the strait between Trinidad and Paria, raising Columbus's flagship to what seemed an immense height, then dropping her suddenly. This alarming experience and some measurements of the North Star's altitude more inaccurate than usual led Columbus to revise his cosmology. Formerly he had believed the Earth to be a perfect sphere, but now he inclined to the medieval view described by Dante. 'I found that it was not round ... but pear-shaped, round except where it has a nipple, for there it is taller, or as if one had a round ball and, on one side, it should be like a woman's breast, and this nipple part is the highest and closest to heaven.' The violent tidal wave, thought Columbus, had been caused by the traditional four rivers of Paradise rushing down the side of the nipple, on top of which must lie the Earthly Paradise. So even the most foresighted thinkers can revert to fantasy; in just such a way Leonardo believed that the movement of the waters of Earth to the tops of mountains was analogous to the movement of blood in the head, both being produced by internal heat lower down.

On his fourth voyage Columbus became marooned in Jamaica. After a time the Indians declined to provide food by barter. Columbus had with him a copy of the *Ephemerides*, by a German astronomer, Regio-montanus, the first booklet of its kind, which gave the position of the heavenly bodies for every day from 1475 to 1506, and foretold a total eclipse of the moon for 29 February 1504. On that day Columbus summoned the chiefs on board, told them that God desired the Indians to supply food and would presently give them a token of his displeasure at their failure to do so. At moonrise the eclipse began. As they saw the Earth's shadow begin to spread over the disc the Indians begged Columbus to stop it. But he emerged from his cabin only when he knew from the booklet that the eclipse was diminishing, to announce that if they would provide food God would take the shadow away. The trick worked and thereafter the Spaniards lacked for nothing. Western man had advanced in knowledge if not in truthfulness since Nicias's disastrous reaction to the eclipse of 413BC!

Even in Jamaica Columbus continued to cling to his dream of delivering Jerusalem, but before he could win backing for such an expedition he died, still unaware of the true nature of the new world he had discovered.

*

The configuration of the Earth, which had eluded Columbus, was presently revealed by the Magellan-Elcano expedition. Sailing from Seville in 1519, Ferdinand Magellan rounded South America through the strait now called after him, and entered a great new sea which he named Pacific. South of the Equator Magellan had of course lost the North Star and from then on had practically nothing but common sense to guide him. He noted that the deviation of the compass, caused by the southern pole, was greater than that caused in the northern hemisphere by the Arctic pole. Also he observed 'many small stars clustered together' in two conspicuous luminous clouds; they were named Magellanic Clouds and are now known to be galaxies.

Tortured by thirst and scurvy, feeding on rat-fouled biscuits, finally reduced to eating the leather off the yardarms, he and his men made their great crossing of the Pacific. On Mactan Island, in the Philippines, Magellan was killed in a fight with natives, but Elcano continued. For bringing home, on 8 September 1522, the leaking but spice-laden *Vitoria*, with only 17 other European survivors and four Indians, 'weaker than men have ever been before', Elcano received from the Emperor an augmentation to his coat of arms: a globe with the inscription 'Primus circumdedisti me.'

Circumnavigation added new space to the globe. It showed the Earth to be about a quarter larger than had hitherto been believed, and established the Americas as continents. Apart from Australasia, the Earth's essential configuration was now known. In a painting, The Battle of Alexander the Great against Darius, Altdorfer marvellously depicted this new world of endless horizons, and seven years later, in 1536, Battista Agnese drew a beautiful map of the whole Earth, showing the track of the Vitoria.

*

Mention has been made of how the Hispaniolans believed Columbus had come from the sky, and as new lands were opened up, so more discoveries came to be made about other people's cosmologies. None was more startling than that of Hernan Cortes, who in the year Magellan set sail arrived in the Mexican coastal town of Cempoalan. Tall, broad-shouldered and deep-chested, Cortes had a pale complexion, serious eyes and a dark beard which partly hid a scar near his lower lip, caused by a knife-fight over a pretty woman. He loved God and gold about equally; also he was 'fond of cards and dice and over-fond of women'. With just 500 men and 14 cannon he planned to conquer all Mexico.

Cortes remained some weeks in Cempoalan, amid whose whitewashed houses stood a stone temple in the form of a stepped pyramid. Every day priests fetched from a prison a young man, sometimes two or three. Leading him to the top of the pyramid, they laid him on a stone, pinioning his arms and legs, while the chief priest plunged an obsidian knife into his breast and plucked out the palpitating heart. This he placed in a calabash before an image of the god. The priests smeared the walls with the victim's blood and gave his limbs to the people, who cooked and ate them.

Cortes was horrified by this rite, and sickened by the sulphurous stench of blood. He sought, and soon found, an explanation for the killings and also for the puzzling crosses depicted on the temple.

The Mexicans conceived the cosmos as a Greek or Maltese cross, each arm representing one of the cardinal points, and the centre Earth. They envisaged five different spaces, corresponding to the cardinal points and the centre, each with its field of force and distinctive properties, and five separate 'times', each linked to a space. There were 400 stars in the northern half of the sky, 400 in the southern. The sun was the most important god; the moon a goddess associated with abundant crops. Among the planets Venus mattered most, an archer god whose arrows brought illness.

The cosmos was unstable. It hovered perpetually on the brink of disaster. Four earlier ages – 'suns' as they were termed – had ended in disaster, and the present age would end likewise unless appropriate action was taken. That action consisted in nourishing the gods – above all, the sun god – who were always in danger of wilting, not with libations or maize or any earth-grown offering, but with the one food that could give them strength, palpitating human hearts.

Mexican civilization – for in some respects it was an advanced society – revolved round this grisly cosmic ritual. The Mexicans made almost continual war on neighbouring peoples, not killing but taking prisoner, while astronomers compiled accurate calendars which specified the day and hour when each god should receive the prisoners' hearts. On the accession of the ruler some years earlier 20,000 hearts had been plucked bleeding from human captive breasts and offered to Uitzilopochtli, the triumphant noonday sun and god of war. Continuously for twenty days the obsidian knives had struck.

Cortes watched the sacrifices in Cempoalan until he could stand them no longer. Though heavily outnumbered, 'fifty of us soldiers,' wrote one of the Spaniards, 'clambered up the pyramid and threw over the images which came rolling down the steps, shattered to pieces. Some looked like fearful dragons as big as calves, others were half-men and half-dogs, hideous to look at. The lords and priests wept, covered their eyes, and prayed for pardon, saying it was not their fault.' Cortes cleaned the temple of blood, whitewashed it, put up a cross and had Mass celebrated there.

One of the gods, Quetzalcoatl, Plumed Serpent, was associated with the setting sun, the wind and the planet Venus. He had taken the form of a white-skinned, black-bearded man, preached against human sacrifice and been driven out by Tezcatlipoca, Smoking Mirror, a great god identified with the constellation known in Europe as the Great Bear. Quetzalcoatl had sailed eastward on a raft, promising to return in a One Reed year and destroy Smoking Mirror. Now 1519 happened to be a One Reed year and Montezuma, ruler of Mexico, felt sure that

Cortes was really Quetzalcoatl. So Montezuma did not oppose the Spaniard's march inland and when he arrived treated him with the reverence due to a god.

Cortes took as his mistress a beautiful, intelligent Mexican girl. But his open-mindedness did not extend to Mexican cosmology and rites. He felt sure these were directly inspired by the Devil. As he brought Mexico into subjection, so the friars with him destroyed books of astronomy and cosmology. Art treasures were shipped home – Dürer, visiting Antwerp, marvelled at the goldwork – but only a few picture-written books were saved. Mexico's alternative cosmology was little studied until quite recently, when it has obliged us to realize that, left to his own devices, man may enrol even sun and moon and stars as accomplices in his fears and cruelty.

*

In order to get the feel of an educated Renaissance man's world-picture let us turn from the great pioneers to a versatile Italian who happens to have left a revealing autobiography. Gerolamo Cardano was born in Pavia in 1501, the illegitimate son of a lawyer. He became a physician and eventually professor of medicine in Pavia. He describes himself as having a fair complexion, yellowish hair, small intent eyes, projecting underlip and a forked beard. He spoke in a shrill voice, had thin arms and unequal hands, the left being elegantly formed with shapely nails, the right clumsy and ill-shapen. He was nervous, dreamy, solitary, accident-prone and very often ill. He liked wearing old clothes and going off for a day's fishing. He married a girl as penniless as himself who died young, leaving him to bring up three children, as well as numerous cats, dogs, birds, goats, lambs, rabbits and storks, with which he filled his small house. Yet Cardano managed to write more than a hundred books, including a very important one which first applied algebra to the resolution of geometrical problems.

'Among the extraordinary, though quite natural circumstances of my life,' wrote Cardano aged seventy-two, 'the first and most unusual is that I was born in the century in which the whole world became known; whereas the ancients were familiar with but little more than a third part of it.' 'The conviction grows,' he added, 'that as a result of these discoveries the fine arts will be neglected and but lightly esteemed, and certainties will be replaced by uncertainties. These things may happen, but for the moment we shall rejoice as in a flower-filled meadow.'

The meadow included new stars discovered in the southern hemisphere, and many of these, such as Canopus and the Southern Cross, Cardano included in an up-to-date cosmology. Yet, hypersensitive as he was to unseen influences, he believed absolutely in astrology. He cast horoscopes, including one of Christ, for which in old age he was placed under house arrest and forbidden to teach, and he carried astrology further than anyone before by inventing a new science, Metoposcopy. This purported to read planetary influences on a man's brow. Cardano divided the forehead by imaginary equidistant lines, beginning close over the eyes, to indicate the regions of the moon, Mercury, Venus and so on. The lines were read accordingly. For instance, if a woman had a straight line across her forehead just above the middle – in the region of Mars – she would be fortunate and get the better of her husband, but if the same line were crooked, she would die by violence.

Today we believe our lives to be inordinately complicated, but how much less complicated they are than Cardano's. As well as having to be alert to the lines on people's brows, he had to notice the lines on their hands – equally revealing of character – and their horoscopes. He had to heed a curious ringing in his ears, by the tone of which he could tell what men at a distance were saying of him. On the many occasions when he suffered from insomnia he had to smear on his body either bear's grease or an ointment of poplar – not just anywhere, but in seventeen specific places corresponding to planetary influences. And when at last he did close his eyes life became more complicated than ever, for Cardano was a compulsive dreamer, and his dreams, he believed, foretold his future.

He dreamed once of a cock that spoke to him in human language; and of a beautiful palatial city the name of which no one could tell him until an old woman said it was called Bacchetta, meaning a cane for punishing schoolboys. But in his strangest dream Cardano's soul went to the moon, where the voice of his father told him he was to remain 7000 years, 'and the same number in star after star until you reach the eighth; after the eighth you will enter the kingdom of God.' Cardano, then a struggling physician of thirty-three, foresaw in that dream the spectrum of his future studies and his fame: the moon signified grammar; Mercury, geometry and arithmetic; Venus, music, divination, poetry; and so on. 'The eighth orb stood for the final harvesting of all understanding, for natural science and various studies. And after these studies I shall rest serenely with my prince, the Lord.'

Cardano's dream calls to mind Ariosto's famous description in *Orlando Furioso* of how Astolfo flew to the moon in Elias's chariot. Astolfo found the moon almost as big as the Earth. It was like untarnished steel and had spacious forests where nymphs were forever hunting game. Most curious of all, the moon was a repository for all that men had lost on Earth – the tears and sighs of lovers, the useless time spent gambling, the chronic idleness of ignorant men, the empty plans which know no rest, even the wits men have lost during periods of foolishness or madness. Astolfo actually found on the moon mad Orlando's wits and brought them back to him. Ariosto's fantasy, like Cardano's dream, typifies Renaissance man's desire to break out of the confining space he had inherited from his medieval past.

Returning to Cardano, we find him in London in 1552, now a famous polymath, having an audience of fifteen-year-old King Edward VI. The King had just recovered from measles and smallpox and is described by Cardano as 'pale, with grey eyes and serious-looking'. Cardano had been given permission to dedicate his latest book to Edward, and because the first part of it offered a new explanation of comets, the King questioned Cardano in Latin about the theory. A comet, Cardano explained, is not a collection of fiery particles; it is light reflected from the planets. Edward asked why the comet did not move at the slow speed of the planets and was told this was due to the angle at which the planets' light was caught, comparable to the changing position of a rainbow on a wall when the sun shines through a crystal. Edward objected that there was nothing similar to a wall in space and Cardano could reply only by repeating himself: 'As in the Milky Way, a comet is a reflection of lights. When many candles are lighted near one another they produce between themselves a certain lucid and white medium.'

Cardano's explanation, though it strikes us today as far-fetched, was advanced for its day. Yet the Italian continued to hold that comets portend such evils as great scarcity of fish and a change in the laws. And, asked by Sir John Cheke, the royal tutor, to cast the King's horoscope, Cardano did so. The King, he said, would always suffer from low vitality, but he would live long. Edward's death the following year in no way weakened Cardano's belief in astrology, could its principles be applied correctly. He had cast his own horoscope, according to which he would die before the age of forty-five, and in fact he lived to be seventy-five, still eccentric, still dreaming, but to the very end a believer in the stars' influence over every aspect of man's body and the length of his life.

*

Also a polymath but in character very different from Cardano is the man who made the most important contribution to Renaissance cosmology. Nicholas Copernicus was born in the free city of Torun, north-central Poland, in 1473, the youngest child of a prosperous merchant. At ten, when his father died, he was adopted by his maternal uncle, a priest and future bishop. Uncle Lucas was an intelligent but sombre man, never known to smile; however, he greatly helped young Nicholas by securing for him a well-paid job as an administrator of Church property. In preparation for this Nicholas for ten years attended Italian universities, becoming proficient in canon law, medicine, mathematics and astronomy. He also learned to paint in the new style, which cleared away 'Gothic' convolutions and irregularities in favour of 'Greek' simplicity and economy of line.

Copernicus returned to Poland aged thirty-three. His self-portrait shows a strong face, the broad brow tapering to an assertive chin, shaggy dark hair, big eyes, a rather thick straight nose, and a mouth small, tight and suggestive of secrecy. He was appointed a lay Canon of Frauenburg Cathedral; that post left him free to marry, but Copernicus always remained a bachelor. The only clue to his motive is that in one of the books on his shelves he marked a passage praising celibacy. He habitually declined mundane invitations, but was not anti-social and always had at least one close male friend. When the Reformation came he adopted a notably tolerant stance to Protestants.

Copernicus's job was the administration of property on behalf of his cathedral chapter. He rode long distances, inspected farms and forests, kept ledgers and made records of all transactions. It was a demanding job and he brought to it energy and honesty. Sometimes, also, he was called in as a physician to treat local dignitaries. Once he was consulted about post-war inflation and the advice he gave is characteristic of his taste, acquired in Italy, for simplicity and economy: he proposed that a single Polish mint should replace the various municipal mints, and that from one pound of fine silver no more than 20 marks should be coined.

The Renaissance was an age of moonlighters, men who did a second job after-hours. Machiavelli, for instance, returning home from his lowly desk job in Florence, would put on a fine scarlet gown, open his copy of Livy, and seek to unravel the political lessons of ancient Rome. Copernicus was another moonlighter. When he returned to his tower house after a busy day and evening service in the cathedral, he would go to a platform on the tower and there, with the simple instrument Ptolemy had used 1400 years before, an armillary astrolabe, he would take measurements of the stars, or retire to his study to calculate planetary orbits. Copernicus might earn his living as a businessman but his real vocation was astronomy.

Using the rules of trigonometry, which German mathematicians were developing on the basis of Ptolemy and Arab authors, Copernicus made more accurate predictions of the movements of sun and planets than had been possible before, and gave a more accurate value for the distance of the moon: 60·30 Earth's radii, which is close to the modern figure of 60·27. But, even more important than this, Copernicus considered the often misty Baltic sky with the eyes of a man trained in Italy. According to the accepted Ptolemaic theory, the heavenly bodies moved not round the Earth's centre, but round the equant, an imaginary point in space in the vicinity of the Earth, and they did so in a complicated series of circular movements, comparable to swinging cabins revolving on a constantly turning Ferris wheel. Altogether there were 40 of these circular movements. This system Copernicus found 'unreasonable', by which he meant that it lacked the neat simplicity and economy he had come to prize in Italy. 'I often considered,' he wrote, 'whether there could perhaps be found a more reasonable arrangement of circles ... in which everything would move uniformly about its proper centre, as the rule of absolute motion requires.' That statement is characteristic of the man. He who advocated a single mint for all Poland wanted a single law – uniform motion around one centre (like the Greeks) – to apply throughout the cosmos.

Was there an alternative theory available? It so happened that there was, and Leonardo for one had probably heard about it. Rediscovered Greek books stated that certain early astronomers, notably Aristarchus of Samos, had 'placed the sun among the fixed stars, and held that the Earth revolves round the sun.' But Aristarchus's theory had never been more than a fringe movement in antiquity because it seemed to contradict everyday experience and because, if the Earth did revolve on its axis at great speed, how was it that objects and people did not fly off into space?

Copernicus believed he could answer these difficulties, provisionally adopted the theory and began to see whether known observations of the heavenly bodies could be made to accord with it. He found that they could. And the alternative theory had one great advantage over Ptolemy's: it offered a simple explanation of the apparently retrograde movements of planets: when the Earth overtakes one of the slower-moving outer planets, that is why the planet appears for a time to recede; when the Earth is overtaken by one of the faster-moving inner planets, an apparent reversal of direction again results.

Copernicus at first hoped to show that Mercury, Venus, Earth, Mars, Jupiter and Saturn revolve, in that order, round the centre of the sun. That would get rid of the untidy equant, and, he believed, reduce the number of circular movements in the solar system to 34. But because planets move in ellipses and Copernicus still believed, as astronomers always had, that they move in circles, he was obliged to reintroduce the hated equant: Earth and the planets, he said, revolve not round the sun but round an imaginary point *near* the sun; and he ended up with 48 circles, eight more than Ptolemy.

Why did Copernicus not abandon his theory when he saw that it failed to provide the hoped-for 'more reasonable arrangement of circles'? We do not know for sure, but probably there were three reasons. First, Copernicus lived in an age of startling discoveries. Ptolemy, for instance, had claimed in his *Geography* that the north temperate zone was separated from a similar temperate zone in the south by an impassable area of deadly heat: a statement which Vasco da Gama's voyage round the Cape had proved untrue. He lived also in an age of startling theories. Down in Bavaria Paracelsus was asserting that darkness is produced by *black stars*, and that man goes through life with *two* bodies: one of flesh and blood, the other an 'astral body' that soars to the stars when he dies. Finally, Copernicus lived on the edge of Europe, far from the modish academics who would have laughed his theory to scorn, and he lived alone, with no wife, no children, no close family, nothing to cherish but his extraordinary theory. And cherish it Copernicus did, for more than thirty years. He wrote it down in book form, but did not publish, evidently hoping to find some way of reducing the number of circles. From time to time he added new material based on more accurate observations. He seems to have been a perfectionist and somewhat secretive, as Leonardo was, and his book would probably have remained for long unknown, like Leonardo's Notebooks, had not a pushing young German admirer, Rheticus, who had heard about Copernicus's theory, come to Frauenburg, read the 212 small folio pages of manuscript, urged him to publish and offered to see the book through the press.

Revolutions of the Heavenly Spheres appeared in 1543. Near the beginning occur Copernicus's two momentous claims: that the apparent daily revolution of the heavenly bodies is due to the Earth's rotation on its own axis, and the apparent annual motion of the sun is due to the fact that the Earth, like the other planets, revolves round the sun, or rather a point near it. Copernicus then defends his claims. To the objection that objects would be whizzed off a moving earth and falling bodies left behind, Copernicus replies that these belong to the Earth and share its motion, which, being 'natural', cannot produce violent or centrifugal effects. To the objection that if the Earth moves it is no longer the centre of the universe, and why therefore should heavy objects gravitate towards it, Copernicus answers that each of the heavenly bodies has its own gravity.

Copernicus went on to show that hundreds of known observations, plus 27 made by himself, tallied with his proposed theory. He gave measurements for the orbits of the planets, and calculated a very accurate length for the year, which was to fulfil Copernicus's hopes by being adopted as the basis of the revised Gregorian calendar of 1582. He drew up tables to foretell the future positions of all the heavenly bodies. In short, the claim that the Earth revolves on its axis and round the sun was backed up by a wealth of first-rate mathematical computation.

Copernicus's book is one of the most important works of cosmology ever published, yet it is part fantasy. The heliocentric theory we now know to be true, the system of circular movements round a point near the sun erroneous. Copernicus has been criticized for having clung to circular movements. If only, writes Arthur Koestler, Copernicus had 'shaken off the authority of Aristotelian physics to which he remained a lifelong slave!' But such criticism is perhaps beside the mark. Copernicus valued the circle not as a doctrinaire Aristotelian but as one who believed that God acts in the most simple, direct and indeed labour-saving way. Of course, he was doubtless glad to have the authority

With the revival of astrology in the Middle Ages men believed that each part of the body is influenced by a sign of the zodiac, from head (Aries) to toe (Pisces).

For men of the Dark Ages heaven was physically very close. In a 5th-century ivory (left), Christ, at the moment of his Ascension, grasps the hand of God.

In a 9th-century manuscript illustration (below) some constellations have become hazy and inaccurate through constant copying unsupported by direct observation. The white snake at the centre represents the Dragon, twisting round the Little Bear and, below, the Great Bear. Near the bottom are Orion and his dogs; at the left-hand edge is the Centaur. The thin white circle represents the Milky Way.

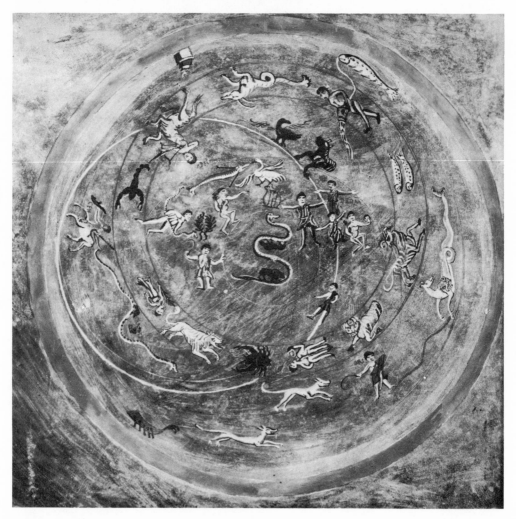

At the end of the Dark Ages men still had a closed-in view of space,
expressed in low-roofed buildings, one of which appears in a pen and ink
drawing of the Presentation (above).

Below: Englishmen point in awe to the comet of 1066, news of which is
brought to Harold. Beneath are ghostly invasion-ships presaged by
the omen.

In a Greek vase painting of Dawn (above), Helios, the sun god, rises from the sea in a chariot drawn by winged horses, while four twinkling stars are personified as boys diving and swimming.

The Romans' belief in astrology is expressed (right) in a statue of the giant Atlas upholding the sky, symbolized by Jupiter and the signs of the zodiac.

(11)

Truths, which Eternity lets fall on Man
With double Weight, through thefe revolving Spheres,
This Death-deep Silence, and incumbent Shade.
Thoughts, fuch as fhall revifit your laft Hour;
Vifit uncall'd, and live when Life expires;
And thy dark Pencil, *Midnight!* darker ftill
In Melancholy dipt, embrowns the whole.

  Yet this, even this, my Laughter-loving Friends! &*c.*
Lorenzo! and thy Brothers of the Smile!
If what imports you moft, can moft engage,
Shall fteal your Ear, and chain you to my Song.
Or if you fail me, know, the wife fhall tafte
The Truths I fing; The Truths I fing fhall feel,
And feeling give Affent, and their Affent
Is ample Recompence, is more than Praife.
But chiefly Thine, O *Litchfield!* nor miftake;
Think not un-introduc'd I force my Way;
Narciffa, not unknown, not unally'd,
By Virtue, or by Blood, illuftrious Youth!
                B 2   To

The planets' regular movements, as explained by Newton, were seen by
the poet Edward Young as evidence for the existence of God. In William
Blake's illustration to Young's *Night Thoughts* Midnight reminds man of
religious truths implicit in a gravitational cosmos.

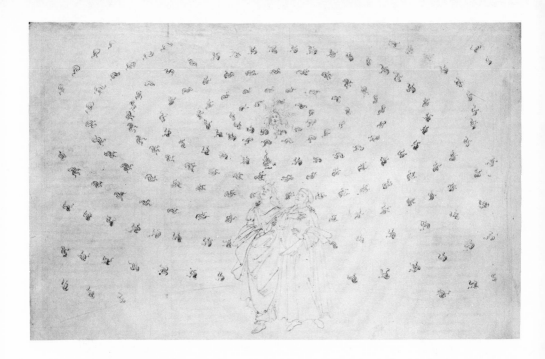

Botticelli portrayed Dante and Beatrice gazing at the flame-like souls of the blessed (**above**), circling a light in which shines the glorified person of Christ, and singing joyfully. Above Beatrice hovers the soul of St Peter, marked 'piero'.

Below: Islamic scientists safeguarded Greek thought during the Dark Ages and improved on some Greek observations. Here an Islamic astronomer observes a meteor with a quadrant.

Pre-Copernican cosmology influenced political theory. Queen Elizabeth
rules by divine right (**above**), while her virtues oversee an immovable
Justice at the heart of her realm.
Below: in the late sixteenth century better star-gazing equipment and the
Copernican controversy made astronomy a popular occupation for
gentlemen.

Medieval man believed the Earth to be surrounded by nine revolving
spheres, each possessing a spiritual character. Giovanni di Paolo's
illustrations to *The Divine Comedy*, painted about 1440, show Dante and
Beatrice in the fifth sphere, that of the planet Mars (**above**); Dante's
ancestor Cacciaguida explains to them that here dwell warrior saints, in
cross formation.
Below: From the eighth sphere, that of the fixed stars, Dante gazes down
on the seven inner spheres, at the centre of which is Earth personified as a
woman holding the Twins, Dante's natal sign.

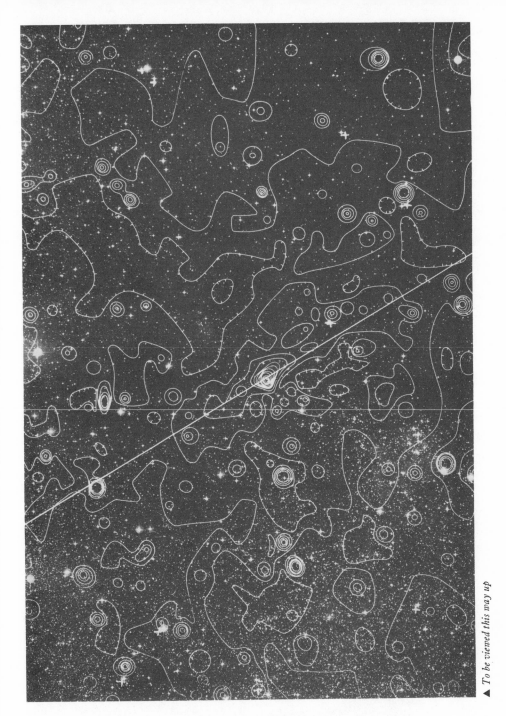

A photograph of the centre of the Galaxy, 30,000 light years away, superimposed on which is a contour map showing the distribution of the intensity of infrared radiation. The white diagonal line is the Galactic equator, the centre-line of our Galaxy: actually a plane, but we see it edge-on. In the middle of the picture is the very intense source of infrared radiation which marks the centre of our rotating Galaxy. Probably it is putting out radiation at all wavelengths, including visible light, but only the radio and infrared radiation can penetrate the dense dust clouds to reach us.

In the first half of the 19th century Englishmen felt themselves to be in serene harmony with the cosmos, a mood expressed in Samuel Palmer's *Cornfield by Moonlight*, (**above**) painted about 1830.
Below: Van Gogh's *The Starry Night*, 1889, was inspired by Victor Hugo's lines describing the stars as symbols of hope. In 1887 an English astronomer had taken the first photographs of a spiral galaxy.

Contemporary man lives in a cosmos of many galaxies, here suggested
(**above**) by Alexander Calder in his ceiling for the Aula Magna,
University of Caracas.
Below: in Eileen Agar's *My Room with a View of the Moon*, the moon,
losing its mystery and remoteness, appears on the left as a pitted object
on a television screen.

Planetary systems like our own extend indefinitely into space: a view widely held on the Continent in the 18th century and depicted in a German engraving of 1744. The black ellipse around our sun is a comet's path.

of the Greek behind him, just as Leonardo was glad to have the authority of Euclid and Archimedes, and Columbus that of the *Book of Esdras*, for men of the early Renaissance were extremely conscious of their inferiority to the golden ages of history – this was one of their mainsprings of action – and without at least a foothold there they would have been unable, psychologically, to launch out into experimental and creative thinking.

While his book was at the printers' Copernicus suffered a cerebral haemorrhage and had to take to his bed, partly paralysed and with his memory impaired. A copy was placed in his hands on 24 May 1543, and later on that day he died. The coincidence is symbolic, for the book itself, and the theory it propounds, were to languish for many years, as we shall see in the next chapter.

One reason for this neglect is that the state of knowledge provided not one shred of evidence to suggest that Copernicus's system was superior to Ptolemy's. Each explained the movement of the heavenly bodies satisfactorily, and predicted their future movements with considerable accuracy. The one certain way of proving Copernicus's theory would have been to discover an apparent shift in the position of one or more of the nearer stars, caused by the movement of the Earth round the sun. This shift or parallax could be detected only with sophisticated instruments and was first to be discovered in 1838.

Another reason for the coolness to Copernicus's theory was expressed by Martin Luther. Copernicus's claim to prove that the Earth goes round, and not the heavens, was, Luther declared, 'as if someone sitting in a moving wagon or ship were to suppose that he was at rest, and that the Earth and the trees were moving past him.'

There were really two Copernican Revolutions. The first declared that the Earth goes round the sun; the second declared that our experience of seeing the sun and stars rotate is erroneous, and the second, as Luther realized, is even more revolutionary than the first. After Copernicus man would continue to watch the sun rise and the stars set, and to describe the happenings in those words, but he would now have to correct himself inwardly: 'My senses mislead me. It is the Earth that is rotating; the sun and stars are stationary.' This was the beginning of a clash between sense impressions and reason, which was to grow, with Descartes for example, into a harmful separation of body and mind.

The Renaissance had begun with man's efforts to peck his way out of the shell of medieval cosmology. Leonardo had sought to soar into the air, while the discoverers had sailed beyond the known horizon to measure the true dimensions of an Earth larger than anyone had supposed, and to bring news of an alternative cosmology. While some, like Cardano, clung to the old desire to link man's destiny with the stars, Copernicus re-thought the whole question of the Earth's passive, static role. What Leonardo had sought to do for man, Copernicus did for the whole Earth: propelled it into space.

CHAPTER SIX

The Blemished Sun

Late Renaissance cosmology illustrated by Shakespeare's plays – Tycho Brahe finds a new star and a comet – Johannes Kepler breaks the circle – Galileo's discoveries with the telescope – his arraignment before the Inquisition – doubts of John Donne and other English writers – Giordano Bruno's vision of infinite worlds – Pascal's anguish in face of infinite space and his response.

In England during the reign of Queen Elizabeth I the Renaissance world-picture reached its most sophisticated form. From newly available classical texts, notably those of Plato, Aristotle and Cicero, Englishmen culled fresh details to extend or improve the world-picture known to Leonardo and Cardano. Confidence in themselves, therefore in the truth of their picture, was fed by national prosperity. Copernicus's theory in distant Poland, a hypothesis unsubstantiated by evidence, as yet did not show above the horizon of the average Englishman, who stood four-square on an Earth firmly at the centre of the cosmos. His world-picture gained additional solidity, and for succeeding generations permanent interest, as it was deepened imaginatively in words of unsurpassed beauty by William Shakespeare.

At the centre of Shakespeare's world stood man. He was made of earth, air, fire and water, the same stuff as other creatures and earthly objects. He possessed four humours, whose even balance ensured health. He dwelt in a world that was not alien, since he and it were linked by invisible strands, so that man's moments of crisis would be echoed or seem to be echoed, as stricken Lear declares on the heath:

> Blow, winds, and crack your cheeks! rage! blow!
> You cataracts and hurricanoes, spout
> Till you have drench'd our steeples, drown'd the cocks!
> You sulphurous and thought-executing fires,

> Vaunt-couriers to oak-cleaving thunderbolts,
> Singe my white head! And thou, all-shaking thunder,
> Smite flat the thick rotundity o' the world!

Englishmen considered themselves fortunate to live in a garden-like island, like that where unfallen man had dwelt, and which Columbus thought he had found in Trinidad: hence John of Gaunt's description of England as 'this other Eden, demi-paradise'.

Above the Earth lay a sphere of air, then a sphere of fire, then the moon, planets, sun and fixed stars.

> The skies are painted with unnumbered sparks;
> They are all fire, and every one doth shine.

Unnumbered though they were, all had been fashioned on the First Day, had never changed in number or brightness and never could change, since everything above the moon was immutable, perfect, an emblem of the divine.

The turning stars, as Pythagoras had taught, made sweet music:

> There's not the smallest orb which thou behold'st
> But in his motion like an angel sings,
> Still quiring to the young-eyed cherubins ...

Man loved the heavens for their beauty, because they reassured him that he was not alone in the world and because they were visible evidence of that unchangingness and perfection to which he aspired. Nor any longer did he fear them. With the hammer of successful enterprise he had begun to break the fetters of astrology. 'The fault, dear Brutus, lies not in our stars / But in ourselves that we are underlings.'

Sometimes, in the lower sky, God allowed strange phenomena to occur as warnings:

> Comets, importing change of times and states,
> Brandish your crystal tresses in the sky ... !

But such occurrences were rare and served to underline the unchangingness of the upper sky.

The blood in man's body rose and fell like sap in trees, or springs and streams in the Earth. Man felt himself to be a microcosm. The kingdom in which he lived, that too was a microcosm. The Elizabethans envisaged the macrocosm as eight circles round a central Earth, with God enthroned beyond the outer circle, and the political microcosm also as eight circles, each representing one of the royal virtues, around a central point, immutable Justice; God's position beyond the virtues being filled by Queen Elizabeth.

Because planets and stars were believed to move in circles, a circle was held to be the perfect shape. It was in a 'wooden O' that Elizabethans watched man the microcosm play out his dramas against a macrocosm which Thomas Heywood compared to the place of performance:

> If then the World a Theater present,
> As by the roundness it appears most fit,
> Built with starre-galleries of hye ascent,
> In which Jehove doth as spectator sit ...

In this cosmic O the Earth was pre-eminent, but the sun enjoyed 'noble eminence' too because it lay between the three outer planets and the three inner ones, Venus, Mercury and the moon. All these heavenly bodies moved in good order, and in periods of darkness became a visible comfort to man. So they were to Ulysses, in *Troilus and Cressida*, when he feared the rise of appetite, lust and anarchy:

> The heavens themselves, the planets and this centre,
> Observe degree, priority and place,
> Insisture, course, proportion, season, form,
> Office and custom, in all line of order;
> And therefore is the glorious planet Sol
> In noble eminence enthroned and sphered
> Amidst the other ...

In this cosmos where everything was in place, man felt at home, finding in its wonderful arrangement a guarantee of order in his own life. Yet he did not feel himself bound by the orderly circles, whether in regard to his humours, his social life or the politics of the State. He was free, freer even than the Athenian, for there were no Olympians to thwart his will. And that will, in a circular cosmos, might move in a straight line, or even zigzag. Man was his own master now, to the point where, in another play, Prince Hamlet dared take into his own hands the divine prerogative of justice.

Whether Shakespeare knew of Copernicus and if so what he thought of him we can only conjecture. Perhaps like his near-contemporary Montaigne he shrugged and said we should not bother whether Ptolemy or Copernicus was right, since 'the sky and the stars have been moving for 3000 years.' Perhaps he did mind but kept his thoughts to himself. At all events, in his plays, nothing disturbs the certainty of his world-picture, where the doings of man, because at its centre, are of paramount importance.

Yet even while Shakespeare was celebrating this rich and satisfying world-picture and that picture appeared most stable, astronomers were growing dissatisfied with its foundations.

The first, Tycho Brahe, came from Hamlet's country. He was born in 1546, eighteen years before Shakespeare, the eldest son of Otto Brahe, a distinguished Danish nobleman and later governor of Helsingborg Castle, opposite Elsinore. When he was fourteen Tycho saw a partial eclipse of the sun, and there and then decided to become an astronomer, an unheard-of career for a nobleman. Strong of body and strong in will, Tycho brushed aside family opposition and studied his chosen subject at no less than six universities. He saw that astronomy required continuous observations of given orbs with improved instruments over long periods of time, and decided to devote his life to making them. But first, at a wedding party, he fought a duel from which he emerged with part of his nose cut off. Thereafter he wore on his nose a metal covering. Contemporaries said it was an alloy of gold and silver, but probably it was of copper, for when his grave was opened in 1901, a greenish trickle was found near the base of the nose. According to the gossipy traveller Thomas Coryate, Tycho regarded his disfigurement as a bar to a socially correct marriage; so he took a peasant girl, Kirstine, as his common-law wife.

At the age of twenty-five Brahe was living near Helsingborg with his uncle Steen. On the evening of 11 November 1572 he happened to be returning to the house for supper from the outhouse laboratory where he and his uncle dabbled in alchemy when he spotted directly overhead near the constellation Cassiopeia a star he had never seen before, bright as Venus at its brightest. Very suprised, Brahe turned to his servants and asked whether they saw it. They said they did. Then, with characteristic thoroughness, Brahe obtained further confirmation from some peasants driving by.

Much excited, Brahe fetched a sextant which he himself had constructed of walnut, with bronze hinges, its arms five and a half feet long, its gradations as accurate as skill could then achieve. With this fine instrument he measured the distance of the bright star from the nine principal stars which compose the W-shape of Cassiopeia.

Every night Brahe observed his star as it followed a course right round the pole. Its position in relation to Cassiopeia remained constant, but over a period of eighteen months it gradually diminished in brightness and changed from white to yellow, to reddish, then to lead colour. At the end of March 1574 Brahe could see it no more.

The phenomenon was tracked by other observers too. Some, with inferior instruments, claimed that it slowly moved, that it was a comet or a planet. But because it twinkled and because his painstaking tracking showed its relation to Cassiopeia was constant Brahe asserted that it was a new star or *nova* – it was he who gave that word currency.

Brahe's claim was an astounding one, for most men of science deemed the eighth sphere, that of the stars, immutable. No new star had been sighted in modern times and the new star which, according to Pliny, Hipparchus had seen about 125 BC was usually assumed not to have been a star, but a comet travelling earthwards, hence with its tail invisible.

Brahe's *nova*, which he estimated to be more than a hundred times as large as the Earth, caused much consternation. It showed that the sphere of fixed stars was not after all immune from change. And what did it portend? Brahe, official astrologer to the Danish Court, assumed it had come into being at the time of the full moon, 5 November, when Mars was the ruling planet, and he predicted a period of wars, seditions, pestilence and venomous snakes. Curiously, 1572 was the year of St Bartholomew's Massacre, which led to twenty-six years of civil and religious war in France. Others, such as the astronomer Landgrave of Hesse, claimed that as the star seen by the Magi foretold the birth of Christ, the new one announced his second coming and the world's end.

Five years later, the world still in being, again in the month of November, Brahe was fishing at one of his fishponds in the gardens of a magnificent observatory presented to him by the King of Denmark when, a little before sunset, he noticed a very bright star in the West. Soon after dusk a spendid tail, twenty-two degrees in length, revealed itself and Brahe realized that a new comet had appeared. He at once began, and methodically continued, a series of measurements with sextant and quadrant. These revealed that the comet had no perceptible parallax, that is, showed no shift against the starry background when viewed by two observers many miles apart; it must therefore be far above the sublunary sphere. Brahe estimated that it moved round the sun in an orbit outside that of Venus and in the direction opposite to that of the planets.

Now, on the authority of Aristotle, astronomers had long held that the planets were attached to solid crystalline spheres or circles, nesting one in another. Brahe realized that if that were so, the comet would have been unable to pass on its course. *Ergo*, there were no solid spheres or circles. Even Copernicus had not questioned their reality; Brahe was the first to do so. 'There are,' he asserted, 'no solid spheres in the heavens.' Only space. Brahe was the man who discovered that above our heads, as far as the most distant stars, extends free and open space.

The Dane's discovery was a momentous one. Unlike Copernicus's theory, it was based on regularly continued observation, but even so it could not prevail overnight. Only gradually, like the comet which had inspired it, did the statement reveal its gleaming tail of metaphysical implications.

In 1600 Brahe took as his assistant a young German astronomer named Johannes Kepler, and the coming together of these two was like a conjunction of Jupiter and Mercury. Brahe looked the famous man he was, burly of figure, plump face sporting six-inch-long drooping moustaches, a fine lace ruff, lace at his wrists and on his white fingers jewelled rings; his manner was bluff, matter-of-fact, assured. Kepler was a poor innkeeper's son and former theology student, wearing threadbare clothes, thin, tense, timid, poetical and as accident-prone as Cardano. To these indifferences was added a major difference of opinion. Unable to detect stellar parallax, Brahe rejected Copernicus's heliocentric theory. Instead he published his own alternative theory, according to which the planets moved round the sun, but the sun, planets and moon revolved round the Earth. Kepler, on the other hand, believed Copernicus's theory to be true. Despite these differences and several quarrels the two men collaborated fruitfully, Brahe the patient observer setting Kepler, the imaginative theorist, to study the orbit of Mars which, among the outer planets, appeared to deviate most from a circle. Before he died in 1601 Brahe bequeathed to Kepler his painstaking observations of thirty-eight years.

Kepler had much in common with Columbus. He was deeply religious and felt a sense of mission to make God's secrets known to man. One winter's day in Prague Kepler observed a few snowflakes that fell on his coat and marvelled at their apparently perfect hexagonal form. Why six sides, he wondered, and never five or seven? Although he lacked the means we have today to answer that hexagonality is the most stable, therefore dominant form, given their atomic architecture, Kepler saw the snowflakes as examples of a general formative tendency in nature. He felt sure that God had created the cosmos according to geometrical figures and mathematical proportions, for otherwise how would man be able to make sense of the variety around him? So Kepler searched the heavens for regular geometrical figures as Columbus had searched the Ocean for Asia, and when Brahe's observations and his own study of Mars convinced him that planets move round the sun in ellipses, of which one of the foci is the sun – the great astronomical discovery known as Kepler's First Law – no one was more disappointed than he, for in his opinion ellipses had nothing to recommend them in the eyes of God or man, and he wrote that though he had to some extent cleaned the stables of astronomy, he was guilty of bringing in 'a cartload of dung', meaning ellipses.

Undaunted, Kepler re-examined the planets in the light of Pythagoras's belief that they emit harmonious music. He wrote a long book, *Harmonies of the World*, to try to show that each planet possesses a distinctive range of voice – bass, tenor, contralto and so on – and that the music it emits is determined by its distance from the common centre of revolution, the sun, as the length of a string on an instrument determines its sound. The main theme of the book is pure fantasy, but in the course of it Kepler stumbled on another important truth: the time a planet takes to go round the sun is mathematically related to its distance from the sun. Kepler expressed it thus: the squares of the period of revolution of the planets are to one another as the cubes of their mean distances from the sun. Though Kepler never arrived at a satisfactory explanation of how the planets move through space – he supposed that the sun swept them along almost as though with an invisible broom – this discovery, later in the seventeenth century, was to help make such an explanation possible.

Kepler made one other important statement about space. He conducted a simple optical experiment to show that if there were a layer of fire between Earth and the moon, as Aristotle supposed, considerable refraction of the rays emitted from the stars should be evident. Since such refraction did not occur, there could be no layer of fire. This, and his assertion that planetary motion was elliptical, were Kepler's two major contributions to cosmology.

*

The third dismantler of the old world-picture, Galileo Galilei, was born most appropriately in Florence, where Greek mathematical texts had been rediscovered, and worked in Venice, where they were printed. In the winter of 1609–10 Galileo, then aged forty-five, constructed a telescope with a magnifying power of 33 diameters. Venice was a glass-making city and Galileo himself saw to the grinding of the lenses. With this superb instrument Galileo explored the night sky. He found a bewildering number of stars, 'more than ten times as many' as were visible to the naked eye; in the Pleiades thirty-six stars instead of the seven so long sung by poets. He found four planets or satellites revolving round Jupiter, and he found that the moon was not a perfectly smooth sphere after all but 'rough and uneven, covered everywhere, just like the Earth's surface, with huge prominences, deep valleys, and chasms'.

Galileo published these discoveries in March 1610 in a booklet called *Messenger of the Stars*. The ensuing excitement may be imagined. From Wales the astronomer Lower wrote to his friend Thomas Hariot: 'Me thinkes my diligent Galileus hath done more in his three fold discoverie than Magellane in opening the streightes to the South sea or the dutch that weare eaten by beares in Nova Zembla.'

The instrument with which these discoveries had been made was hailed by Kepler as 'more precious than any sceptre! Is not he who holds thee in his hand made king and lord of the works of God?' Others had reservations, sensing that the human eye had been proved inadequate, a faltering organ without the crutch of a telescope, while for the Bible-reading physician Thomas Browne of Norwich it posed a question which he confided to the privacy of his Commonplace Book: 'When god commanded Abraham to looke up to heaven to number the starres thereof, that hee extraordinarily enlarged his sight to behold the host of heaven & the innumerable heape of starres wch Telescopes now shoe unto us, some men might be persuaded to beleeve.'

Galileo went on to discover that Venus had phases, like the moon. These could be satisfactorily explained only if Venus revolved round the sun, either in accordance with Copernicus's world-picture or with Brahe's. Galileo opted for Copernicus's, without Kepler's elliptical orbits however. Other cosmologists opposed Copernicus because they could detect no stellar parallax, and a majority preferred to believe with Brahe that the planets moved round the sun, but the sun, planets and moon revolved round the Earth. For the ordinary educated person the important point was that no agreement had been reached. Galileo's discoveries did not diminish doubt, they increased it.

Even Galileo's discovery that the moon was a body not unlike the Earth raised all manner of doubts. Galileo said that for his part he was very glad to find that one at least of the heavenly bodies was not immutable. Immutability meant deadness and he preferred an orb that could change and produce. He foresaw tangerines being planted on the moon and bearing fruit, while Kepler, imaginative as always, believed the crater-like hollows on the moon were circular walled cities built by moon people nineteen times the size of man. But few could take either so cool or so imaginative a view, and men were deeply perturbed by the moon's new status.

In spring 1611 Galileo made a new discovery with his telescope: dark marks on the body of the sun. At first he thought they were hitherto unknown stars revolving about the sun, but on further observation he saw that they were very irregular, some being very dark and others less so; that one would often divide into two or three and at other times two or three would unite into one; and that they were all carried round with the sun, which turned on its axis in a little less than a lunar month. Galileo concluded that the marks were attached to the surface of the sun, and called them *macchie solari*, sun-spots.

Fiercely contested at first, the discovery that the 'perfect' sun, the emblem of divinity, was blemished caused disarray. It was as though truth itself had been found wanting, and Galileo's sun-spots came to stand for all the uncertainty raised in men's minds by the telescope.

The greatly increased number of stars also caused heart-searching. Kepler recognized that 'the entire planetary system practically disappears when compared with the fixed star system', and therefore man would seem to belong to 'the absolute trifles'. He tried to console himself by saying that size is not everything, otherwise God would prefer the crocodile or elephant to man.

*

We are able to follow in detail the reaction of one typical Englishman to the dismantling of the old cosmology. John Donne was born in London in 1572, and so was eight years younger than Shakespeare. His mother's family, Heywood, had close links with that of Sir Thomas More, and she brought John up a Catholic. After Oxford, Cambridge and Lincoln's Inn, Donne fought in Spain, became a courtier, wrote witty poems and made a runaway marriage for which, briefly, he was imprisoned. Good-looking, with a lean face, sensuous lips and intense, disturbing eyes, Donne had much of Hamlet. He was brilliant, complex, melancholy, charming and at times unstable.

The early Donne specialized in conceits taken from the sky. Brahe's *nova* having been followed, in 1604, by a second *nova*, Donne wrote:

> We have added to the world Virginia, and sent
> Two new starres lately to the firmament

while in another poem he described himself as 'Meteor-like, of stuff and forme perplext, / Whose *what*, and *where*, in disputation is.'

In 1610 Donne was living in a rather damp gabled cottage in Mitcham, Surrey, with his wife and six children. He was aged thirty-eight. For reasons unknown he had given up his Catholicism, was trying unsuccessfully to get State employment, and felt a lack of direction. At that moment Galileo published his discoveries. They came as a profound shock to Donne, sweeping away as they did the reassuringly familiar cosmos he had explored in words. They coincided, moreover, with the death of Elizabeth Drury, the fifteen-year-old daughter of Donne's friend and patron. Donne picked up his pen and wrote one of the great poems of doubt, *An Anatomie of the World*, in which he associated the loss of Elizabeth with the loss of geocentric certainty, and hinted too that the world itself might be drawing to its end.

> The sun is lost, and th'earth, and no mans wit
> Can well direct him where to looke for it.
> And freely men confesse that this world's spent
> When in the Planets, and the Firmament
> They seeke so many new; then see that this
> Is crumbled out againe to his Atomies.

Donne went on to draw the kind of conclusion that would have appalled Shakespeare's Ulysses:

'Tis all in peeces, all cohaerence gone;
All just supply, and all Relation:
Prince, Subject, Father, Sonne, are things forgot,
For every man alone thinkes he hath got
To be a Phoenix, and that then can bee
None of that kinde, of which he is, but hee.

And Donne summed up his feelings by declaring 'the new Philosophy calls all in doubt.'

Between the time of Galileo's discoveries and 1615 Donne underwent a period of spiritual anguish, at the end of which he entered the Church of England. What part the new cosmology played in this we do not know, but a passage from one of his sermons may throw light:

Admiration, wonder, stands in the midst, betweene knowledge and faith, and hath an eye towards both. If I know a thing, or beleeve a thing, I do no longer wonder: but when I finde that I have reason to stop upon the consideration of a thing, so as that I see enough to induce admiration, to make me wonder, I come by that step, and God leads me by that hand, to a knowledge, if it be of a naturall or civill thing, or to a faith, if it be a supernaturall, and spirituall thing.

With the years, particularly after his appointment as Dean of St Paul's, Donne found a measure of peace but still he would associate the new cosmology with doubt and instability. Recovering from an illness but still subject to giddiness, he jested: 'I am up, and I seem to stand, and I go round; and I am a new argument of the new philosophy, that the Earth moves round.' 'Copernicism in the mathematics,' he wrote to a friend, 'hath carried Earth farther up, from the stupid centre; and yet not honoured it, nor advantaged it, because for the necessity of appearances, it hath carried heaven so much higher from it.' Most movingly perhaps, when his wife Ann, who had borne him twelve children, died at the early age of thirty-four Donne felt his grief transcending the bounds of the old cosmos:

You which beyond that heaven which was most high
Have found new sphears, and of new lands can write,
Powre new seas in mine eyes, that so I might
Drowne my world with my weeping earnestly.

[115]

Though Donne was unique in fusing passion and thought about the changing cosmology, other English poets reacted strongly in their different ways. Francis Quarles, a more pedestrian writer than Donne, shied away from the newly discovered immense distances. In 'A Cupid trying in vain to grasp a Globe in his arms' he saw the scientist's quest for space as a form of greed:

> We gape, we grasp, we gripe, add store to store;
> Enough requires too much; too much craves more ...
> Thus we, poor little worlds! with blood and sweat,
> In vain attempt to comprehend the great.

As for George Herbert, like Donne a priest, he too turned away from what he termed 'some fourtie heav'ns, or more':

> O rack me not to such a vast extent;
> Those distances belong to thee:
> The world's too little for thy tent,
>   A grave too big for me.

The Platonic tradition continued strong, and one prose writer who belonged to it, Robert Burton, stated in *The Anatomy of Melancholy* his dismay at not finding the stars arranged more harmoniously: 'Why are some big, some little, why are they so confusedly, unequally situated in the heavens, and set so much out of order? In all other things nature is equal, proportionable, and constant ... Why are the heavens so irregular? ... and whence is this difference?'

The general consternation was summed up by a Scot, William Drummond, in 1623. His first sentence, which echoes a line by Donne, calls for explanation. Gerolamo Cardano had redefined the word 'element' to mean something productive of life and hence declared that fire was not an element, a position which had been reinforced by Kepler's proof that there was no layer of fire between Earth and moon.

'The element of fire is quite put out ...' wrote Drummond. 'The air is but water rarefied, the Earth is found to move, and is no more the centre of the universe ... ; stars are not fixed, but swim in the ethereal spaces, comets are mounted above the planets ... Thus sciences, by the diverse motion of the globe of the brain of man, are become opinions, nay, errors, and leave the imagination in a thousand labyrinths. What is all we know, compared with what we know not?'

These doubts were sufficient in themselves to weigh heavily upon men of the late Renaissance, but to them in 1633 were added a new cluster of doubts of a different but no less troubling kind. They flowed from a dispute in distant Rome which had its origin in the character of Galileo. The Florentine had the quick temper which often goes with red hair, and as a boy had been made much of by three doting sisters. He was already pugnacious, assertive and quick to ridicule opponents when at forty he became world-famous. Adulation went to his head and Galileo came to believe that he alone could pronounce on celestial matters. When the Jesuit priest Horatio Grassi, of the Collegium Romanum, published a lecture on comets which failed to mention Galileo's views, the Florentine mounted an attack on Grassi, culminating in *The Assayer*, which held the Jesuit up to ridicule. This, and un-provoked attacks on two other Jesuits, brought him the enmity of the Order, so that when Galileo publicly championed Copernicus's theory in his *Dialogue on the Great World Systems*, the Jesuits used their influence against him. In 1633, by now nearly seventy, Galileo was summoned to Rome and put on trial before the Inquisition. The charges were that he had publicly declared to be probable two propositions which, according to the Inquisition, were 'absurd and false philosophi-cally', namely that the sun is the centre of the world and that the Earth moves.

The trial lasted two months. At the end of it the judges found Galileo guilty. They prohibited his *Dialogue*, forbade him to teach or write further on the subject, and ordered him to abjure before a plenary assembly of the Inquisition.

With unexpected docility Galileo submitted. 'I abjure, curse and detest the aforesaid errors and heresies,' he declared, 'and I swear that in future I will never again say or assert, verbally or in writing, anything that might furnish occasion for a similar suspicion regarding me.' For his remaining eight years he lived quietly at home, writing no more on cosmology and sinking slowly into blindness.

The condemnation of Galileo raised three grave doubts, one immediate, the others later but more far-reaching. The first doubt was whether by publishing some new and contentious theory a Catholic might not lay himself open to trial for heresy. Fear of such action caused Descartes, for one, to hide his conviction that Copernicanism was true, and probably also played a part in his decision to move home from France to Holland, where he would be less subject to Church authority.

The second doubt grew more slowly. When Newton eventually proved Copernicus's theory true, the Church's condemnation of Galileo began to look extremely foolish. If the Church had committed itself to an out-of-date cosmic theory, perhaps the religious truths it taught were equally out of date? Doubt arose in many people's minds about the Church's whole claim to be champion and sole interpreter of God's truth.

The third doubt was most pernicious of all. The Church had condemned two scientific statements as heretical on the grounds that they contradicted a literal interpretation of certain passages of Scripture. By so doing the Church quite unnecessarily brought into opposition cosmology on the one hand and Scripture on the other, forgetting that both are aspects of a single truth. The opposition between the two was to continue well into the twentieth century, causing much heart-searching among generations of scientists belonging to the Catholic Church and creating, in cosmology, a mental climate akin to schizophrenia.

We saw earlier how Brahe's observations dismantled the old crystalline spheres, replacing them with open space. To this innovation one man in particular reacted with extraordinary enthusiasm.

Giordano Bruno was born near the south Italian town of Nola in 1584, the son of a soldier. South Italy was then occupied by the Spaniards and their harsh rule may have given Bruno his dislike of authority. Heir like many south Italians to a long tradition of animism, as a child Bruno believed he saw spirits on hills where beeches and laurels grew. One day he was taken to nearby Vesuvius. From his home it had seemed bare and forbidding, but now he saw it had trees and lush vegetation, whereas his distant home, set amid vines, had become dark and dim. 'For the first time I became aware that sight could deceive,' he later recalled, and he might have added that already he trusted those spirits seen by the inner eye more than visible externals.

At seventeen Bruno joined the Dominicans, exchanging his baptismal name of Felipe for Giordano, and seven years later was ordained priest. He was a frank, excitable, gesticulating, indiscreet young man, untidy – buttons forever missing – but loving life, and full of enthusiasms and curiosity. He could not bear to be hemmed in, either in body or mind.

He loved the boundless soaring of intuition and correspondingly distrusted our limited world. Hence he came to favour the Arian heresy, that Christ is man only, not God – for how could God be hemmed in in a human body? After a warning from his superiors about his Arian tendencies, Bruno left his monastery and began a life of wandering.

Five years later he arrived in Paris, where he lectured on what might be called How to Improve your Memory and Start Living. He taught Henri III a method of memorizing *Genesis* and was rewarded with royal patronage. He also published a book, *The Shadows of Ideas*, which held that man and visible phenomena are very imperfect shadows of the Ineffable One, who is to be reached not through reasoning but through mystical union.

In 1583 Bruno came to England, and there probably encountered the Copernican theory as developed by a leading English cosmographer. Thomas Digges published in 1576, as part of a new edition of a book called *Prognostication*, an account of Copernicus's cosmology and illustrated it with a diagram showing the orbits of the planets round a central sun and, beyond, numerous scattered stars, captioned thus: 'This orb of stars fixed infinitely up extendeth itself in altitude spherically and therefore immovable. The palace of felicity garnished with perpetual shining glorious lights innumerable ...'

The key words were '*infinitely up*', for Digges, going beyond Brahe, held the view that space is infinite. The Roman poet Lucretius had imagined as much, and the fifteenth-century Platonist, Nicholas of Cusa, had proposed the same view on metaphysical grounds, but Digges was one of the first to say so on scientific grounds.

Bruno seized joyfully on Digges's assertion that space is infinite, for he saw in it evidence of God's unboundedness. He lectured on Copernicanism at Oxford where, he said, the dons 'knew much more about beer than about Greek,' joined Philip Sidney's circle and wrote two important books, *The Ash Wednesday Supper* and *The Infinite Universe and its Worlds*, in which he carried Digges's notion to the furthest limits then imaginable. Not only was the universe infinite, said Bruno, but there were an infinite number of worlds. 'There are countless suns and an infinity of planets which circle round their suns as our seven planets circle round our sun.' Furthermore they were inhabited by people, some better, some worse than ourselves. Bruno held that the universe as a whole is eternal; later he was to revise this, and assert that its worlds decay and perish, their constituent parts entering into fresh combinations. Intoxicated with his new-found notion of infinity, Bruno extended it downwards too. 'A countless multitude of creatures,' he declared, 'inhabit men's bodies and all composite things.'

In this infinite universe there could be no up and down, for all things were constantly changing position and form. Good and evil were not a matter of personal responsibility but oscillations of the soul in a flux of light and darkness. Bruno was never far from pantheism and the blurring of right and wrong which accompanies it.

After two and a half years in England Bruno returned to the Continent. Accident-prone like so many dreamers, including Cardano, he had his baggage and money stolen on the journey. He resumed his wandering and published more books, in one of which he claimed that there are undiscovered planets revolving round our sun. In all of them he conveyed his joyful, confident vision of an ever-living, infinite Universe.

In 1591 Bruno arrived in Venice, then the most liberal city in Italy, hoping to be reconciled with the Church and to live again as a priest, though not with the Dominicans. He claimed to have done valuable work: 'The Nolan has given freedom to the human spirit and made its knowledge free. It was suffocating in the close air of a narrow prison-house, whence, but only through chinks, it gazed at the far-off stars. Its wings were clipped, so that it was unable to cleave the veiled cloud and reach the reality beyond.' By a tragic irony it was Bruno who now came to be denounced for heresy by a Venetian acquaintance and thrown into a narrow prison-house.

Under questioning by Venice's Inquisition Bruno declared that he held 'the universe to be infinite as a result of the infinite divine power; for I think it unworthy of divine goodness and power to have produced merely one finite world when it was able to bring into being an infinity of worlds.' He admitted that he had grave doubts whether the Father, Son and Spirit were distinct Persons, but that he had never made his doubts public. After weeks of interrogation Bruno was told to abjure. He did so, and falling on his knees begged for mercy. But by then the Inquisition at Rome had demanded that Bruno be handed over to them.

The Roman Inquisition was more severe than the Venetian. In 1567 and 1570 it had put two Italians to death for heresy. It held Bruno in prison for seven years before bringing him to trial. Though the documents of the trial have been lost or destroyed, there is enough to show that eight new charges were brought, including one that he denied transubstantiation. This time Bruno declined to retract. At the end, as at the beginning, of his life he could not believe in a hemming-in of the Infinite, whether in the flesh of Christ or in a disc of unleavened bread. He was found guilty by eight cardinals, unfrocked, excommunicated and with the usual hollow recommendation to mercy handed over to the secular arm. On 17 February in the jubilee year 1600, aged fifty-two, Bruno was led barefoot to the Field of Flowers wearing a garment which, for those with eyes to see, perhaps recalled Mexican ritual: a white sheet emblazoned at the corners with the cross of St Andrew, and bestrewn with devils and red flames. Bruno died by burning at the stake. According to one account he was gagged, according to another he called out that his soul would ascend in the smoke to Paradise.

To his judges Bruno had said, 'Perhaps your fear in passing judgment on me is greater than mine in receiving it.' It was both a brave and perceptive remark. Future generations have endorsed Bruno's belief that no authority should be allowed to hem in the human spirit, and condemned those who sent Bruno to the stake for the theological consequences of a cosmology.

Bruno gloried in the infinite and his cosmological writings in that vein were to prove influential. But awareness of the infinite did not affect all men of this age in the same way. One in particular reacted very differently from Bruno.

Blaise Pascal was born in 1623 in Paris, where his father, a lawyer, held a senior post in the Board of Excise. When he was three his mother died, and this probably left Blaise with an abiding sense of insecurity, mitigated to some extent by the care with which he was looked after by his two sisters. Blaise's health was never robust and at the age of four he suffered from an eye infection. It is a curious sidelight on medicine of the day that two cats were put to death as part of the cure.

Blaise's father was a pious man but no bigot and his circle of cultivated friends included Le Pailleur, who wrote an essay on 'Galileo's opinion concerning the movement of the Earth', an ironic eulogy of ignorance as the surest haven against heresy. Pascal *père* carefully supervised his son's education and noted signs of brilliance. At sixteen the boy wrote an original Treatise on Conic Sections, and at eighteen constructed the first calculating machine. On leaving school Blaise began various researches, notably into the nature of a vacuum and into mathematical calculus. To make a living he took part in a scheme for draining marshes in Poitou and for running a regular stage coach service.

This energetic young man, widely regarded as a mathematical genius, had a disability. The soft areas of cartilage between the bones of the skull usually harden by the age of eighteen months, but in Pascal's case five of them failed to close and harden. As a result, he often suffered from migraines so severe they halted his scientific thinking. During such periods of prostration Pascal would reflect on religion, and in particular on certain new doubts about the status of reason.

They had arisen initially from the discovery, in the New World, of cannibals. That man should eat his fellow men was doubtless regrettable, but that he should claim he was right to eat them, and invoke reason to support the usage was profoundly disturbing to many in Europe, for it seemed to call in doubt both absolute right and reason. Montaigne was one who thought out the implications. Casting about in time present and past he counted no less than 280 sovereign goods, each supported by reason. Unable to choose between them Montaigne lost faith in reason's powers and adopted a position of extreme scepticism.

Blaise Pascal was extremely disturbed by Montaigne's arguments, and in his *Pensées* carried them a stage further: 'The historians of Mexico, of the five suns, the last of which is only 800 years old. Difference between a book accepted by a nation and one that moulds a nation.' This cryptic note suggests that Pascal realized that man's religion, no less than his reason, could endorse cannibalism.

Pascal then pursued a slightly different tack. Man, he claimed, is indeed a wretched creature, even more wretched than Montaigne believed. A mere nothing suffices to distort his judgment, the veering of a weathercock, too much or too little wine. A mere nothing starts him off on mad adventures. 'Custom' prejudices our understandings and transforms accidental associations into necessary links. Our 'imagination', without our being aware of it, substitutes its own fantasies for the reality of things. What good are all the riches of the world to the man who imagines himself poor? If the preacher has not shaved properly, can the gravest judge listen to the most edifying sermons without a secret inclination to laugh? Thus imagination is the queen of this world; her laws insidiously substitute themselves for the laws of reason.

This was dismaying in the extreme. But there was something even worse, known only to Pascal the mathematician, who had explored the new territory of calculus and found no logical contradiction in the concept of an infinitely large number. We know, he wrote, that an infinite number exists, but 'we do not know what it is: it is not equal, nor is it unequal, for if you add one to it, you do not change it.'

Man's reason might be unable to grasp infinity, yet infinity existed, up there in the night sky. Whereas Bruno had embraced infinite space with naïve joy, Pascal, a mathematician who knew precisely that infinity entailed, and how powerless the mind was to comprehend it, looked at the heavens and trembled with fear. Repeatedly in the *Pensées* Pascal alluded to his fear: 'I see these frightening spaces in the universe, which enclose men, and I find myself attached to one corner of this vast expanse, without knowing why I have been placed here rather than there, nor why I have been destined to live my short life at this point of eternity rather than at another. On all sides I see only infinities, which enclose me like an atom, and like a momentary shadow.' Elsewhere he penned the lapidary phrase: 'The eternal silence of these infinite spaces makes me afraid.' Why afraid? Since reason has made man *aware* of infinity, may not reason ultimately enable him to comprehend it? Pascal believed not. Finite reason, operating in a finite lifespan, can never comprehend the infinite. So again, on the most dramatic scale of all – the cosmic – man was proved most wretched.

Pascal asked, How can I find a way out of these terrifying doubts? After much searching he fastened on that very wretchedness which so dismayed him. Man is aware of his wretchedness, and there lies his hope, for such awareness presupposes that he was made for better things. Man senses in himself possibilities of independence, even of greatness. He is a reed, but a thinking reed. He contains in himself both marks of extreme weakness and marks of the sublime. Why such paradoxes? What can explain them? And Pascal answers, only the Fall and the Redemption, Adam and Jesus Christ.

The philosophers, thought Pascal, were wrong in their conception of God. God is not a mover, or a geometer, towards whom we can reason by way of clear ideas. He is a hidden God. We have only glimmerings of Him. And that is understandable, for those glimmerings come not from reason but from the heart. In saying this Pascal was not making a retreat from reason, only from the then fashionable limited view of reason as a detached manipulator of clear ideas. For the heart, as conceived by Pascal, was very far from being mere sentiment, it was an essential part of the knowing self. In an important shorter work, *The Mind of the Mathematician*, Pascal claimed that the heart is an integral element even of mathematics, the supposedly perfect science, for through the intuitions of the heart we know first principles, such as space, time, motion and number, and them reason cannot impugn. Pascal was in fact re-affirming the unity of the knowing self, claiming that all our philosophical ideas, even the most abstract, have an existential quality; we can claim validity for them only as they ring true to that which is most deeply personal.

Pascal's plea for the unity of the knowing self was to go unheeded for the rest of the seventeenth century, and longer. Only in recent years has it again come to the fore as a perhaps more satisfactory explanation of the value judgments which appear to transfuse even our most professedly impersonal thinking.

In Pascal's own life the conclusions he arrived at brought immediate decisive changes. Through his sister Jacqueline, a nun at Port Royal, Pascal became attracted to the austere piety of Jansenism. In 1646, when he was twenty-three, he underwent a first 'conversion', as the result of which he led a less worldly life and adopted pious Jansenist practices. In 1654, between half past ten in the evening of 23 November and half past midnight he underwent a second profound religious experience, which he described thus:

> Fire.
> 'God of Abraham, God of Isaac, God of Jacob' not of the philosophers and men of learning. Certainty. Certainty. Feelings. Joy. Peace ...

Pascal thereupon withdrew to Port Royal, where he led a life of austerity in a plain room from which he had the tapestry removed. For a year he worked at an apologetic of Christianity designed to answer the doubts of his contemporaries, not least those arising from new views of the cosmos. Then ill health made him stop, and that is why of perhaps the most brilliant work of Christian philosophy ever conceived we have only the fragments known as *Pensées*. His pain increased, and he realized he was near the end. Did the silence of the eternal spaces still frighten him? Perhaps, for his last words were, 'May God never abandon me.' Shortly afterwards, aged only thirty-nine, Pascal died.

CHAPTER SEVEN

The Apple and
the Comet

Speculation about the moon – John Wilkins's Discovery of a World in the
Moon – *Englishmen's adaptation to a larger cosmos – John Milton and*
Paradise Lost – *Christopher Wren expresses the new concept of space in St
Paul's – Descartes's view of a liquid heaven where stars turn into planets
– the gardens of Versailles – the French Royal Observatory's cosmological
discoveries – Isaac Newton, the* Principia *and the law of universal gravitation – Newton's Scholium about God – reaction to the* Principia *in England
and abroad – the return of Halley's comet.*

Injustice sometimes attends discovery. As the continent found by
Columbus was named after a later mariner, Amerigo Vespucci, so
the lunar craters and seas discovered by Galileo were named by
later opponents, the Jesuits Grimaldi and Riccioli. These early
selenographers called craters and ridges after eminent philosophers and
astronomers, such as Plato, Aristotle, Ptolemy, Grimaldi (a big crater),
and Riccioli (a medium-sized crater), but they awarded only a small
crater to Galileo. They called the dark expanses Sea of Rains, Sea of
Serenity, Sea of Tranquillity, Bay of Rainbows and so on. Naturally
such delicious names both expressed and stimulated intense speculation
about the nature of the moon.

The most influential of these speculators, John Wilkins, was born
in Northamptonshire in 1614, eldest son of a goldsmith who conducted
experiments in perpetual motion. After going to Oxford he took Holy
Orders. A chance conversation while coursing a hare brought him
the post of chaplain to Lord Say and Sele. In 1648 he became Warden
of Wadham College, Oxford. According to John Aubrey he was 'a
lustie, strong-grown, well-set, broad-shouldered person, cheerful and

hospitable; no great-read man, but one of much and deepe thinking, and of a working head; and a prudent man as well as ingeniose.'

When the Puritan revolution broke out, Wilkins, a moderate, looked likely to lose his job, and it was from prudence, we are told, rather than love, that he married the Protector's youngest sister, a widow. At the Restoration Wilkins again showed prudence, came to terms with the new order and obtained an important London parish, later becoming a bishop.

Wilkins took an interest in every branch of science. He invented a universal language of 3000 words, a forerunner of Esperanto, which was written in shorthand and could be learned, he claimed, in a month. He helped to found the Royal Society of London for the Improvement of Natural Knowledge, and became its first secretary.

Wilkins was twenty-four when he published *Discovery of a World in the Moon*, to which he added, for the third edition of 1640, a *Discourse concerning the Possibility of a Passage thither*. In contrast to fantastical Voyages such as Cyrano de Bergerac's *Comic History of the Lunar States*, the Englishman's approach was scientific.

Wilkins took the dark patches on the moon to be sea, the brighter parts land. He thought the absence of dark patches near the circumference was caused by the intervention of luminous vapours, and hence that the moon had an atmosphere. 'Since the sun is in their zenith every month, and doth tarry there so long before he leaves it,' would the heat be intolerable? Wilkins answered that 'the equality of their nights doth much temper the scorching of the day; and the extreme cold that comes from the one requires some space before it can be dispelled by the other; so that the heat spending a great while before it can get the victory hath not afterwards much time to rage in.'

If the moon had a tolerable atmosphere, hills and valleys, since God had made nothing in vain, Wilkins thought it was probably inhabited, but declined to speculate about the inhabitants other than to guess that they might be midway between men and angels.

Could man fly to the moon on artificial wings? Wilkins forsaw a number of difficulties. Fast-running birds such as pheasant and partridge fly less well than the swallow and swift. Man was more like the first group; he would be a slow flier and tire quickly. What about the temperature in space? While admitting that mountain-tops were extremely cold, Wilkins believed that this cold layer might have been made for the purpose of condensing clouds, and above it the air might be warmer. As for the thinness of the air in space, this might make breathing difficult, but Wilkins ingeniously suggested taking along moistened sponges which would thicken the air.

Wilkins reckoned the atmosphere to be twenty miles high. Once above that height man 'may move with far greater swiftness than any living creatures here below, because then he is without all gravity.' Wilkins supposed man might fly a thousand miles a day, and it would take him half a year to cover the estimated 179,712 miles. But without gravity he would need to exert himself less and therefore would require less food. He might enter a kind of somnolent state akin to hibernation, but even so he would require some means of sustenance.

Wilkins considered the opinion of a first-century Alexandrian, Philo the Jew, that if we could hear the music of the spheres we could live by 'feeding on the ear only,' and receiving no other nourishment, 'and for this very reason [says he] was Moses enabled to tarry forty days and forty nights in the Mount without eating anything, because he there heard the melody of the heavens.' Wilkins preferred what seems to us an equally implausible idea, that the pure air of space contained some sort of nourishment, sufficient to keep a traveller going.

' 'Tis the opinion of Kepler,' wrote Wilkins, 'that as soon as the art of flying is found out, some of their German nation will make one of the first colonies, that shall transplant into that other world.' On which he commented drily, 'I suppose his appropriating this pre-eminence to his own countrymen may arise from an overpartial affection to them.'

Though he did not speculate who would be first on the moon, Wilkins had quite clear ideas about how to get there. Either man would devise big bird-like wings or, more likely, he would build 'a flying chariot, in which a man may sit, and give such a motion unto it, as shall convey him through the air. And this perhaps might be made large enough to carry divers men at the same time, together with food for their viaticum, and commodities for traffic. It is not the bigness of any thing in this kind that can hinder its motion, if the motive faculty be answerable thereunto. We see a great ship swims as well as a small cork, and an eagle flies in the air as well as a little gnat.'

'Notwithstanding all these seeming impossibilities,' Wilkins concluded, ' 'tis likely enough that there may be a means invented of journeying to the moon; and how happy shall they be, that are first successful in this attempt!'

When Wilkins's book was published, the Duchess of Newcastle, a writer of what were then called 'fancies', questioned him about his space travellers.

'Where, Sir, shall they be lodged, since you confess there are no inns on the Way?'

With characteristic cheerfulness Wilkins replied, 'Surely, Madame, you who have written so many romances will not refuse my mariners rest and refreshment – in one of your many castles in the air.'

So futuristic an author was bound to be laughed at by some. In an Oxford oration against the Royal Society Robert South ridiculed Wilkins: 'and when he should obtain a domine in the world of the moon, he could make an Archbishop of Cuckoo.' To this it was answered 'that the seat was not vacant, There was a Bishop upon [it] already, called the Man in the Moon ...', for Wilkins in that year had been given the see of Chester.

While he was Warden of Wadham College, Wilkins pursued flying experiments, helped by his brilliant but bilious polymath friend, Robert Hooke. Hooke even constructed a model flying chariot which, with the help of weights and springs, 'rised and sustain'd itself in the Air', but unfortunately we have no details of it.

Talk of flying to the moon made Englishmen quite moonstruck. Samuel Pepys, 8 August 1666: 'It being a mighty fine bright night, and so upon my leads, though very sleepy, till one in the morning, looking on the moon and Jupiter, with this twelve-foote glasse and another of six foote, that he [Reeves, a glass-grinder] hath brought with him tonight, and the sights mighty pleasant, and one of the glasses I will buy, it being very usefull. So to bed mighty sleepy, but with much pleasure.' Pepys, of course, was merely indulging his curiosity about the moon, not trying to situate it in a whole, for there was then still no one satisfactory world-picture available.

Some thought this moon-craze went too far. In *Hudibras* Samuel Butler was to picture the excitement of learned members of the Royal Society who thought they had discovered an elephant in the moon but were chagrined by finding that a fly had lodged in the tube of their telescope!

*

How did Wilkins and his generation react to the dilemmas described in the last chapter? As regards heliocentricism, they were weaned on it, and therefore untroubled by it; Henry Power, physician and naturalist, could write in 1664: 'As for the Earth being the centre of the world, 'tis now an opinion so generally exploded that I need not trouble you nor myself with it.' As regards the issues raised by Galileo's trial, churchmen explained that in Scripture the Holy Spirit sometimes 'conformed his expression to the error of our conceits', and spoke of sun and moon not as they are but as they appear to us.

The disorderly arrangement of the stars, of which Burton had complained, was justified now by the Cambridge Platonist, Henry More:

> The meaner mind works with more nicetie,
> As Spiders wont to weave their idle web,
> But braver spirits do all things gallantly,
> Of lesser failings nought at all affred:
>> So Natures carelesse pencil dipt in light
>> With sprinkled starres hath spattered the Night.

As for the infinite spaces that had made Pascal tremble, Englishmen, reluctant to follow any line of thought to its extreme point, settled back on the comforting view that the cosmos, whether finite or infinite, made in time or eternal, was an expression of God. We have an example in Henry Vaughan's 'The World':

> I saw Eternity the other night
> Like a great Ring of pure and endless light,
>> All calm, as it was bright
> And round beneath it, Time in hours, days, years
>> Driv'n by the spheres
> Like a vast shadow mov'd ...

It was now admitted, as Henry Power put it, that 'the world was not made primarily, nor solely for the use of man, nor in subserviency unto him and his faculties.' But Englishmen lost no sleep over this, and in 1672 John Wilkins's dying words were in marked contrast to Pascal's. 'I am prepared,' said Bishop Wilkins, 'for the great experiment.'

*

If Englishmen of Wilkins's generation attained an enviable serenity regarding nature, it was largely because a great poet forged a world-picture in which the new discoveries of the telescope cohered with the old truths of the Bible.

John Milton was born in London in 1608, the only son of a prosperous scrivener who also composed music. At St Paul's School and Cambridge he became proficient in Greek, Latin and Italian; later, living in the country at his father's expense, he made a name with *L'Allegro* and *Il Penseroso*. When he was twenty-nine he travelled to Italy. He had a warm reception in various academies, especially in Florence, where he recited a Latin poem. 'There it was,' he later recalled, 'that I found and visited the famous Galileo grown old, a prisoner to Inquisition, for thinking in Astronomy otherwise than the Franciscan and Dominican licencers thought.'

Galileo was then aged seventy-four. In the past few months he had lost the sight of his right eye, then of his left: as he wrote to a friend, 'this heaven, this Earth, this universe, which I by my marvellous discoveries and clear demonstrations had enlarged a hundred thousand times beyond the belief of the wise men of past ages, henceforward for me is shrunk into such a small space as is filled by my own bodily sensations.'

The young poet and the aged scientist had points in common: boundless curiosity, a sense of their own importance, a love of music and the countryside. But as yet Milton had shown little interest in cosmology, and in his verse he had treated the sky conventionally.

Galileo and the telescope were so much associated that it would have been natural for Milton, visiting Galileo's villa, to have used the instrument to observe those stars which his host could no longer see; certainly when he came to write *Paradise Lost* he specifically alluded to the 'glass of Galileo'. Whether or not he actually used a telescope on this occasion, his meeting with the blind astronomer could hardly have failed to stir the young Englishman and to turn his imagination to the new cosmology.

Either in Florence or Milan Milton saw a comedy by Andreini called *Adam, or Original Sin*. The actors were God the Father, the devils, the angels, Adam, Eve, the serpent, Death and the Seven Deadly Sins. It opened with a speech by the archangel Michael to a chorus of angels: 'Let the rainbow be the bow of the violin of the heavens; let the seven planets be the seven notes of our music; let Time beat out an exact measure ...'

This information comes from Voltaire, in his essay on epic poetry. 'Milton,' continues Voltaire, 'discovered beneath the absurdity of the work the hidden sublimity of the subject ... He conceived the idea of making a tragedy of Andreini's farce; he even wrote one act and a half. I was assured of this fact by some men of letters who had it from his daughter, who died while I was in London.'

In 1639 Milton returned to England and was caught up in the Civil War. He married in haste; six weeks later his wife left him, whether because of his chilly, distant and domineering character is not known. Milton wrote four tracts demanding legalization of divorce for incompatibility, but later his wife returned and they resumed life together. In 1641 he made notes of 98 possible subjects for tragedy, sixty of them Biblical, including the Fall, but for many years his duties allowed him no time to write the great work for which he felt destined. Then in 1652, at the age of forty-three, Milton became totally blind. Three years later he began to meditate and, in 1658, to write *Paradise Lost*.

Milton's professed subject is the same as Andreini's play: the fall of Adam, and Milton's purpose to 'justify the ways of God to men'. But *Paradise Lost* is really about God the omnipotent lord of the thousand, thousand stars newly discovered, of the sun, the moon, the Earth and interstellar space. Milton is discovering, and describing, a new kind of God in a new kind of universe, whose omnipotence is manifest in immensity, and in scene after scene Milton shows the reader huge distances, vast perspectives, dizzying heights and unfathomable abysses. He does so in a language exactly attuned to his subject: very long sentences containing the many abstract words one would expect from a Latin scholar, using few colours and much contrast of light and dark – this last characteristic perhaps an effect of Milton's blindness.

One example must stand for many. Satan, flying through space towards this world:

> Round he surveys, (and well might, where he stood
> So high above the circling canopy
> Of Night's extended shade) from eastern point
> Of Libra to the fleecy star that bears
> Andromeda far off Atlantic seas
> Beyond the horizon; then from pole to pole
> He views his breadth, – and, without longer pause,
> Down right into the World's first region throws
> His flight precipitant, and winds with ease
> Through the pure marble air his oblique way
> Amongst innumerable stars that shone,
> Stars distant, but nigh-hand seemed other worlds.

In the garden of Eden the Angel tells Adam that knowledge of the true astronomical hypothesis is not essential to man. For the purpose of his epic Milton sets the Earth at the centre of the cosmos, presumably because this tallied better with Scripture. But he included many of Galileo's discoveries, including sun-spots, the true nature of the Milky Way and the Earth-like surface of the moon: 'new lands, Rivers or mountains in her spotty globe'.

Was the moon inhabited? Perhaps: 'Her spots thou seest / As clouds, and clouds may rain, and rain produce / Fruits in her softened soil for some to eat, / Allotted there.' Satan, flying through space, even speculates that there may be other inhabited planets, 'But who dwelt happy there / He staid not to inquire.'

Though more conscious of space than any poet before or since, Milton did not believe space was infinite. God has circumscribed the universe: 'This is thy just circumference, O World!' So space is not terrifying, as it was for Pascal. It may contain other inhabited planets, it may even in future, as Satan suggests upon the lake of Hell, produce new worlds, but it is never infinite. Always it is bounded and contained by God, who alone can say: 'I am who fill Infinitude.'

Milton's epic did indeed 'justify the ways of God to men'. Published in 1667, for the next two hundred years it was to provide readers in England and far beyond with an unforgettably powerful picture of the immensity of space and time, where man, however, was not ultimately lost, but redeemed. It became for many an indispensable corollary to the Bible, and a cosmology whose beauty of language emulated the beauty of the stars.

The cosmology which Milton put into verse made its mark in other spheres of civilized life, and this may be seen in the work of an Englishman of the generation after Milton. Christopher Wren was born in 1632, the son of a rector who became Chaplain to the Forces of Charles I during the Civil War. A delicate child, he received his early education privately from his brother-in-law William Holder, an able mathematician. Then he attended Westminster School, where he invented a 'weather clock' with a revolving cylinder, 'by means of which a record can be kept through the night'. He went on to Wadham College, where he conducted experiments and devised inventions with the Warden, John Wilkins. After a brilliant undergraduate career – 'that miracle of a youth' the diarist Evelyn called him – at twenty-four Wren was appointed Professor of Astronomy at Gresham College.

Wren was physically unprepossessing: round-faced, short and slight of build. But he was modest, cheerful and got on well with almost everyone. At Oxford he took part in amateur theatricals; in London at St Bartholomew's Fair he tried stilt-walking; he enjoyed his pipe and drinking coffee with friends. He taught himself sufficient medicine to look after his delicate health and once, according to Hooke, 'cured his Lady of the thrush by hanging a bag of live bog-lice about her neck.'

In breadth of interest Wren is comparable to Leonardo. During the next ten years, as well as doing important astronomical work on comets and on Saturn, Wren invented an artificial eye, several new ways of engraving and etching, a technique for weaving many ribbons at once by turning a single wheel, 'a way of embroidery for beds cheap and fair', pneumatic engines, a speaking organ, easier methods of whale-fishing, and a way 'to measure the height of a mountain only by journeying over it'.

Wren had a strong artistic side. He made beautifully exact drawings of brain dissections, and drawings of a louse, a flea and a nit, done with a magnifying glass, which Charles II hung in his private apartments. This artistic side gradually got the upper hand, and although for some years yet he was to remain a professor of astronomy, at the age of twenty-nine Wren turned to architecture. He accepted the post of Deputy Surveyor of the King's Works, with the understanding that he would soon have to do something about Old St Paul's in the City of London, which had fallen into disrepair.

Wren's architectural ideas flowed from his astronomy, as set out in his inaugural Gresham lecture. God was 'the finest and greatest of the geometers', and 'mathematical demonstrations built upon the impregnable foundations of Geometry and Arithmetick are the only Truth that can sink into the Mind of Man void of all uncertainty.' It followed that a building in praise of God should be based on geometrical shapes, of which Wren preferred the square and circle.

Through the family library – his father had been an excellent amateur architect – Wren would early have become familiar with Italian Renaissance architecture. This had grown out of the Platonist cosmology prevailing in Florence, which saw God no longer as a Being outside the universe, but as a Presence immanent in it. Church architects tried to express this by reintroducing the dome, which the Romans, notably in the Pantheon, had used as a symbol of the heavens, and by disposing space harmoniously in a well-lit, regularly proportioned building.

In 1665, the year of the Plague, Wren spent eight months in France, and there for the first time saw specimens of this new kind of church, such as the Sorbonne and Val de Grâce. Whereas the low dark roundness of Romanesque suggests a close Heaven, but one owing more to apprehension than comprehension, and the height of Gothic architecture is a reaching up to an external God, its spires piercing through the crystalline spheres, Renaissance churches expressed in their domes the omnipresence of God.

Wren liked these domes, and Louis Le Vau's domed salon at Le Raincy. One of his Oxford friends, Wallis, invented the mathematical symbol for infinity – two circles joined – and Wren doubtless saw in the dome's circular lines an architectural symbol of the same. Visiting Versailles he noted, 'Building certainly ought to have the Attribute of eternal.'

Wren returned home in February 1666. In September the Great Fire destroyed much of London and seriously damaged the already decrepit St Paul's. The cathedral had been begun about 1088 and consecrated under Edward I. It was a long cruciform building with a tower which had once carried the tallest steeple in Christendom, but the steeple had been destroyed by lightning in 1561. During the Civil War it had been used for stabling horses, and by the time the flames of the 1666 fire were extinguished the fabric was no longer safe.

Wren designed a new cathedral in the shape of a Greek cross, four equal arms suggestive of space receding equally in all directions, crowned by a large dome resting on eight piers surrounded by an ambulatory of eight lesser domes. He constructed a wooden model of this design, which still exists. King Charles II liked it, for during his exile he had grown accustomed to French architecture, but the conservative Church authorities found it too modern and to Wren's keen disappointment turned it down. In 1675 Wren submitted a new design, for a cruciform Gothic building with Roman trimmings, its cupola topped by a tall spire. This hybrid was approved, but Wren took care to secure from the King full liberty to make 'variations' during the work 'as from time to time he should see proper'.

The foundation stone was laid in 1675. 47,000 cartloads of rubble had to be removed, while in Dorset's Isle of Portland 700 quarrymen toiled to produce the hard pearly limestone Wren had stipulated for the exterior. During the thirty-six years of building Wren took full advantage of his concession from the King to make radical 'variations'. He soon rejected the cruciform shape for a nave of the same length as the choir, gave greater emphasis to the central space at the crossing and, abandoning the spire altogether, replaced it by a dome resting on a drum pierced by sixteen oval windows.

In October 1708 the last stone was added to the lantern by Wren's elder son, born in the year work was begun. Wren was then aged seventy-six and a tradition holds that he witnessed the ceremony from a basket slung from a giant crane. He was to live another fourteen years, and though in his will he asked for a burial 'without pomp', he was laid to rest with all the honours due to him in the crypt of his cathedral, a tablet above the tomb declaring, 'Reader, if you seek a monument, look about you.'

St Paul's is not only a supremely beautiful church, it is in some respects the seventeenth-century equivalent of a planetarium. It embodies that cosmos Wren had observed through his telescope and the immensity of which he described in his inaugural Gresham address: 'every nebulous star appearing as if it were the Firmament of some other World – bury'd in the vast abyss of intermundious vacuum.' Light is everywhere in the new St Paul's, and space; round-arched bays receding into the distance; saucer domes in the transepts, their spare ornamentation picked out with gold paint, like so many 'other worlds' clustered around the great dome representing the heaven above this world, of central importance now no longer cosmologically but because in it God chose to become man.

*

In France a different world-picture prevailed. It originated with René Descartes. Born in Touraine in 1596 of well-to-do parents, Descartes had studied and become proficient in many sciences, including mechanics, medicine, geometry and optics. After service as a peacetime soldier he settled in Holland, where he could publish more freely on controversial subjects than in Catholic France. He was invited to teach philosophy to the blue-stocking Queen of Sweden, but that attempt to fashion a wise ruler ended even less happily than Plato's. Queen Christina was soon to abdicate and to be involved in a sordid murder, while Descartes, required to begin his lessons at 5 a.m. even in winter, succumbed to pneumonia.

Descartes approached the natural world through mechanics. He declared that the process by which a seed develops is purely mechanical. Dissecting animals, he compared the action of their muscles to weights on pulleys and, more generally, gave a mechanical explanation of all animal bodies, including man's. To that extent he was a materialist. But he was also an idealist, in the sense that from his ability to think clear ideas he deduced his own existence as a thinking agent, and from there went on to deduce the existence of others and of God. The net effect was to separate, radically, the body and the mind.

Descartes carried out in the cosmos a similar separation of matter and mind, this time the mind of God. Rejecting the Platonist immanence dear to Englishmen, Descartes set his God outside the cosmos, and explained the latter in mechanical terms. 'The entire universe,' he declared, 'is a machine.' Space is full of matter; there is no such thing as a void. The properties of matter are extension (size) and motion. Whereas the matter of Earth is mainly hard and when it moves tends to continue in a straight line, the matter of space is *liquid*. Endowed by God with motion, it can be likened to a fast-flowing river. In it whirls or vortices have formed (Anaxagoras's vortex had been of aether; Descartes' vortices are genuine liquids). These in time have formed stars, which are liquids containing particles of light. Light particles traverse space in straight lines and reach the viewer instantaneously.

Stars, including the sun, are subject to change: spots can form on them 'as spume swims on a boiling liquid'; the spots can dim, even extinguish stars. Sometimes, too, one vortex is absorbed by another; then the star of the lesser vortex turns into a comet or, if it lacks the solidity to become a comet, into a planet. Earth is one such planet, and is carried round the sun by the vortex of which the sun is the centre.

There was much in this world-picture attractive to Frenchmen: the mind, undistracted by its body, calmly manipulating ideas clear as daylight, dismissing as unreal anything hazy; sky and Earth explicable by the same laws, the stars no more mysterious than a firework show. Pascal, who had seen this cosmology emerge, of course abhorred it. Not so Louis XIV and his courtiers. They eagerly adopted it, for in a period of fragmented knowledge it offered coherence. It was to contribute much to that extraordinary assurance we find throughout the reign, in the ceremonies at Court as in the sermons of Bossuet, and it was the presupposition underlying much French thinking and art.

One example is the garden. As late as the sixteenth century a French garden bore the mark of the pre-Copernican world-picture. It was composed of disunified parts, separated by high yew hedges, the whole lacking space and enclosed by a high wall. Then in 1661 a man of great originality, André Le Nôtre, who had been trained as an architect, completed for a highly intelligent Finance Minister, Nicolas Fouquet, a garden which gave expression to Descartes's world-picture. Light and bodies in space, left to themselves, move in straight lines, so in this garden straight lines predominate, but there are vortices too, and those are represented by circular pools and fountains. Everything in space is linked, so there are no isolated trees or bushes. Matter, as grasped by the mind, has extension but no colour or smell; so the garden is composed of masses of shade and light, rather than of flowers. The immensity of space is conveyed by receding parallel lines, notably a central avenue between convoluted box-hedge parterres, bisected by a canal, leading to a grotto and fountains partly enclosed by woods, and finally to a grassy slope between trees at the end of which stood a gilded statue of Hercules. The distance from the house to the statue was three kilometres, so that the garden not only symbolized the sky, but seemed to merge with it.

When he was summoned to lay out gardens for King Louis XIV, Le Nôtre had to weave in a second theme. Louis's personal emblem was the sun. He had made full use of it as a unifying force during and immediately after the civil war known as the Fronde: in 1655, for instance, he danced the role of Apollo, attended by the nine Muses, in the ballet *The Wedding of Thetis and Peleus*. Thereafter a golden sun was worked into royal tapestries, furniture, even wrought-iron balconies, and it must have its place also in the King's garden.

At Versailles Le Nôtre had even more space and greater resources than at Vaux-le-Vicomte. He extended the gardens north, south and west, creating long vistas in each of these directions, and made abundant use of water in the form of pools, fountains and a mile-long canal. These *pièces d'eau* mirrored the sky, and when their fountains played became like vortices. Rainbow-coloured sprays were blown out beyond the rim of the pools, and at night, during firework displays, the canal reflected rockets zooming through the darkness like comets.

The notion of the gardens as an image of the heavens was heightened by extensive use of statuary. At the centre of the Bassin d'Apollon, half a mile west of the château, the chariot of Apollo rises at dawn from the sea amid tritons and monsters of the deep. The Bath of Apollo and the Fountain of Apollo's mother, Latona, also have statuary suggesting that the gardens are a visible expression of the Sun God's pre-eminence in heaven, and the Sun King's pre-eminence on Earth. Poets said the vista westwards formed a pathway to the heavens so that the Sun God, father of life, could descend to his domain at Versailles.

Louis XIV took seriously the obligations implicit in his cosmic title. He founded a royal observatory in Paris, and here the director, Jean Dominique Cassini, discovered four satellites of Saturn, naming them, predictably, after the King. With Louis's approval Cassini sent Jean Richer to Cayenne, in French Guiana, latitude 5 degrees north, where he found that a pendulum swings more slowly than in Paris, 49 degrees north, which indicated that gravity is less intense near the Equator than in higher latitudes. Since gravity was known to increase as a body came closer to the centre of the Earth, it followed that the Earth is not a perfect sphere, but an oblate spheroid.

In 1675 Olaus Roemer, a Danish astronomer working at the Paris Observatory, investigated eclipses of the satellites of Jupiter. He found that when the Earth was closest to Jupiter the eclipse took place earlier than expected; when farthest from Jupiter, it occurred later. He concluded that time is required for light to travel from one point to another, and calculated that the sun's rays take eleven minutes to travel to the Earth (the modern figure is 8 minutes 19 seconds).

Roemer's discovery dented Cartesian cosmology, for Descartes, as we have seen, held that light moves instantaneously. But Cartesianism had such a grip on the Continent that it could not easily be dislodged. It continued to provide a framework for writers and artists in France, Italy and Germany; even in England thinkers such as Thomas Burnet, author of *Sacred Theory of the Earth*, found themselves captivated by so unitary a picture. Some English astronomers might be sceptical of how exactly a vortex could cause elliptical orbits, but until the eighth decade of the seventeenth century no one possessed the vision and mathematical expertise to proffer an alternative theory.

*

Isaac Newton was born in the Lincolnshire hamlet of Woolsthorpe on Christmas Day 1642. He was a premature child, so tiny, his mother would often tell him, that he could have fitted into a quart jug. His father, a farmer, died before he was born, and within two years of his birth his mother remarried, leaving Isaac in the charge of his grandmother. The effect of this separation was to impart a deep sense of insecurity.

Isaac Newton was a silent boy, near-sighted, secretive, but capable with his hands. He constructed wooden models, including one of a mill activated by a mouse, and paper kites, which he flew at night with lanterns attached to their tails, frightening the country people who thought they were comets. He liked to observe the motion of the sun, and would mark on the ground with pegs the hours and half-hours cast by its shadow.

When he was fourteen his mother tried to make a farmer out of him but after a few months, on the advice of his former headmaster, sent him to prepare for university. At eighteen he went up to Trinity College, Cambridge. In due course he became a Fellow and Professor of Mathematics.

In appearance Newton was short, and the back of his head protruded noticeably. His craggy face was marked by a strong jaw and a fine nose. His eyes had the absent look produced by near-sightedness, and his hair went grey at thirty. He cared little about clothes or food: in London 'his gruel, or milk and eggs, that was carried to him warm for supper, he would often eat cold for breakfast.' As a person he was locked up in himself. He treated people politely but kept his distance, and he never married. About criticism, however slight, he was exceedingly touchy. According to an astronomer who knew him well, Flamsteed, 'he was of the most fearful, cautious, and suspicious temper, that I ever knew.'

That early separation of a fatherless child from its mother left Newton all his life longing for the union, for the oneness that had escaped him in childhood and was to elude him altogether when his mother died. It is perhaps no accident that much of Newton's thinking shows the same search for oneness. In religion he became a Unitarian; in optics he showed that one white light produced the rainbow; in cosmology, as we shall see, he unified forces previously believed separate.

During his early years at Cambridge Newton pondered the question, How do planets move in space? In order to begin answering he assumed that Copernicus was right about the sun being the centre of our planetary system, and that Kepler's laws were correct. He also had at his disposal Galileo's discovery that a projectile fired from a cannon describes a parabola composed of two straight lines – the impetus forward of the projectile, and the force downward of gravity. Newton took Kepler's third law, that the squares of the period of revolution of the planets are to one another as the cubes of their mean distances from the sun. Simplifying in order to facilitate calculation, he assumed that the planets describe circular paths with constant speeds. He then asked himself what force would be required to restrain the tendency of a planet to move in a straight line by pulling it back just sufficiently to make its orbit circular. He found that the force would equal the mass of one body multiplied by the mass of the other, then divided by the square of the distance between them. This is called the Inverse Square Law.

The plague which had driven Wren to Paris drove Newton to Lincolnshire. He was then aged twenty-three. There, according to Newton himself, the 'notion of gravitation' came to his mind 'as he sat in a comtemplative mood' and 'was occasioned by the fall of an apple'. The moon spins in space near the Earth but without ever falling on to the Earth; an apple falls with a thud to the ground: between the forces at work in these two cases there would seem to be no possible connection. It was Newton's genius to assume a connection. Suppose, he said, that the two forces are really forms of one and the same force, what then? The pull of the Earth on the apple and on the moon will follow the same inverse square law as the pull of the sun on the planets.

The moon is 60 times as far away from the centre of the Earth as the apple, so the Earth should attract the moon with 1/3600th of the pull it exerts on the apple. It was known that a body on the Earth's surface falls in such a way that its speed increases by 32 feet per second every second: in the first second it falls 16 feet, during the next second 48 feet, and so on. In order to calculate the rate of the moon's fall to Earth, the exact distance from the surface to the centre of the Earth was required, and this in turn depended on the size of a degree of latitude. As he was away from Cambridge and his books, Newton used a current estimate of 60 miles to one degree of latitude, which was about one-sixth too little. As a result Newton would have calculated that the rate of the moon's fall to Earth was such that it would fall 12 feet during the first hour, whereas if the inverse square law held true the figure should have been 16 feet in the first hour. Newton could not be satisfied with such a discrepancy. He concluded with characteristic caution that some other factor, perhaps the Cartesian vortices, was at work and put the problem aside for nearly twenty years.

The appearance of a comet in 1680, and of a second comet in 1682 that moved in a direction opposite to the planets, aroused new interest in the paths of these conspicuous but short-lived phenomena, and turned Newton's mind back to astronomy. In June 1682, at a meeting of the Royal Society, Newton heard about the work of Monsieur Jean Picard, who was mapping France with sophisticated instruments, and had found the length of a degree to be 69·1 miles. Newton repeated his former calculations using Picard's value and found that the rate of the moon's fall to Earth exactly corresponded to the inverse square law.

Newton then got to grips with the elliptical motion of planets, using his own brilliant method, called infinitesimals: that is, studying each curve not as something fixed and dead, cut and dried, but as something that varies continuously, like flowing water. Newton considered the motion along an ellipse from one point to another during an infinitely small interval of time, and evaluated the deflection from the tangent during that interval, assuming the deflection to be proportional to the inverse square of the distance from a focus. In this way he established, mathematically, the curve a body will follow under the sun's gravitational attraction.

Newton, it might be thought, would have rushed into print, but that was not his way. He had no particular desire to share his discovery, and no desire at all to publish it and so lay himself open to possible criticism. Instead of proclaiming his calculation, he put it with a mass of other papers, and so things might have remained but for the entry into the quiet of Newton's Cambridge rooms of a dynamic young man.

Edmund Halley was fourteen years younger than Newton, and his antithesis. He was ebullient, had a sense of humour and enjoyed life to the full. He had shown precocious gifts as an astronomer and, requiring observations of Jupiter and Saturn in the southern hemisphere, he coolly left his undergraduate studies at Oxford and sailed to St Helena, where he not only obtained his observations but determined for the first time accurate positions of 341 stars. In 1684 he was working on the orbits of comets and, more generally, on planetary motion. Learning from Christopher Wren that Newton had discussed the second subject some years previously with the famous architect, Halley travelled to Cambridge to see Newton. Halley was then twenty-seven, Newton forty-one.

Halley 'at once indicated the object of his visit by asking Newton what would be the curve described by the planets on the supposition that gravity diminished as the square of the distance. Newton immediately answered, *an ellipse*. Struck with joy and amazement, Halley asked him how he knew it? Why, replied he, I have calculated it; and being asked for the calculation, he could not find it . . .'

Halley managed to convey to Newton his excitement about the mislaid calculation and its implications, and before the young man left Newton had promised to look it out and send it to him. However, he could not find it and had to work out a new calculation.

When this reached him in London, Halley at once perceived its importance and again travelled to Cambridge. Newton was quite content merely to speak of the calculation in his sparsely-attended lectures, but Halley urged him to write up the whole system of gravity in book form. At first Newton refused. But Halley possessed quite exceptional gifts of persuasion. As Rheticus had pressed Copernicus, so he pressed Newton, using just the right mixture of discreet praise, urgency and tact, and promising to look after all the practical arrangements. Newton finally acquiesced, and in fifteen months wrote his *Philosophiae Naturalis Principia Mathematica*.

Robert Hooke, the querulous secretary of the Royal Society, saw the manuscript and claimed that he had made some of the discoveries in it before Newton. The claim was unjustified but Newton, deeply hurt, wished to hold back the whole of Book III, for fear of public criticism. Finally Halley calmed this latest alarm. In 1687 the *Principia* was published, at Halley's expense, and dedicated to the Royal Society.

Book I discusses orbital motion, and shows that gravitation is the only force necessary to account for the elliptic orbits of the planets and their satellites. Book II discusses frictional forces of the sun's vortex as posited in Descartes's highly-considered cosmology, and shows that they fail to explain adequately the movement of heavenly bodies. Book III is introduced with the magisterial claim, 'I now demonstrate the frame of the System of the World.' In it Newton shows how the diverse orbits of the planets can be explained by gravitation proportional to mass, and that they move round the sun inversely as the square of their distance from it. The moon's motion, including all the slight variations in it, is similarly explained, and also the courses of comets.

The effect of gravity on a sphere revolving on its axis is to flatten the poles and to make the equator protrude. So gravity explained what Richer's measurements had suggested, that the Earth is an oblate spheroid.

As the strongest proof of his theory Newton offered an explanation of the ebb and flow of the tides. Until then, the tides were supposed to be caused by the moon only, but no one quite knew how; Descartes, for instance, thought the moon exerted pressure downward on the seas. Newton explained that both moon and sun exert a *pull upward* on the seas.

When do we find the highest tides? Twice each month: at new moon, when sun and moon both pull on the same side of the earth, and at full moon, when the sun and moon are opposite. The reason is that each body creates two tidal bulges on opposite sides of the Earth (altogether the sun's effect is only half that of the moon). These bulges arise as the Earth and the other body (say, the moon) travel around their common centre of gravity: this revolution produces a centrifugal force, exactly balanced at the Earth's centre by the moon's gravity. On the side away from the moon, its gravitational pull is weaker, and the unbalanced centrifugal force pulls the oceans out into a bulge; underneath the moon, the stronger effect of gravity creates the bulge. As the Earth spins each day, the tidal bulges pass any seaport twice; but the interval is stretched from 12 hours to 12 hours 25 minutes as the moon surges ahead in its orbit about the Earth.

The tide rise to be expected from the pull of moon and sun is between 10 and 12 feet; what, then asked Newton, of the bays 'at Plymouth and Chepstow Bridge in England, at the mountains of St Michael, and the town of Avranches, in Normandy, and at Cambaia and Pegu in the East Indies', where the shores are sometimes left dry for many miles, and the tide rise may be 30, 40 or 50 feet and above? Newton attributed the excess to nearby shallow channels which 'hurry' the sea in and out with exceptional violence.

By providing an explanation of many such details as well as of other puzzling phenomena Newton's *Principia* made out an almost un-challengeable case for this law of gravitation, according to which every body in the universe attracts every other body with a force proportional to the product of their masses and inversely proportional to the square of the distance between them.

*

The first thing to be said about Newton's book is that it fulfilled the promise of its title: *Mathematical Principles of Natural Philosophy*. It provided a world-picture based on the measurement of masses and distances, angles and velocities. It was thus in the line of numerical cosmologies initiated by Pythagoras and Archimedes, but it asked and answered a bigger question than they: not only what courses bodies follow in space, but *how* they move.

The last generally accepted world-picture had been that of Shakespeare, with a four-element Earth at the centre of a cosmos of circles. Newton now provided a picture in terms of mass, ellipses and a single law of gravity. Copernicus and Kepler, he said, had been right, though wrong on details. The Earth and planets do go round the sun, the moon does go round the Earth. But there was no need for Copernicus's epicycles nor for Kepler's sweeping motion of the sun, as of a broom. The one force of gravity could explain all movement in orbits. Gravity explained too why things and people were not whisked off a rotating globe, and so silenced one of the chief objections to heliocentrism.

Newton's world-picture offered, in three ways, a satisfying coherence. It fused recent disconnected discoveries, laws and insights into a coherent whole. By explaining how bodies remain close in space without colliding or separating, it provided what may be called coherence in the solar system. Thirdly, by showing that the same law governs a falling apple and a nomad comet it made the formerly distinct sublunary and superlunary worlds cohere.

How did God fit into this world-picture? 'When I wrote my treatise [the *Principia*] ...' Newton declared in 1692, 'I had an eye upon such principles as might work with considering men, for the belief of a Deity.' But true to his opinion that science should limit itself to experimentation and to formulation of laws, in the *Principia* Newton made no explicit mention of God. The furthest he went in the way of speculation was to say briefly that the speed of the planets was declining, and that unless the universe was to perish they would have to be re-activated or re-made (*refacti*).

Newton was therefore hurt when the famous German mathematician-philosopher, Gottfried Leibniz, expressed disapproval of the *Principia* on the ground that its philosophy was materialistic and subversive of the Christian religion. Newton's God, claimed Leibniz, was merely a super-mechanic who could not even create a satisfactory universe, but to keep it going must constantly repair its worn parts.

With the help of a brilliant young divine, Richard Bentley, Newton explored the metaphysical implications of his cosmology and came to see that the transverse impulse needed for each planet to go into a nearly circular orbit could be accounted for only by 'the divine arm'. To the second edition of the *Principia*, in 1713, Newton added a Scholium, in which he stated that no mere mechanical causes could give birth to so many regular motions: 'This most beautiful system of the sun, planets, and comets, could only proceed from the counsel and domination of an intelligent and powerful Being.'

What status did man have in Newton's world-picture? In one sense he was smaller, for his planet had been moved from the centre to the fringe. But Newton's contemporaries liked to think that man had more than righted the balance: in the person of Newton he had so grown in stature that he could fathom the workings of the whole solar system. So Alexander Pope's tribute did not at the time seem exaggerated:

> Nature, and Nature's Laws lay hid in Night.
> God said, *Let Newton be!* and *All was Light*.

Like any profoundly new set of truths, Newton's were slow to win general acceptance. Yet such was the abundance of evidence he offered that by the time of his death in 1727 Newton's principle of universal gravitation was being taught in all English universities.

*

On the Continent it was another story. Frenchmen, following Descartes, understood by force something that pushes. The pull of gravity savoured to them of mystification, on a par with Aristotle's stone falling to Earth because it has a 'longing' for rest. Hence the *Journal des Savants'* opinion that the mechanics of the *Principia* 'do not fulfil the necessary requirements of rendering the universe intelligible'.

Furthermore, in 1696 Fontenelle had published *Conversations on the Plurality of Worlds*. In it he adopted many of Wilkins's arguments as his own and prophesied that communication with other inhabited worlds would one day be possible. But he explained the plurality of worlds exclusively in terms of Descartes's vortices. The book became immensely popular and for long proved a barrier to Newtonianism.

During his youthful visit to England Voltaire, that most Anglophile of Frenchmen, became a convert to Newtonianism. On his return home he encouraged his mistress, Madame du Châtelet, to translate the *Principia* and he himself in 1737–8, as we shall see in the next chapter, was to publish so lucid an account of gravitation that by mid-century educated people in France, as in England, had come to envisage the cosmos in terms of the mutual attraction of bodies.

In Germany most people still preferred Leibniz's view of the cosmos. Leibniz held that there is no empty space. Everything is full of monads, tiny particles possessing both spatial magnitude and thought, pushing one another about but always in such perfect harmony that 'All is for the best in the best of all possible worlds,' a belief that Voltaire was to pillory in *Candide*. Though he believed in the basic, underlying unity of the human spirit, and had tried to reunite the Protestant and Catholic Churches, Leibniz was unable to work out a compromise between his own conscious monads and Newton's lifeless matter spinning in space.

The final triumph of Newton occurred in a most dramatic and appropriate manner. Using Newton's principles and techniques, Edmund Halley had plotted the course of the 1682 comet. He found evidence that a notable comet had been reported at intervals of 75 years as far back as 1066. He assumed that all those reports referred to the comet seen in 1682, then on that assumption calculated the comet's orbit, and predicted that it was due to reappear in August 1757. He added, however, that it would be retarded by the attraction of Jupiter, and therefore might not be visible before the end of 1758 or the beginning of 1759.

In 1758 the curious situation arose of astronomers actually waiting to welcome a comet, yet without knowing what a comet was. The English public held it to be an exhalation of dry air, portending a long spell of hot weather and possibly infectious disease; Cartesians believed it to be a displaced star, while Newton had declared it to be ignited vaporous matter which would eventually fall into, and rejuvenate, a star. Particularly puzzling was why its tail should sometimes precede it.

On Christmas Day 1758 one of the watching astronomers sighted the comet. It had journeyed 5300 million kilometres away from the sun and was completing its vast elongated ellipse, yet its movement then and now was as predictable as that of the Earth and planets, or of a falling apple. The great silvery-tailed light, 12 degrees in length, passed over England and over the grave in Westminster Abbey of one who as a small boy had frightened country people by flying a paper kite with a lantern attached to its tail. It was now known as Halley's comet, but the cosmology it proved was Newton's.

CHAPTER EIGHT

Who Bowled these Flaming Globes?

From Newton's teaching is deduced a new Christian world-picture – Edward Young and his poem Night Thoughts *– James Ferguson and his orreries – Benjamin Franklin and James Madison – Voltaire uses Newtonianism to attack the Church – Rousseau, Bernardin de Saint-Pierre and the sentimental picture of Mother Nature – French Revolutionary cults – Herschel, Uranus and nebulae – Laplace perfects Newtonianism but dispenses with Newton's God – Napoleon's world-picture.*

In his *Principia* Newton had offered an exclusively scientific explanation of the universe. God and Revelation he put on one side as different, complementary and equally valid subjects of enquiry. To his followers was left the task of deducing from his equations religious (or atheistic) propositions.

The most influential of those followers was Richard Bentley. A brilliant classical scholar, Bentley was a warm-hearted, imperious man who enjoyed his glass of port but said of claret that 'it would be port if it could.' He held the post of chaplain to the Bishop of Worcester when he was invited in 1692 to deliver the first Robert Boyle lectures in St Mary-le-Bow church, London. He chose as his theme the Confutation of Atheism, by which he meant the updated Epicureanism held by Thomas Hobbes and his admirers, according to which atoms had combined by chance to produce cosmos out of chaos.

Bentley allowed for the sake of argument that planets could form by gravitation alone, but he considered it preposterous that each should have formed at a point of its actual orbit. If, however, the formation had occurred at a point further away from the sun, or closer to it, gravitation alone could not have brought them to the correct distance. Furthermore, the amount of spin observable in their movements (what

[148]

astronomers call angular momentum) could in no way be accounted for by gravitation.

Bentley went on to consider several other surprising facts. One was the sufficiently wide spacing between planets, which eliminated undue mutual disturbances; this, he held, could be accounted for by 'no natural and necessary cause', only by 'divine art and conduct'.

Perhaps Bentley's most telling argument was a statistical one. He computed that 'the empty space of our solar region ... is 8575 hundred thousand million million times more ample than all the corporeal substance in it.' We may fairly suppose the same proportion to hold throughout the Universe. If, as Epicureans claimed, matter in the original chaos was evenly distributed, it meant that every particle had to be 'above nine million times its own length from any other particle'. Picture two ships fitted with durable timber and rigging, but without pilot or mariners, set adrift from opposite Poles; they might not collide for thousands of years, yet they were ten thousand times more likely to collide than two atoms. Bentley computed that the odds against one atom striking another by chance as it moved through the solar system was a hundred million million to one. It followed that the chances of the sun, moon and planets having been formed by random linking of atoms were practically zero.

That Bentley's sermons went through five editions in five years tells us something about the mood in England. Teachers of science at the universities were as a rule obliged to be ordained ministers of the Anglican Church, and this had laid the basis of what has been called a 'holy' alliance between science and Christianity. Hence the God called for by Newtonianism was generally identified with the Christian God. According to this view the cosmos was infinite in space but finite in time, having been created 6000 years ago, as the Old Testament said. God's interventions to correct the planets' loss of speed implied that only man had fallen, not all nature as had been believed in the seventeenth century. Everything in space was for the best. If Jupiter had four moons, that was because, being so much farther than the Earth from the sun, the inhabitants of Jupiter, if any, needed more light at night. The Flood, as Halley suggested, had been caused by the near passage of a comet, and the world would end either by the near passage of another comet or by the sun being darkened by its spots, which were likened to craters throwing out lava. But little attention was paid to such catastrophes. What fascinated that age was the smooth working of the solar system as an expression of almighty power. Joseph Addison, who did much to popularize Newtonianism in *The Spectator*, was one of many to express that feeling in his Ode, 'The spacious Firmament on high' (1712), which is still sung in English churches. In it Addison hails the radiant orbs

> For ever Singing as they Shine,
> 'The Hand that made us is Divine.'

It was a view that brought great comfort, if we are to judge by Addison's last words: 'See in what peace a Christian can die.'

An argument from the stars to God could, and sometimes did, lead to Deism – Addison's Ode has a tinge of Deism. Some, notably Dissenters, were uneasy about this, hence the note of qualification in a hymn by Isaac Watts:

> Thy voice produc'd the seas and spheres;
> Bid the waves roar, and planets shine;
> But nothing like thy Self appears,
> Through all these spacious works of thine.

What may be called the typically English Christian version of Newtonianism was expressed in a long poem which became one of the most widely read and influential books of the eighteenth century. Its author, Edward Young, was born in 1683, only son of the rector of Upham in Hampshire. A big, cheerful, warm-hearted man, he got on well with people and, being witty, easily made friends. After taking a law degree at Oxford he became a Fellow of All Souls. He published a series of Odes, then two moderately successful tragedies, *Busiris, King of Egypt*, and *The Revenge*. He wrote only at night – or, occasionally, on horseback – whereupon his friend the Duke of Wharton sent him a skull with a candle fixed in it as the most appropriate lamp to write tragedy by. Young chose that genre not because he was gloomy or preoccupied with death but because 'We love to be at once miserable and unhurt; so are we made.'

Young next turned to satirizing the rakes and fops of his day. He had his eye on a political career but, like Donne, was too honest for politics. At forty-four he took Holy Orders, and presently became rector of Welwyn in Hertfordshire, where he was to spend the rest of his life.

At forty-eight he married a widow, Lady Elizabeth Lee, who gave him one son and ten years of happiness. After her death he interested himself in his many friends, and his garden, where he planted a fine avenue of limes. There 'I consider myself as unrivalled Sultan, I am just now going to take a walk in my seraglio, and which will be the happy daisy I cannot yet tell.' He suffered much from rheumatism, treating it at Tunbridge Wells, but remained even into his eighty-third year a cheerful, witty, sought-after friend.

'There is but one objection against marriage,' Young wrote, 'and that is one which the wise world amongst its ten thousand objections never makes; I mean that the husband and wife seldom die in one day and then the survivor must be necessarily miserable.' It was during his misery after Elizabeth's death that Young began *Night Thoughts*. It is a long, frankly didactic blank verse poem in which a devout Christian who has read Newton looks upward for assurance.

> Who turns his eye on Nature's midnight face
> But must inquire – 'What hand behind the scene,
> What arm almighty, put these wheeling globes
> In motion, and wound-up the vast machine? ...
> Who bowl'd them flaming through the dark profound,
> Numerous as glittering gems of morning dew ...?'

Young is in no doubt that the stars are:

> The temple and the preacher! O how loud
> It calls Devotion, genuine growth of Night!
> Devotion! daughter of Astronomy!
> An undevout astronomer is mad.
> True, all things speak a GOD; but, in the small,
> Men trace out Him; in great, He seizes man;
> Seizes, and elevates, and rapts, and fills
> With new inquiries, 'mid associates new.

From so orderly a cosmos Young even learns an ethic:

> The planets of each system represent
> Kind neighbours; mutual amity prevails;
> Sweet interchange of rays, received, return'd;
> Enlightening, and enlighten'd! All, at once,
> Attracting, and attracted! Patriot-like,
> None sins against the welfare of the whole;

> But their reciprocal, unselfish aid
> Affords an emblem of millennial love ...
> Wilt thou not feel the bias Nature gave?
> Canst thou descend from converse with the Skies,
> And seize thy brother's throat? For what? a clod?
> An inch of earth? The Planets cry, 'Forbear.'

But this is no mere Deist's cosmos.

> Nature is Christian; preaches to mankind,
> And bids dead matter aid us in our creed.
> Hast thou ne'er seen the comet's flaming flight?
> The illustrious stranger, passing, terror sheds
> On gazing nations, from his fiery train
> Of length enormous; takes his ample round
> Through depths of ether; coasts unnumber'd worlds
> Of more than solar glory; doubles wide
> Heaven's mighty cape; and then revisits earth,
> From the long travel of a thousand years.
> Thus, at the destined period, shall return
> HE, once on earth, who bids the comet blaze;
> And, with Him, all our triumph o'er the tomb.

Though the poem contains much bombast and a few lines bad enough to make us laugh today, the intensity of Young's belief and his richly inventive applications of his main theme caused *Night Thoughts* to run through edition after edition. Though it also had readers abroad, Englishmen in particular responded to its special compound of level-headedness and mystery, epitomized in these two non-consecutive lines:

> Religion, what? – The proof of common sense ...

And

> Could we conceive Him, God he could not be.

No wonder Charles Wesley was to say that, Scripture excepted, no piece of writing was so useful to him in preaching Methodism.

*

Another Briton from quite a different background and in quite a different medium harnessed Newtonianism to Christianity. James Ferguson was born in Rothiemay, a poor part of the Scottish Highlands, in 1710, the second son of a crofter. One day part of the cottage roof collapsed and the seven-year-old boy watched his father, single-handed, heave it back into position with a stout pole. He was impressed by the feat and began some simple experimenting himself; soon he was fashioning levers, pulleys, model mills and spinning-wheels. He was taught by his father to read and write but the family was so poor that after only three months' schooling James had to become a shepherd boy. At night, beside his sheep, the mechanically-minded Scots lad would watch the planets' zodiacal paths, which he fancied resembled 'the narrow ruts made by cart-wheels, sometimes on one side of the plain road, sometimes on the other, crossing the road at small angles, but never going far from either side of it.'

It happened that the self-taught butler of a neighbouring gentleman knew Latin, Greek and French and was a good mathematician, and this unusual man kindly taught the shepherd boy decimals and algebra. Then young Ferguson worked for a miller, but the man drank and almost starved the boy, who after a year returned home. Ferguson then made a wooden clock, a broken bottle serving as a makeshift bell, and on the strength of this was taken into the house of Sir James Dunbar, where he mended clocks and for the local ladies drew needlework patterns. There were two round stones atop the gateway to Dunbar's house, and these Ferguson painted, one as a terrestrial globe, the other as a celestial globe, correctly aligned on the Pole. He was taken to Edinburgh by Dunbar's sister, Lady Dipple, where he earned his keep by painting portraits and for the first time in his life saw an orrery.

An orrery is a scale model of the solar system, in which a clock-work mechanism moves planets round the sun at their correct relative distances and speeds. One of the first had been constructed in 1700 for an amateur astronomer, the Earl of Orrery, and it had since become a familiar item of furniture in noblemen's houses: a sturdy illustration of Newton's coherent universe.

Ferguson was fascinated by the orrery, made a simple one himself out of wood and knew he had found his vocation.

In 1743, by now married, Ferguson moved to London and embarked on a career as inventor of cosmological models and popularizer of

Newtonian astronomy. He was a man of sober habits and sedate manner, with a lean, careworn face and a big nose. He usually wore a large full stuff wig, which made him look older than his years, knee breeches of black velvet or plush, fastened at the knee with silver buckles, and black stockings. He had one foible: a terror of being destitute. He kept salting away small sums in Edinburgh in case his wife and family should overspend, but neglected to keep account of them or of interest due on the capital. When he was elected to the Royal Society in 1763, he was excused payment of all fees on the ground of poverty, though he had about £1500 deposited in Edinburgh. His wife Isabella bore him four children but she was not robust and died before her husband of consumption.

In 1745 Ferguson constructed a new, more elaborate orrery and published a booklet about its use. He also invented and made an Eclipsareon, for 'Exhibiting the Time, Duration, and Quantity of Solar Eclipses at all places of the Earth', the sun being represented by a candle. He invented and made a Cometarium, 'which shows the motion of a comet moving round the sun, describing equal areas in equal times, and may be so contrived as to show such a motion for any degree of eccentricity.' Finally he invented and made a Planetarium, which accurately represented the revolutions of the Earth and all the other planets. The wheel turning Saturn had no less than 206 teeth and the whole intermeshed as cunningly as clockwork.

With this precision-made apparatus Ferguson lectured up and down the country. On 9 February 1767 the *Bath Chronicle* carried this advertisement:

Mr Ferguson will begin a lecture on the Orrery at the Lamb Inn, in Stall St, at six o'clock this evening; in which all vicissitudes of seasons and the times, causes, and return of all the eclipses of the Sun and Moon will be explained and demonstrated on the principles of Nature, together with the phenomena of Saturn's ring. The year of our Saviour's crucifixion will be astronomically ascertained, and the darkness at the time of the crucifixion proved to have been out of the Common Course of Nature. NB. Each subscriber who attends this Lecture is to pay Half-a-Crown; and no Gold will be changed at the Lecture Room.

According to Ferguson, the crucifixion took place on 3 April, AD 33, 'on which day, my orrery shews that the Moon was Full, and consequently could not then eclipse the Sun.' It followed that the darkening of the Earth was miraculous, that Christ was above the Newtonian laws.

Ferguson's audiences included ladies – 'Some nymphs prefer astronomy to love: Elope from mortal man, and range above.' They crowded to see his models of the solar system driven by a single piece of clockwork which the Wheelwright of the Heavens, as Ferguson was called, would wind up, just as God had wound up the universe. It was a visual equivalent of Young's *Night Thoughts*.

On his travels Ferguson visited the great Silk Mill in Derby, where 26,586 wheels were driven by one great water-wheel, revolving three times a minute. Unlike today, in the eighteenth century an Englishman had only to see machinery to think of God and Ferguson copied into a notebook these lines from a poem about the Mill:

> Now turn from Earth to Heaven thy doubting eyes,
> And read th'amazing Glories of the Skies!
> Worlds without number roll in different Spheres,
> Keep to their seasons, and complete their years ...
> STUPENDOUS POWER and THOUGHT! Enquire no more:
> Own the FIRST MOVER: and, convinc'd, ADORE.

It is pleasing to find the self-taught former shepherd boy being invited to Kew 'to converse on Philosophical and Mechanical topics' with George III, as well as on the turning of wood and ivory, for the King, it was said by a turner who knew his work, 'with industry, might have made 40 shillings, or 50 shillings a week as a hard-wood and ivory turner.'

Ferguson doubtless discussed with the King one of the favourite theories: 'ten thousand ten thousand Worlds, all in rapid motion, yet calm, regular and harmonious ... and these Worlds peopled with myriads of intelligent beings, formed for endless progression in perfection and felicity ... How good must HE be, who made and governs the Whole.'

In his cosmic optimism Ferguson joined hands with Edward Young, as indeed with most Englishmen. From about 1750 to 1800 a comfortable world-picture prevailed in Britain. A universe that ran as smoothly as a clock had been wound up and set ticking by the same God who had, in a different and more specific way, revealed himself in Jesus Christ.

*

A comparison between James Ferguson and Benjamin Franklin is of interest in itself, both being largely self-taught and inventive, but still more so as showing how an American reacted to Newton's world-picture. Franklin was born in Boston in 1706 and after two years' schooling assisted his father, a tallow-chandler and soap-boiler. At twelve he was apprenticed to a printer, and as a printer came to London, where he met the secretary of the Royal Society, Dr Pemberton, at a coffee-house and read Pemberton's exposition of Newtonianism. Settling in Philadelphia, he formed a common-law union with his landlady's daughter, whose husband had deserted her, made a living as a printer and a name as a 'natural philosopher'.

Pennsylvania had been founded by an Englishman, but was still wild and untamed. There were no vicarage and Fellows' gardens as in 'Old England' to encourage contemplation, few privately-owned telescopes. So Franklin was inclined not to commune with nature, but rather to turn it to man's account. His string of useful inventions include the rocking-chair, bifocal spectacles, the Franklin stove (with an open fireplace), and a clock which told the hours, minutes and seconds with only three wheels and two pinions in the movement, which was improved by James Ferguson, and known as Ferguson's clock. His greatest discovery, of course, was that clouds are electrified and that the lightning discharge is a rapid release from the clouds of electric 'fluid'. This too he turned to practical use by inventing the lightning-rod.

The same practical note is evident in Franklin's moral life. Influenced by the Nonconformist background of Pennsylvania and by reading the Deism of Shaftesbury, at twenty-two he drafted a personal creed and code of conduct. It stated his belief in one God, who governs the world through his providence, and who is best worshipped by men doing good to their fellows. There is no mention of Christ. Franklin listed thirteen virtues, headed by Temperance, to each of which he gave four weeks' strict attention a year, as a means to self-improvement.

Franklin knew Young's *Night Thoughts*. But whereas Ferguson preferred the contemplative passages, in his *Poor Richard's Almanac* Franklin cited one of its few pieces of practical advice, this warning against inaction:

> Of Man's miraculous Mistakes, this bears
> The Palm, 'That all Men are *about to live*,'
> For ever on the Brink of being born.
> How excellent the Life they *mean* to lead!

Franklin went on to play a leading part in his country's struggle for independence, and at eighty-one found himself a member of the Convention called to frame a Constitution. If those who had the largest hand in drafting the Constitution were younger than he, many of them, such as James Madison, shared his view that the Newtonian universe proved the existence of God, but that the acceptance of any brand of Christian orthodoxy was intellectually impossible.

Now when men of this background gather to devise a new system of government they will draw their principles from human nature and, consciously or subconsciously, from nature itself, in this case Newtonianism. Around 1786 James Madison wrote an unfinished essay, 'The Symmetry of Nature', advocating a system of human laws akin to those regulating the planetary system, and he was doubtless thinking of Newtonian checks and balances when, in No. 51 of *The Federalist*, he proposed so to contrive 'the interior structure of the government as that its several constituent parts may, by their mutual relations, be the means of keeping each other in their proper places.'

The framers of the Constitution wished the smaller states to retain their independence in internal affairs, not get 'swallowed up'. They spoke much of a 'balancing influence'. Finally they devised a federal system like no other political system, yet very like Newton's. Instead of specifying central government power, they proclaimed a loose definition of the means by which that power could be exercised. Sovereignty resided no place in particular: power was not quite anywhere.

These men pictured the dome of the heavens controlled throughout by immutable laws, which are reducible to writing; complete uniformity – the sun, the moon and an apple being subject to the same laws; no hierarchy of elements, where fire would be 'superior' to earth or water; instead bodies have importance according to mass, i.e. quantitative factors. The Constitution as finally agreed established that all men are free and equal before the law, which is consigned to writing, but that the majority rule. Furthermore, just as the regular flow of the tides follows from the pull of more than one body, each in a different direction, so the Constitution established three distinct bodies – a Presidency, a Legislature and a Supreme Court – to regulate the tide of human affairs.

The influence of the current world-picture on the Constitution-framers is arguable, for, being at best a presupposition, it was not put into words. But look at early Washington. Streets ran, as they still do, east and west, north and south; the federal government building had a low, flattened dome, like a celestial arena, and the Hall of Representatives, as built to Thornton's design, was a combination of ellipses with different curvatures.

*

Turning now to France, we find a world-picture different from the American and the cosily English. In eighteenth-century France a man's success in society depended upon good birth, good looks and good nature. These qualities had been denied François Arouet, who called himself Voltaire. Though he possessed a superb intelligence, he was only a lawyer's son, short of stature, plain of face and lacking in manly panache. This embittered him early in life, so that Young, who met Voltaire, could describe him thus: 'Thou art so witty, profligate, and thin / At once we think thee Milton's Death and Sin.' Shortly after his return from England the young man whom society failed to appreciate began a one-man war against that society, in particular against its corner-stone, the Catholic Church. First, Voltaire published *Lettres Philosophiques*, which vaunted England's freedom of conscience and political liberty, and four years later *Eléments de la Philosophie de Newton*. It says much about conditions in France that because Cartesianism was the 'official' cosmology, Voltaire could not obtain a licence to publish in Paris, and was obliged to issue his book in Amsterdam. On the other hand it would be ingenuous to think that Voltaire wrote the book to purvey certain truths of physics.

Newton, it will be remembered, believed that the scientist's job was to describe the forces operating in Nature, not to offer metaphysical explanations. He had replied to Leibniz's charge of materialism by adding a Scholium to the second edition of the *Principia*, but the Scholium was no more than a postscript. This did not suit the Frenchman, whose main concern was his war against the Church, so he decided to stand Newton's method on its head. Voltaire treats the nature of light in Book II of the *Eléments*, the theory of gravitation in Book III, but Book I he devotes to a discussion of God, Man and the Soul! Indeed, on one of the first pages Voltaire declares: 'Newton's whole philosophy necessarily leads us to the knowledge of a supreme Being, who has freely created and arranged all things.'

Newton's upbringing, education and training, the whole framework of science in his day, were based on belief in the Lord God of the Bible, who cares for man his servant and wishes man to come to know him. Voltaire discards this altogether in favour of Newton the timeless sage. 'Newton,' he writes, 'found no argument for the existence of God more convincing or more beautiful than Plato's.' Newton's unconcern about food becomes under the Frenchman's pen a Brahmin-like reluctance to eat meat, and, adds Voltaire in a sentence more redolent of French than of English *mores*, it was from compassion that Newton never allowed animals to be cooked alive over a slow fire, 'in order to make their flesh more succulent'. 'This compassion for animals turned to real charity towards his fellow men. In fact, without a humane attitude to other men, a virtue which embraces all virtues, a man would scarcely deserve to be called a philosopher.' Forgetting that Newton had been for most of his life a recluse, Voltaire announced: 'Newton thought that the urge we all feel to live in society is the basis of natural law.'

With clever touches such as this Voltaire depicted a sage of no particular land or epoch, living in accordance with natural law, and discovering other natural laws about the cosmos which prove that the universe was made by God – not the Christian God but the kind of wise God Indian and Chinese sages believe in – 'naturally', without any help from the Church.

The importance of Voltaire's deceitful popularization was very great. It linked the latest cosmology inextricably with Deism, although in fact Newton had been at pains to keep the two apart and although in the Scholium he specifically rejects the Deists' God, 'a being, however perfect, without dominion'.

In 1755 an earthquake destroyed the centre of Lisbon, killing 20,000 people. It was the worst earthquake in recorded history, and caused consternation throughout Europe. Voltaire wrote a poem to drive home his point that whereas the Christians' God could never have permitted such carnage, it was quite compatible with the Deists' God, Creator of the cosmos but not directly concerned with the welfare of man.

While Voltaire, with the stars as chalk-marks on his blackboard, taught the scientific reasonableness of Deism, Jean-Jacques Rousseau was proclaiming a sentimental approach to Nature. Marked by the death of his mother a few days after his birth, he turned for security to the Earth as a kind of mother, and shied away from the stars, at once challenging and disturbing. Perhaps because he was short-sighted, we find in his books astonishingly few accurate descriptions of flora and fauna. What we do find is a mixture of haziness and self-centredness. He particularly values bountifulness – hence his predilection for exotic lands where banana and coconut trees provide abundantly for 'noble savages'. There, indeed in any 'primitive' setting, man's intrinsic goodness comes to the fore: he responds to the beauties as well as to the bounties of Nature, and these fill him with a melting feeling between joy and sadness which Rousseau chooses to call religious adoration.

Bernardin de Saint-Pierre, a disciple of Rousseau, was a more observant naturalist than he and, since he had lived in Mauritius, an expert on noble savages. He published his *Etudes de la Nature* in 1784, at Rousseau's prompting: its 900 pages are the fullest statement in French of the sentimental view of Nature. Here are some of its claims:

Brambles have thorns to prevent sheep disturbing birds that roost there, while catching enough wool to be useful for nest-building. Birches and larches shed their foliage in winter to provide animals with bedding, but the pine remains evergreen in order to afford them shelter when it snows. Birches are specially large near lakes, so that man can use them as dug-out canoes. Trees bearing cinnamon, nutmeg and cloves grow near swamps, so that their fragrance may dissipate the stench. Aconites are hideously coloured and smell bad to warn man that they are poisonous. There is white foam on black rocks to warn the sailor against reefs. Plums and peaches ripen in summer because it is then that man wishes to quench his thirst. A melon has ridges so that it may be divided and eaten at the family table. The reds and yellows of parrots and toucans are Nature's way of pleasing man, who might otherwise find the green jungle monotonous to his eye.

Earthquakes are not really bad: they purify the air, and since they usually occur in a given region, men there can build houses of wood – as the survivors of Lisbon have done. Some have said that Benjamin Franklin's lightning-conductor has 'disarmed God'. That is impious and false. Lightning is not an instrument of God's anger, it is necessary for cooling the air during the heat of summer.

In this good world man too is naturally good. Bernardin offers as proof the fact that man takes pleasure in the sight of a well-proportioned animal, and this pleasure increases when he sees it in its natural habitat: for example, he takes more pleasure in seeing a turtle-dove nesting in the wild than when it is caged in an aviary.

This glib finding of purposes in Nature was to be taken up in England by William Paley and, as we shall see, with disastrous consequences. For the moment, in France, it had serious political effects. Since the Author of the world designed such abundance and harmony, if men are unhappy, then the fault lies with society. 'Man is born free and is everywhere in chains.' This argument contributed to the French Revolution and through the Revolution to the attainment of Voltaire's original aim, the humiliation of the Catholic Church.

In 1793 the Mass was superseded by Voltaireanism. On 10 November that year, in the *ci-devant* cathedral of Notre Dame, now a Temple of Reason, a stage mountain was constructed, with a round temple on top, inscribed *A la Philosophie*. From it stepped an actress, dressed in white, with a red bonnet, carrying a pike. She sat on a mound of greenery and was hailed with a hymn beginning:

Descends, O Liberté, fille de la Nature ...

In another ceremony, at the Bastille, a statue of Nature was unveiled, inscribed, 'We are all her children.' From Nature's breasts two streams of water gushed into a pool. Water played an important part in Revolutionary imagery, a feminine symbol to replace the male symbol of kinghood, and as many as possible of the new *départements* were named after rivers. Sun, moon and stars were conspicuous by their absence, for the *fille de la Nature* very quickly became an *enfant de la Patrie*, and imagery must reinforce those aspects of Nature peculiar to France, not the heavenly bodies which shed their light on all men without distinction.

In 1794 this cult was changed by Robespierre to Rousseauism. Robespierre had met Jean-Jacques in the last year of his life, found 'his face darkened with the sorrows inflicted by men's injustice', and embraced his view that the world was made for man by a Supreme Being, revealed not of course in Christ but by the stirrings of a pure heart. Robespierre fostered his Deism by chosen passages of Young's *Night Thoughts*, a translation of which he kept under his pillow. For the cult of Reason he substituted the Festival of the Supreme Being, held in the Champ de Mars on 6 June 1794. Here men assembled bearing oak branches, women bunches of roses. A decorated ox-cart rolled up, laden with farm tools, emblems of honest toil. Robespierre burned a representation of Atheism, which was immediately replaced by one of Wisdom. 'Frenchmen,' he declared to the crowd, 'you are at war with Kings, you are therefore worthy to honour the God-head.'

The Revolutionary cults are important because they show the logical conclusion of the teachings of Voltaire and Rousseau. Although short-lived, they illustrate a climate of opinion under which many lost their faith, including scientists such as Laplace.

Pierre Simon Laplace was born in 1749 in the lush cheese-making pastureland of Lower Normandy. He came of a very poor peasant family but won a bursary to the local military school and stayed on as a teacher. Then he went to Paris where d'Alembert got him the job of mathematics master at the Ecole Militaire.

Laplace was of medium build, pleasant-looking, with a precise manner, and despite a weak voice he rose rapidly. At the Revolution he declared himself a republican, discarded his Catholic faith and helped Napoleon assemble scientists for the Egyptian campaign. When Napoleon came to power Laplace was appointed Minister of the Interior. But the mathematician-astronomer looked for subtleties every-where and, according to Napoleon, brought to his job 'the method of infinitesimal calculus'. After six weeks he was relieved of his duties and made a Senator. He married and had at least one son, but he seems not to have been a very strong character. As soon as Louis XVIII returned, he offered the King his services with what colleagues con-sidered unseemly haste. He was created a marquis, after which he took pains to hush up his humble birth.

Laplace began his most important work following the discovery of a new planet, in 1781, by William Herschel, a Hanover-born amateur astronomer working in England. Scarcely visible to the naked eye, the planet revolves round the sun at about twice the distance of Saturn. Herschel called it Georgian, after King George III, but the name which finally stuck was Uranus, the Greek word for 'Heaven'. Herschel also discovered two of its moons, which he named Oberon and Titania. Keats was to take up the discovery of Uranus in his sonnet on dis-covering Chapman's translation of Homer:

> Then felt I like some watcher of the skies
> When a new planet swims into his ken ...

Laplace was one of the first to calculate Uranus's orbit. He found that when it came close to Jupiter and Saturn the orbit showed perturbations. Helped by another brilliant mathematician, Joseph Louis Lagrange, Laplace calculated that these, and other slight perturbations in the movements of other planets, could all be explained by Newtonian gravity. Many astronomers had been puzzled by these and other seeming instabilities and took them as evidence that the universe was 'running down'. Not at all, said Laplace. Any perturbations could all be explained by mutual attraction and would eventually right themselves, though it might take hundred of years. The solar system, Laplace declared, 'will always oscillate about a mean state from which it will deviate but by very small quantities'; it was in effect a perpetual-motion machine.

In 1796 Laplace published an *Essay on the System of the World*. He started from what he called the 'astonishing' fact that all the planets move round the sun from west to east, and nearly in the same plane, the satellites move round their planets in the same direction and in approximately the same plane as the planets (this is not true of Uranus's satellites, but that fact had not yet emerged), and finally the sun, the planets and their satellites all rotate on their axis in the direction of their movements of projection, and almost in the same plane. 'Envisaged in this way the planetary system offers us, therefore, 42 movements the planes of which are inclined to the sun's equator by, at most, ninety degrees. The odds against this happening by chance are more than four thousand billion.'

How then did it happen? Laplace again drew on Herschel's work. Using powerful telescopes he made himself, Herschel had discovered more than two thousand nebulae, that is, luminous patches, some composed of clusters of very distant stars, others apparently made of 'a shining fluid, of a nature totally unknown to us'. Some nebulae even appeared to change shape.

Laplace argued that the planets had at one time been fluid, for only so could they have become, in the course of long spinning, flattened at the poles. Both their fluid state and the direction of their movement could be explained if they had all originated from an immense, incandescent nebula, rotating from west to east, of which our sun is a relic. As it cooled, the nebula contracted, and rings of matter detached themselves and were left behind, the rings so formed all lying in the same plane. The rings being unstable, however, each broke up into rotating masses which eventually coalesced to form a separate planet.

Contraction of the planets produced satellites, and such phenomena as Saturn's ring, which is a satellite in the making.

In an age when Cavendish, Priestley and Lavoisier were making spectacular progress in the study of gases, Laplace's *System* offered a perfected restatement of Newton's mechanical cosmos, and instead of the Englishman's 'divine arm' impulse, a chemical explanation of its origin. Gaseous matter, expanding or contracting solely according to changes in temperature, could of itself have formed a solar system. No external movement would have been required, said Laplace, and therefore no unmoved Mover.

Laplace presented Napoleon with a copy of *Celestial Mechanics*, in five volumes.

'You have written this huge book,' said Napoleon, 'without once mentioning the author of the universe.'

'Sire, I had no need of that *hypothesis*,' replied Laplace.

Napoleon's world-pictures – for he held several in turn – are of interest and to some extent typical of the France of his day. As a boy in Ajaccio he held the usual Catholic view of the heavens. At military academy in Burgundy he learned a lot of mathematics but less science than an English schoolboy learned, and nothing about astronomy – in a Catholic country now an awkward topic. As a subaltern he became keen on Rousseau. 'Go down to the sea shore:' he wrote at this time, 'look at the morning star as it sets majestically on the breast of the infinite: melancholy will overcome you ... None can resist the melancholy in Nature.'

Napoleon dropped this emotional response to landscape and sky under the influence of the cult of Reason, and as a leading general of the Republican armies he counted among his enemies the Catholic Church both as a temporal power and in the kind of role it had adopted to Galileo. He would have described himself then as an agnostic. Yet, sailing to Egypt, he had lain on deck, asking his scientists whether the planets were inhabited, how old the Earth was, and whether it would perish by fire or by flood. Many, like his friend Gaspard Monge, the first man to liquefy a gas, were atheists. Then Napoleon would point to the sky. 'You may say what you like, but who made all those?'

When he became First Consul and started reconstructing Napoleon saw that belief in the Christian God was a fact of life in France and, though he could not share the popular belief, he did restore the Catholic Church. But he had no time for an unscientific interpretation of nature, whether clerical or lay. When Bernardin de Saint-Pierre complained that scientists scoffed at his sentimental reading of nature, Napoleon asked sharply, 'Do you understand differential calculus, Monsier Bernardin? – Well then, go and learn it.'

Yet when Laplace produced his nebular hypothesis, Napoleon found its atheism difficult to swallow. It was more than esteem for his devoutly Catholic mother, more than his experience of how men can behave when they throw God overboard. Napoleon had pondered the perennial human perplexities: how the spirit functions through the flesh which at the same time seems to burden its activity, how heroism based on sinew and muscle may take on the aspect of divinity. Yet any incipient belief in God did not deter Napoleon, in 1814, from trying to commit suicide by swallowing poison.

A prisoner on the rainswept rock of St Helena, in December 1817 Napoleon said to his companion Gourgaud: 'The intelligence watching over the movement of the stars – only a property of matter, you know – do you really believe that that watches over men's actions too, and keeps account of them?' 'Sire, I believe in God, and would be very unhappy to be an atheist.' 'Bah! look at Monge and Laplace. Vanity of vanities!'

Napoleon enjoyed getting a rise out of Gourgaud, just as he did out of the atheist General Bertrand by praising the Catholic Church. The truth is, Napoleon was still divided. He felt sympathy with the scientists, but what then of courage and glory and France – all those values which became mere vapour if life had no meaning? And divided Napoleon died – sending for a Catholic priest, but still unsure whether man was just a spin-off from a nebula.

CHAPTER NINE

From Detachment
to Involvement

Kant asserts that space is something the mind imposes on reality and denies the possibility of natural theology – Goethe starts a reaction against a clockwork universe explained by mathematics – William Blake turns from a scientific to a Biblically poetic view of Nature – in The Ancient Mariner Coleridge *shows Divine justice working through cosmic phenomena – in the favoured Lake District Wordsworth finds spiritual peace – Shelley, the aetherial atheist – Charlotte Brontë draws strength from the Northern Lights.*

In eighteenth-century Germany cosmology was being approached in a radically different way by means of a radically different method. Immanuel Kant was born in 1724 in Königsberg, fourth child of a poor saddle-maker. His mother, who belonged to the devout sect of Pietists, gave her son a religious upbringing that stressed right-doing more than dogma. Kant studied for a doctorate in philosophy at Königsberg University – where his one recreation was billiards – and for nine years taught as a tutor in rich families. In his spare time he studied astronomy. Perhaps under the influence of Young, a favourite poet, he took 'ecstatic delight' in the starry sky. In 1755 he became tutor at the University, and in 1770 professor of logic and metaphysics, a post he held until retirement.

Kant was about five foot tall, thin, narrow-chested, his right shoulder higher than his left. His cast of face resembled a monkey's save that his brow was broad and he had attractive blue eyes. He was so sensitive that the scratching of pens disturbed him when he lectured and a newspaper damp from the press could give him a cold. He fussed over his health and kept a close eye on the room thermometers but he rarely took medicine, citing the epitaph of one who had over-indulged in that respect: 'So-and-so was well, but because he wanted to be better than well he is here.'

Kant considered women decorative and emotional and incapable of principled action. He liked conversing with them, but never on serious subjects, to which their minds were not attuned. 'They need not know more about the universe than is required to make a view of the heavens, on a beautiful evening, affecting.' Thrice he thought of marrying but by the time he had calculated how far an emotional life would interfere with his philosophizing the lady in question had gone.

Kant mastered all his appetites save a craving for English cheese and French crystallized fruit. He was indifferent to music and painting, the only picture in his scantily furnished bachelor house being one of Rousseau. His life was as regular as the cathedral clock, and the neighbours knew it was exactly half past three when Kant, in his grey coat and holding a Spanish cane, stepped out of his door and walked towards the small avenue of limes, still called after him 'The Philosopher's Walk'. Eight times up, eight times back, slowly in summer, so as not to perspire.

Kant enjoyed serious conversation and often invited a few friends to dinner. In his younger days he was witty and made jokes but from middle age onward even this spring of feeling closed and he rarely laughed. He disliked replying to letters, and those he did write were devoid of intimacy and affection. He believed in God, but rarely attended church and never prayed. Though he lived in the same city with his sisters, who had been servant-girls and married men of that class, not once in twenty-five years did he speak to them.

'He was so occupied with his own thought,' wrote one who knew him, 'that he found it difficult to appreciate the spirit of another philosophy; in old age he found it altogether impossible.' In 1798, when Napoleon sailed from Toulon, Kant asserted that his destination was Portugal, since that would most harm England. So satisfactorily did he demonstrate to himself this supposed stratagem of Napoleon, that even after the French had landed in Egypt, and the Government had announced the fact to all Europe, Kant still asserted that the expedition was against Portugal, and that the announcement was a pretext to mislead the English.

As a young man, more open to others' opinions, Kant had followed Leibniz in believing that all was for the best in the best of worlds: even the Lisbon earthquake, according to this view, had beneficial results, since shocks emanating from it were believed to have caused medicinal springs to appear in Toplita, benefiting the people of that town.

Kant's predilection for regularity and logic and his youthful optimism are to be found in his *Universal Natural History and Theory of the Heavens*, published at the age of thirty-one. Drawing on the work of Thomas Wright of Durham, who claimed that the rings of Saturn were in process of coalescing into moons, and that this was a new solar system being formed, Kant asserted that the universe is composed of an infinite number of solar systems, being born, dying, and phoenix-like being born again. Most such systems have inhabited planets. As distance from a sun increases, the inhabitants compensate for lack of heat by possessing very elastic, supple bodies, and quick, right-thinking minds. The inhabitants of Jupiter will be so superior to us as to view Newton as we view a monkey, while the inhabitants of hot Mars will be in every way dense. Kant, a great man for laying down laws, actually called this view a law of the universe.

What signs are there of God? 'Air, water, heat ... and all those useful consequences without which Nature could not but remain desolate, waste and unfruitful ... if their natures were necessary by themselves and independent, what an impossibility would it be that they should so exactly fit in to each other with their natural activities and tendencies, just as if a reflective prudent choice had combined them!' Kant concluded: 'There is a God, just because Nature, *even in chaos*, cannot proceed otherwise than regularly and according to order.'

Were God to act immediately in the universe, all planetary orbits would be circular and in exactly the same plane; evidently God created the universe and the laws of movement, then allowed them free play. In this manner the German set a mechanistic explanation in the framework of a providential universe.

Leibniz had said there was no space – everything tiny particles; while Newton had declared space to be something real, existing even if there were no bodies in it. As he grew into middle age, Kant felt increasingly doubtful about both views. Then he began reading the Scottish empiricist, David Hume. Hume asserted that when we speak of a cause, this is not something occurring in the real world, it is a connection our mind makes between two events. That pronouncement, says Kant, awoke him from his 'dogmatic slumber'. He began to re-model his whole philosophy and in so doing made it an expression of his aloof, self-centred life in which mind, devoid of emotion, analysed its own thinking.

The result, published in a series of books from the age of fifty-seven, reversed Kant's earlier position. The world as it really is is closed to man. All man can do is apprehend it through his mind, and in so doing he imposes on it a grid. Part of this grid is space, part is time, part is the notion of causality. When we think we see the stars in space on a clear night, what really happens is that our mind coordinates sense-data of the stars according to unchanging laws, and part of this coordination we call space.

This was a profoundly original view and from it followed important consequences. Because we cannot know the world as it is, only our picture of the world, we cannot argue from the orderly movement of the planets, say, to the existence of an Orderer. Metaphysics is a sham; natural theology a lie.

Kant, however, did not stop there. He went on to argue that man possesses a conscience, conscience presupposes freedom, and can lead to virtue. Virtue is not always rewarded in this life, and since nowhere in living Nature do we find purposeless urges, we are led to believe that virtue will find its reward in another life. The God who cannot be found in Nature is resurrected by Kant upon practical, ethical grounds.

Kant's effect on cosmological thinking was not unlike Picasso's, a century and a half later, on representational painting. Both claimed that a hallowed tradition was no more than a sum of conventions, and sought to establish a new, more valid reality built up from the nature of the self. But whereas Picasso's self embraced a rich wide spectrum of activities, Kant's was limited to the pure white light of a lonely mind reflecting austerely on the nature of thought. He retracted all his earlier, warmer writings, believing them to be part of the sham. What remained was an isolated 'pure' intelligence, cut off from feeling and from people, buoyed up by a residue of early Pietism. His was a cold, stern, aloof view of the cosmos, yet not without its attraction for some. Lines from one of his books, inscribed on his tomb in Königsberg Cathedral, convey the man's quintessence: 'Two things there are which, the oftener and more steadfastly we contemplate them, fill the mind with an ever new, an ever rising admiration and reverence: the starry heavens above, the moral law within.' But only the moral law led to God.

*

It would be difficult to imagine a man more different from Kant than Wolfgang Goethe. He was born in 1749 in Frankfurt, eldest and only surviving son of a prosperous lawyer. His mother, half her husband's age and only eighteen when she gave birth to her son, played a tenderly influential role in Wolfgang's childhood. She taught him to love the Bible and imparted to him her animism, that curious feature of the German temperament which had so struck Tacitus. 'Air, fire, earth and water I represented under the forms of princesses ... As we thought of paths which led from star to star, and that we should one day inhabit the stars, and thought of the great spirits we should meet there, I was as eager for the hours of story-telling as the children themselves.' Young Goethe made an altar to the princesses from a music-stand, decked it with minerals and flowers, and crowned it with a flame lit by a burning-glass from the rays of the newly-risen sun.

Goethe had been blessed with a strong body and striking good looks, keen intelligence and a poet's sensibility, a gift for friendship and a gift for love. But what he concentrated on were his weaknesses. While reading for a degree in law, he grappled with these. To rid himself of a terror of ghosts and demons, he would walk through graveyards at night; to rid himself of his horror of diseased and dead bodies, he worked in a dissecting-room; to cure a fear of heights, he would climb to the lantern of Strasbourg Cathedral, sit there for a quarter of an hour, then step out on a ledge and force himself to look down on the tiny figures below.

Goethe mastered his emotions, not in order to escape feeling, as Kant had done, but in order to feel more fully and fruitfully. As a leader of the *Sturm und Drang* movement he plunged into Nature, swimming and skating, quoting Shakespeare, many times falling in love, eager to experience everything. At twenty-five he published *The Sorrows of Young Werther,* a novel about an idealist dreamer who falls in love with a married woman and in despair shoots himself. The novel brought him fame. Two years later he had the great good fortune to win the friendship of young Prince Karl August, and was invited to help administer and reform his pocket-sized State of Weimar.

Now began a new, even fuller life. After spending a whole winter night drinking and dancing, at dawn Goethe would be out on a hunt. No ditch was too broad for him, no hedge too high, no rocky bridlepath too hard. Next morning he would be in the Prince's council chamber, drawing up plans for new roads, or be out inspecting mines and irrigation canals. He wrote and produced plays for the local theatre. He fell in love with Charlotte von Stein, wife of the Master of the Horse, and began what the Germans call a 'soul-friendship', which, while observing the conventions, enriched and refined him.

At thirty-seven Goethe felt his administrative duties making him narrow. Off he dashed to Italy, there discovering physical love, classical art and the classical view of Nature, which reinforced his boyhood animism. This journey too was an enrichment which allowed him on his return to complete his great play *Iphigenia*. Then he married and had a son. In *Faust* he expressed man's desire to know all things, an ambition he pursued in real life. Forestry and agriculture led him to study botany, mining led him to mineralogy and geology.

'The most human of men', so the poet Wieland described young Goethe, and Napoleon's tribute when he met him in middle age, '*Voilà un homme!*' emphasizes how fully Goethe had developed his many gifts. Involved with friends and affairs, making himself useful in a hundred ways, Goethe stands at the far end of the human spectrum from Kant, and his view of the natural world was correspondingly different.

Goethe started from the conviction that man is inseparably part of the natural world. Not just of one aspect of that world, but of the whole of it. He has reciprocal relations with everything else in the cosmos, with sun and moon, hill and sea, tree and flower. Futhermore man has three needs: to worship, to produce and to contemplate. Any interpretation of Nature must satisfy these.

Unwaveringly involved in the business of living, Goethe naturally rejected Kant's view that an insuperable gulf divides man from things as they are. The true gulf, said Goethe, comes from being one-sided, responding to the world with our mind only. 'Everything fragmentary is blameworthy.'

A mathematical interpretation of Nature would be fragmentary, therefore blameworthy. In conversation with his friend Eckermann Goethe declared: 'As if, forsooth! things only exist when they can be mathematically demonstrated. It would be foolish for a man not to believe in his mistress's love because she could not prove it to him mathematically. She can mathematically prove her dowry, but not her love.'

On the grounds that they too were only fragmentary explanations, Goethe rejected the mechanical cosmologies of Newton and Laplace. By isolating linear causal sequences they failed to do justice to Nature's multi-dimensionality. Just as the mechanical theory of sound tells us nothing of the quality of music, so such cosmologies fail to convey the inner life of Nature as a whole. Worse, 'number and measurement in all their baldness destroy form and banish the spirit of living contemplation.'

In Goethe the visual faculty was highly developed and he asked of a cosmology that it should be picturable. This was not an unfair demand, for about half of the fibres that convey sensation to our brains stem from the optic nerves and most men are primarily visualizers. But it led Goethe into an awkward position. Newton had declared that white light is composed of rays, each a different colour, and demonstrated this by passing white light through a prism. Goethe found Newton's theory impossible to square with everyday experience. He could not believe that the pure sensation of white could possibly be caused by anything save a simple, uncompounded substance. He preferred to hold the view that colours arise from a mixture of white light and a turbid medium pertaining to objects which causes them to take on colour in daytime.

Goethe's theory of colour does not fit the facts and is indefensible. Yet he clung to it. This was a quirk and does not detract from the importance of the rest of his world-picture.

Goethe expressed his view of Nature in symbols, believing they tell us more than concepts do. The universe, he said, is unfathomable, because God is within it, flowing forth into activity, directing the flow of every stream, the growth of every leaf. Goethe never quite threw off his youthful animism, so he sometimes gives us the never very convincing notion of a world-soul. More often he speaks as a pantheist. This led him, as it had led Seneca, into difficulties, for if God is in all things, including us, how do we separate ourselves sufficiently to conceive him, and to act as independent persons?

Goethe accepted that man is composed of antinomies, what Pascal termed the Angel and the Beast, but in order to reconcile them he saw no need for Christ's saving grace. God was already in the world, sanctifying.

But the cosmos is not made for man. To claim that would be, again, to isolate some particular feature of Nature at the expense of all the rest. The most we can say is that man in his spiritual aspect is the highest manifestation of the Divine on Earth, and that part of the flow of Nature is directed towards man. 'It would have been for [God] a poor occupation to compose this heavy world out of simple elements and to keep it rolling in the sunbeams from year to year if He had not had the plan of founding upon this material basis a nursery for a world of spirits.'

In offering a cosmology that would satisfy all man's needs, Goethe

joined hands with classical man, who had also known the joy of wholeness. The link took a symbolic form on 4 September 1784, when Goethe
climbed the Brocken and, his spirit stirred mightily, inscribed in the
visitors' book at the summit two lines from the Roman poet Manilius:

> *Quis coelum posset nisi coeli munere nosse*
> *Et reperire Deum, nisi qui pars ipse Deorum est?*
>
> Who could know heaven save by heaven's gift
> And discover God save one who shares himself in the divine?

*

On the Continent during and immediately after the French Revolution
politics crowded out cosmology, the nature of the State the state of
Nature, though there were exceptions, particularly in music. Inspired
perhaps by a visit to Herschel's observatory in Slough, in 1798 Joseph
Haydn composed *The Creation*, a joyful celebration of a universe
abounding in goodness, the most dramatic moment of which occurs when
chaos, suggested by the key of C-minor, gives way to D-flat major, then
to C-major, and the thunderous proclamation, 'Let there be light.'

At this period it was chiefly in the safely-moated castle of England that
poets had the serenity to grapple with the new world-picture, and here
a saying of Kant is applicable: 'The German genius develops into roots,
the Italian into foliage, the French into flowers, the English into fruit.'
That fruit owed something to the new roots put down by Goethe and,
to a lesser extent, the roots put down by Kant.

William Blake was born in London in 1757 of poor parents. He taught
himself to draw and paint, and later taught his young wife to read and
write. He practised the trade of engraving and his first large book of his
own designs illustrated Young's *Night Thoughts*. He was a Christian
Dissenter and a radical, protesting against a society that thwarted the
fulfilment of man, and against the new machinery, 'the dark Satanic
mills', which he believed broke man's spirit. He associated them with the
machinery of Newton's world-picture, and rebelled against that too.

Blake shared Goethe's vision of Nature as a single whole, edged with
something larger than the measurable, and his best-known quatrain is not
far from the German's view:

> To see a World in a Grain of Sand
> And a Heaven in a Wild Flower,
> Hold Infinity in the palm of your hand
> And Eternity in an hour.

Blake shared too Goethe's distaste for any partial explanation of man's cosmic destiny, be it cynical, maudlin or mechanistic:

> Mock on, mock on, Voltaire, Rousseau;
> Mock on, mock on, 'tis all in vain!
> You throw the sand against the wind,
> And the wind throws it back again.
>
> And every sand becomes a Gem
> Reflected in the beams divine;
> Blown back they blind the mocking Eye
> But still in Israel's paths they shine.
>
> The Atoms of Democritus
> And Newton's particles of Light
> Are sands upon the Red Sea Shore
> Where Israel's tents do shine so bright.

As the references to Israel indicate, Blake's view of the cosmos was that of a Christian steeped in the Bible. Here he parts company from Goethe and from such English predecessors as Young. Whereas the author of *Night Thoughts* had argued from the order of the solar system to a loving God, Blake made two sets of etchings on the theme, 'There is No Natural Religion' and held that the book of the sky must be deciphered through the Book of Books.

This led him to adopt the cosmology of the Jews of the Bible:

> The Sky is an immortal Tent built by the sons of Los ...
> And on its verge the Sun rises & sets, the Clouds bow
> To meet the flat Earth & the Sea in such an order'd Space ...

Blake really believed that the Earth is flat, that the sun moves round it and is no bigger than it appears to a child, for the telescope alters 'the ratio of the Spectator's Organs, but leaves Objects untouch'd'.

Though he sometimes attained to that innocent vision of a child, more often Blake projected on to space his deeply felt Christianity. 'Do you not see a round disc of fire somewhat like a Guinea?' Blake asked himself while regarding a sunset. 'O no, no, I see an innumerable company of the heavenly host crying "Holy, holy, holy, is the Lord God Almighty." '

Blake the visionary painter peopled space with titanic hermaphroditic figures, clothed in the ankle-length robes of Old Testament times. Now they are towering, now crouching; sometimes they hover like clouds. Nothing quite like them had been seen before in art. They represent the eternal looming behind the temporal. And over them towers the figure of Christ, 'who appear'd to me as Coming to Judgment among his Saints and throwing off the Temporal that the Eternal might be Establish'd . . .'

But Blake is no idle visionary. He is not content just to descry the eternal in space. 'Our Lord is the word of God & every thing on Earth is the word of God.' For that to be truly so, man's Satanic mills must be cleared away. The heart of Blake's vision lies in his longing to establish a new Jerusalem on 'England's green and pleasant land'.

*

Blake's younger contemporary, Samuel Taylor Coleridge, is more approachable as a person than Blake, indeed he is among the most lovable of the English Romantics. He responded to the natural world with throbbing sensibility and acute intelligence; he had a generous heart and considerable courage; and, though it cost him much in personal unhappiness, he tenaciously sought a picture of the world that would satisfy his whole self, heart as well as head.

Coleridge was born in 1772. A clergyman's son, he nevertheless lacked Blake's convinced belief in Christianity and when as an eight-year-old boy he walked with his father under the Devonshire sky, he saw space and the stars in terms of the genii and magic of a favourite book, *The Arabian Nights*. At Cambridge he wrote a Greek Ode to Astronomy, in which he expressed a wish to visit 'worlds which elder Suns have vivified'. Instead, he gathered a group of friends with the intention of planting a generous-minded colony on the banks of the Susquehannah, in Pennsylvania. He married one of his would-be colonists in haste, and lived to regret it – in England. Coleridge was too impractical to be a colonist and lacked the sexual drive to be a satisfactory husband.

Yet Coleridge possessed superb powers of perception. Nervous by temperament, he would often lie awake at night. What interested him was not the orderly turning of the planets, but phenomena strange and fantastic. On 22 October 1801 he wrote in his Notebook: 'I saw one after the other, nearly in the same place, two perfect Moon Rainbows – the one foot in the field below my garden, the other in the field nearest but two to the Church – It was grey-moonlight-mist-color.' Coleridge complemented such observations by reading books of exploration, particularly passages about ice, frost and snow, imagery which, perhaps because it is sexually neuter, attracted him.

Rather as Blake had fused his reading of the Bible with his experiences of nature, Coleridge fused his exotic reading with his nocturnal observations and a vestigial Christianity to produce *The Ancient Mariner*.

At the cosmological level, the poem is about the Earth, with its climate of torrid heat and groaning ice, set in space between a sun and moon whose strange and wildly fluctuating phenomena manifest a Divine justice. Water plays an important role too, but it is post-Robespierrean water, either frozen solid or unfecund save that 'slimy things did crawl with legs / Upon the slimy sea.'

The moon and moonlight stand, in Coleridge's poem, for the world of spirits lying behind the visible world. They add a daemonic dimension to events on the ship. They are in opposition to the sun, which in one passage is associated with God, perhaps the crucified God: 'Nor dim nor red, like God's own head'. The shooting of the albatross, its consequence, the death of all the crew but one, the mariner's guilt and eventual absolution – all happen between the sun and the moon, which dominate events as surely as the Olympians a Greek tragedy.

The Ancient Mariner is the most extreme example of the Romantic imagination interpreting the cosmos in terms of God's justice. Yet shortly after writing it, Coleridge began to doubt whether Nature is living and Divine. Turning his back on poetry, he travelled to Germany and embraced a modified version of Kant's view that Nature furnishes only the raw material of sensation on which the mind imposes its own forms of thought. Henceforth Coleridge sought to understand the workings of imagination. How far, for instance, were the blood-red sun and icy moon of his poem the actual sun and moon? And if they were not, what kind of reality and validity did they possess?

As he grew older, Coleridge came to resemble Kant in the sense that he sank deep into the complexities of his own thinking, which he studded with neologisms. There is no space here to follow him in detail. But in his last years he came to see that his personal flaws – including opium-taking – could not be put right by human agency. An impersonal God had sufficed the self-contained, self-righteous Kant, but Coleridge, weaker and humbler, felt the need of a Redeemer. He turned to the

Christian religion because he needed the help of a personal God and he remained in it, he said, because he received that needed help. The justification for religion, finally, lay not in the evidences of the cosmos nor the imperatives of a moral law, but in the fact that it produced practical results for the individual near to despair.

*

Coleridge's friend, William Wordsworth, had no taste for the exotic or febrile. He preferred a quiet relationship with familiar English landscape. Like Goethe, he needed to worship, to produce and to contemplate, and for these activities chose the Lake District. It was not, as he liked to think, a typical section of God's Earth, but park-like and, to be honest, almost a 'reserve' for poets, where stones were conveniently cushioned with moss, and should a storm blow up, Wordsworth could count on finding a well-read recluse, who would invite him in to share his fire, his oatcakes, his milk and his maxims.

Also, there were mountains. As recently as in Donne's time mountains had been wens, excrescences, disfiguring the pure spherical lines of Earth and best overlooked; to Wordsworth they were evidence of God's creative power, and at night on their peaks man might view the heavens in all their brilliance.

In *The Excursion* Wordsworth described his early eclipse of faith in the 1790's, then his pantheism – which so upset Blake as to give him stomach-ache! – and, after the death by drowning of his sailor brother, his return to orthodox Christianity, strengthened by mystical communing with nature. More sometimes than communing, for the poet's mind would mingle with what he saw and transmute it:

> between those heights
> And on the top of either pinnacle,
> More keenly than elsewhere in night's blue vault,
> Sparkle the stars, as of their station proud.
> Thoughts are not busier in the mind of man
> Than the mute agents stirring there ...

Like Goethe, Wordsworth was wary of the scientists. Cavendish had recently broken down water into hydrogen and oxygen; of the chemists

[177]

generally and the astronomers who 'weigh the planets in the hollow of their hand' Wordsworth wrote:

> Oh! there is laughter at their work in heaven!
> Enquire of ancient Wisdom; go, demand
> Of mighty Nature, if 'twas ever meant
> That we should pry far off yet be unraised;
> That we should pore, and dwindle as we pore,
> Viewing all objects unremittingly
> In disconnection dead and spiritless ...

Between the scientist's cold fact and a fancy that may narrow or mislead only Christian faith can provide a sure equipoise. Wordsworth put it like this:

> I have seen
> A curious child, who dwelt upon a tract
> Of inland ground, applying to his ear
> The convolutions of a smooth-lipped shell;
> To which, in silence hushed, his very soul
> Listened intensely ...
> Even such a shell the universe itself
> Is to the ear of faith.

Blake and the older Coleridge would have agreed.

But not so Percy Bysshe Shelley. From this cluster of Romantic poets he detached himself and comet-like pursued a lonely path. In his prep-school science lessons what excited the boy Shelley were stars 'so distant that their light has not reached the Earth since creation', inhabited doubtless by 'a higher race than ours': in other words, the invisible and the superior. With this taste went a powerful instinct to idealize: only the perfect could satisfy Shelley. The Christians about him, especially when they opposed his love for Harriet Grove, were plainly far from perfect. Christianity therefore was not true. Nor then were the notion of a First Cause and the Newtonian world-picture on which Christianity in England had come to rely; and Shelley shocked Oxford by declaring as much in *The Necessity of Atheism*.

After so radical a rejection, what remained? The invisible. Only the invisible, or the faintly perceptible, were spirit and perfect. Only they could satisfy. From this standpoint flowed that most original and powerful poem, *To Night*:

> Wrap thy form in a mantle grey,
>   Star-inwrought!
> Blind with thine hair the eyes of day;
> Kiss her until she be wearied out,
> Then wander o'er city, and sea, and land
> Touching all with thine opiate wand –
>   Come, long-sought!

The stars were most to Shelley's liking when least material, when they 'tremble, gleam and disappear'. In the moon he sometimes saw the sister of Spirit, sometimes a body quite alien. The classical Greeks had not connected the moon and madness; it was the late Romans who coined the word 'lunatic': St Jerome, in his Latin translation of the New Testament, rendered the Greek word for 'a man possessed' by 'lunatic'. Thence the connection passed into cliché. But Shelley had so far freed himself from convention that he was able, daringly, to see the moon itself in those terms:

> And like a dying lady, lean and pale,
> Who totters forth, wrapped in a gauzy veil,
> Out of her chamber, led by the insane
> And feeble wanderings of her fading brain,
> The moon arose up in the murky East,
> A white and shapeless mass –

Shelley found support for his rejection of a First Cause in Laplace; in the works of that astronomer and his compatriot Bailly he read too that the axis of the Earth's poles is slowly tending to the vertical. In *Queen Mab*, a prophecy of happier days in the light of a new religion of perfect love, Shelley stated his belief in this curious and erroneous theory, which would extend even to the polar ice-caps and the tropics a bland, fruitful climate suitable for 'regeneration's work'.

Shelley's atheism stemmed from his view of society. When he stood alone with wind and cloud, vaporous mist or a lark heard but unseen, glimpses of a Spirit within Nature, akin to Plato's World-Soul, kept appearing through the bars of his atheism. In *Ode to Heaven* he movingly expressed the resultant conflict of views. A first Spirit hails the sky as 'Presence-chamber, temple, home, / Ever-canopying dome, / Of acts and ages yet to come!' A second Spirit claims that what we see is only the shadow of a dream, and that the truth will appear beyond the portal of the grave. But unsupported by Revelation, such a view

cannot withstand the words of the third Spirit, who decries such pre-
sumption, declares Heaven to be a mere 'globe of dew', and proclaims
a cosmic Spirit:

> What are suns and spheres which flee
> With the instinct of that Spirit
> Of which ye are but a part?
> Drops which Nature's mighty heart
> Drives through thinnest veins! Depart!

Despite such glimpses of some vast power at work within Nature,
Shelley continued to the end to reject a personal God. The least earth-
bound of the Romantic poets was the atheist among them.

*

On 28 December 1817 Charles Lamb, John Keats, William Wordsworth
and Benjamin Haydon dined together. They discussed the merits of
Homer, Shakespeare, Milton and Virgil; and as the evening advanced
Lamb, slightly tipsy, began to abuse Haydon for putting Newton's head
into a picture he had just finished, 'A fellow,' he declared, 'who believed
nothing unless it was as clear as three sides of a triangle!' Then Lamb
and Keats agreed that Newton had destroyed the poetry of the rainbow
by reducing it to its prismatic colours. They all ended with a toast,
'drinking Newton's health and confusion to mathematics'. And in *Lamia*,
published three years later, Keats beautifully stated his case:

> Do not all charms fly
> At the mere touch of cold philosophy?
> There was an awful rainbow once in heaven;
> We know her woof, her texture; she is given
> In the dull catalogue of common things.
> Philosophy will clip an angel's wings,
> Conquer all mysteries by rule and line,
> Empty the haunted air, and gnomed mine –
> Unweave a rainbow.

Keats might protest, but Laplace's application of Newtonianism was
about to score another triumph. Uranus, the planet discovered by
Herschel, showed certain irregularities of movement. In 1846 J. C.
Adams of Cambridge and Le Verrier of Paris, working independently,

indicated the part of the heavens where the perturbing body was to be found. Telescopic search revealed it as predicted and it was given the name Neptune. There were now known to be eight planets, including Earth, revolving gravitationally round the sun, Neptune taking 165 years to make his journey.

What, meanwhile, of women's response to celestial phenomena? The Brontë sisters, living in the lonely parsonage of Haworth on the edge of bleak Yorkshire moors, have left a record of their experiences. With Emily one feels that she needed the stars with every particle of her being, and that that need was distinctive of her sex. It took three forms. The first, sharpened perhaps by her inability to establish a satisfactory relationship with people, was the need for a kind of mystical union: in the poem called 'Stars' she wrote of her feeling of peace as she gazed from her bedroom window: star followed star, 'While one sweet influence near and far / Thrilled through and proved us one.'

In its second form, Emily Brontë needed to see the stars as centres of worlds happier than ours: she may have been enabled to do this through reading in Thomas Dick's popular *The Sidereal Heavens* (1840) that there are 50,000 inhabited planets:

> I'll think there's not one world above,
> Far as these straining eyes can see,
> Where Wisdom ever laughed at Love,
> Or Virtue crouched to Infamy ...

Finally, near the end of her brief life, Emily Brontë turned to the stars for courage. Men had turned in that direction for intellectual reassurance and spiritual solace, but none quite in this way, for the unflinching resolve desperately required by a consumptive girl in order to face death. From that sprang the well-known poem beginning:

> No coward soul is mine
> No trembler in the world's storm-troubled sphere:
> I see Heaven's glories shine,
> And faith shines equal, arming me from fear ...

Emily's sister, Charlotte, as the eldest, had to shoulder the family cares and responsibilities. She sought in Nature not a mystical experience but signposts and courage for the task of living. In Chapter 5 of Charlotte's autobiographical novel *Villette* Lucy Snowe has been to consult an old servant of the family, her only remaining link with her lost connections. She is without money or prospects and finds herself without guidance as to her next move. The old servant cannot advise her; Lucy returns alone through desolate fields to her lodging, in darkness and without human habitations near. It was only 'by the leading of the stars that I traced my path', yet she was not afraid, not afraid even by the sudden apparition of the *aurora borealis* – that 'moving Mystery' as she calls it, which shone that night.

The *aurora borealis*, or Northern Lights, is usually of reddish colour, inclining to yellow, and sends out frequent coruscations of pale light, which seem to rise from the horizon in a pyramidal, undulating form, shooting very fast up to the zenith. It results at times of geo-magnetic disturbances from the glow of atoms in the thin upper atmosphere as they are hit by fast-moving electrons and protons speeding from the sun, but in Charlotte Brontë's day its cause was quite unknown.

This 'solemn stranger', as she calls it, influences Lucy Snowe far otherwise than through her fears. '... Some new power it seemed to bring. I drew in energy with the keen low breeze that blew in its path. A bold thought was sent to my mind; my mind was made strong to receive it ...'

So Lucy decides to risk seeking her living in London. She goes there, and thence to find employment as a governess in Belgium, where her destiny awaits her (as it did the real Charlotte).

In Chapter 15 a more direct cosmic influence is described. Lucy has been ill during the school's long summer holidays and left alone, except for the cook, in the empty premises. She is agonizingly in love, without hope of requital, and physically crushed. Then a change in the elements occurs: after a scorching drought tempestuous rain falls. Lucy is roused; overcoming her weakness, she staggers out to seek comfort in Sainte Gudule and, Protestant as she is, to make confession to the priest. The narrative is instinct with truth, for Charlotte Brontë had done that very thing, '... Twilight was falling, and I deemed its influence pitiful; ... it seemed to me that at this hour there was affection and sorrow in Heaven above for all the pain suffered on Earth beneath ...' Going out, she hears the bells of a church: '... any solemn rite, any spectacle of sincere worship, any opening for appeal to God was as welcome to me as bread to one in want ...'

In Chapter 26 Lucy Snowe buries the packet of her love letters and, with it, all hope of happiness for herself. 'The air of the night was very still, but dim with a peculiar mist, which changed the moonlight into a luminous haze. In this air, or this mist, there was some quality – electrical, perhaps – which acted in strange sort upon me. I felt then as I had felt a year ago in England – on a night when the *aurora borealis* was streaming and sweeping round heaven ... I felt, not happy, far otherwise, but strong with reinforced strength.'

In communing with the cosmos Wordsworth, who already possessed strength, had found peace; Charlotte Brontë, less strong but quite as brave, found physical and moral strength. Her heroine's response to the Northern Lights and to moonlight may be said to mark the culmination of the Romantic movement. She has gone as far as is possible in the direction away from Laplace's nebula.

The English Romantic movement, so indebted to Rousseau and to Goethe, gave back its fruit to the Continent. Around 1850 clusters of poets in France, Germany and Italy were communing with the larger world of Nature. Perhaps never before or since had man felt so close to the rock and tree of his own world, and to the light of distant worlds, the sight of which he felt enlarged his soul.

CHAPTER TEN

The Shadow of the Ape

The Englishman's world-picture in 1850 – serene assurance that the cosmos reflects the goodness of God – Charles Darwin's The Origin of Species *– cruel and wasteful Nature – man descended from apelike ancestors – controversy and pessimism – Thomas Hardy's world-picture – popularity of Omar Khayyam's* Rubáiyát *– Karl Marx – youthful poems about the stars – his atheistic materialism and attitude to Nature – Victor Hugo, Van Gogh and* Starry Night *– Jules Verne and space-fiction.*

The year 1850 saw peace from Berlin to California, and is a convenient point at which to summarize the educated man's cosmology, which had now outgrown an orrery's power to depict. 'Choose any well levelled field or bowling-green,' wrote Sir John Herschel, son of William. 'On it place a globe, two feet in diameter; this will represent the sun; Mercury will be represented by a grain of mustard seed, on the circumference of a circle 164 feet in diameter for its orbit; Venus a pea, on a circle of 430 feet; Mars a rather large pin's head, on a circle of 654 feet ... Jupiter a moderate-sized orange, in a circle nearly half a mile across, Saturn a small orange, on a circle of four-fifths of a mile; Uranus a full-sized cherry, or small plum, upon the circumference of a circle more than a mile and a half, and Neptune a good-sized plum on a circle about two miles and a half in diameter.'

The sun was known to be some 92 million miles from the Earth. What produced its immense heat no one had the least idea. Sun-spots were being studied. John Herschel thought they were the dark, solid body of the sun itself. They were reckoned to occur every ten years – actually eleven – and to disturb the weather.

The moon was believed to be volcanic. No one any longer believed it had seas, or an atmosphere. Its mountains were being measured. In 1850 it was successfully photographed by W. C. Bond at Harvard College Observatory by means of the new daguerrcotype process.

At last someone had calculated the distance of a star from Earth. Friedrich Bessel had measured stellar parallax for star 61 in the constellation of the Swan, and thereby calculated that that star was 600,000 times further away from Earth than the sun (the modern value is 706,000).

Cosmography was moving also towards a very large numbers of stars, graded by brilliancy or magnitude – a system first devised by Hipparchus. The stars of the northern hemisphere, down to the ninth magnitude, were counted and found to total 324,198. The southern hemisphere was expected to yield a comparable harvest. Binary, or double, stars had been discovered, and fourteen nebulae of a new class called 'spiral' because two spiral arms projected from opposite sides of the nucleus.

The two Herschels, William and John, showed convincingly that the arrangement of the stars in the sky, especially those in the circular band of the Milky Way, could best be explained by assuming that they are spread within an immensely wide but fairly thin layer, shaped like a discus or lens, with the sun and the Earth somewhere near the centre (on this last point, as we now know, they were mistaken). If an observer looks along the long diameter of the lens he will see a dense mass of stars – the Milky Way; if he looks along the short diameter of the lens, he will see only nearby stars, with no background luminosity of distant stars.

Perhaps the most surprising recent discovery was that the solar system did not remain in a fixed area of this lens-shaped system. The sun and its planets, including Earth, were moving very fast in the direction of the constellation Hercules.

That the universe was infinite in extent had been believed by Newton and Kant, to name only two, and was a common assumption in the eighteenth century. Then, in 1826, the Dutch astronomer Heinrich Olbers asked, 'Why is the sky dark at night?' If the universe is infinite, and there are an infinite number of stars evenly spread throughout it, the whole sky should be dazzling. Olbers concluded that the universe is not infinite or, if it is, the stars are not infinite in number.

William Herschel, using an 18-inch telescope, counted the stars in sample regions of the sky. If there is an infinity of stars, they ought to increase in number with dimness according to a fixed formula, but Herschel found they did not. On Herschel's evidence and Olbers's reasoning, the nineteenth century believed that the universe, though very large, is finite.

On Earth the number of elements isolated and described had grown to around sixty, and they were no longer viewed in qualitative terms. According to the chemist John Dalton, each element was made up of stable indivisible particles or atoms – he used Democritus's word – of a kind peculiar to itself alone, and the basic difference between one atom and another was weight.

Geologists now knew enough about rock formation to be sure that their planet had been formed over a very long period of time. Similarly, fossil evidence suggested that God had created new groups of animals down the ages. Robert Chambers summed up in *Vestiges of the Natural History of Creation*: 'The inorganic has been thought to have one final comprehensive law, GRAVITATION. The organic, the other great department of mundane things, rests in like manner on one law, and that is – DEVELOPMENT. In France the term evolution was preferred and Jean Baptiste de Lamarck had proposed a theory of evolution which attributed to an animal's nervous system a power to change structure.

Had this array of stars and planets and animals come into being without God through the built-in laws of matter and motion? A few retained Laplace's nebular hypothesis and answered Yes. But others, especially in England, retorted with a firm No. Samuel Vince, Professor of Astronomy at Cambridge, argued from the law of gravity to a providential God, for if gravitational force were to vary inversely as the cube of the distance, or in any inverse ratio higher than the cube, the planets would either have perpetually receded from the sun, or continually approached, and at last have fallen into it.

Vince also cited the existence of an atmosphere just right for vegetable and animal life, and the inclination of the Earth on its axis, without which we should lack the succession of heat and cold necessary for the ripening of crops.

In the very widely read *Natural Theology; or Evidence of the Existence and Attributes of the Deity collected from the Appearances of Nature* (1802) Archdeacon William Paley developed Vince's line of thought and extended the cosmological argument into the sphere of animal life, less recklessly than Bernardin de Saint-Pierre had done but still, as later events were to show, unwisely. Paley considered a mollusc's bivalve shell, compared it to a hinge made by man and concluded that the shell, like the hinge, was made by an intelligent being; he considered the giraffe and asked why its neck was so long. Evidently in order to crop tall branches; this too was prevision by an intelligent being, and so on through the animal kingdom.

In *Vestiges* Robert Chambers added a further touch. Perhaps after eating a hearty dinner followed by a bottle of port, he wrote that man generally has a sense of *well-being*, evident also in animals, and this is just what we would expect from a benevolent Creator.

If we had to choose an actual picture to illustrate this world-picture, we could hardly do better than *Cornfield by Moonlight*, painted about 1830 by a follower of Blake, Samuel Palmer, while living at Shoreham, Sussex. In order to indicate the increased importance of the celestial orbs in proportion to the Earth Palmer shows the moon and stars larger than they would be in life. Like Tintoretto in *The Baptism of Christ*, he uses their light to diminish our sense of the solidity of earthly things, and so leads us into a world more charged with God's presence than is daylight. It is a work in which land and sky harmonize as serenely as do Nature and the supernatural.

Into this cornfield strode the figure of Charles Darwin. His work was intimately bound up with his inner life, and his inner life is exceptionally well documented – a case-history not only of genius in action, but of genius reflecting the pressures of the mid-nineteenth century – so we shall look at both.

Charles Darwin was born in Shrewsbury in 1809, second son and fourth child of Robert Darwin, physician, and of Susannah Wedgwood. Dr Darwin, as depicted in a sketch from life, was abnormally large – eventually weighing 300 lbs – with a bloated face, physically repellent, highly intelligent, opinionated and, according to his son 'very easily made angry'. Charles lost his mother at eight and seems to have suffered from paternal overbearingness. As a schoolboy he would invent and recount tall stories, evidently in order to attract attention.

Dr Darwin was a Freemason, and either an agnostic or an atheist. Yet he urged his son to drop his scientific interests, to choose the respectable career of clergyman and, in preparation for that, to take a degree in Divinity. The feelings of Charles Darwin on receiving such advice may be imagined, but as he himself was a convinced Christian he was able to acquiesce without insincerity.

At Cambridge Darwin was influenced by two books. One was John Herschel's *Study of Natural Philosophy*, in which the astronomer laid down that scientific method is the discovery through induction of causes, leading to the establishment of a single law of general application, such as Newton's law of gravity. The other was William Paley's *Natural Theology*, already described. Darwin could 'almost ... have said it by heart' and found its argument for the existence of God from design in Nature 'conclusive'.

At twenty-two Darwin was offered the post of naturalist on a surveying brig, HMS *Beagle*. Darwin wanted to accept, but his father opposed the idea. Darwin did not feel strong enough to stand up to his father and invoked the help of his uncle, Josiah Wedgwood. This turned the scales and Dr Darwin agreed to pay his son's expenses.

While abroad Darwin was to write often to his sisters, rarely to the father who was making the voyage possible. Once he did begin a letter to his father, then continued it as though to his sisters. More generally – and the importance of this will appear later – he had acquired a strong aversion to authority. Before sailing he watched William IV's coronation procession, and wrote wishfully, 'I can hardly think there will be a Coronation this time fifty years.'

Darwin sailed from Plymouth two days after Christmas 1831. The *Beagle*, 242 tons, trimmed her canvas to the same trade-wind as Columbus's ship, and this voyage by the twenty-two-year-old Englishman was to be quite as momentous as Columbus's.

Darwin landed in Bahia, Brazil. Sensitive to the beauty of landscape, he instantly liked the tropical forest, confiding to his notebook his admiration at the colourful flowers and birds. Later he responded no less fervently to the austere boundless plains of Patagonia.

A self-taught geologist and naturalist, Darwin had very little field experience; indeed he was pitifully ill-equipped. Yet just as Cortes with inadequate forces and much courage had effected the military conquest of all Mexico, so during the next four years Darwin was to effect a scientific conquest of all South America. He studied in detail its rocks and fauna, coming to understand how, as a continent, it had been built, and how its creatures had evolved and become dispersed.

Darwin had two outstandingly important experiences in South America. One was a major earthquake, which destroyed the Chilean town of Concepción. Darwin went to see the damage and got a clear idea of how earthquake, like volcano, could have contributed over millions of years to the formation of the Andes. The other experience was his encounter with Fuegians, naked tribespeople who eked out a miserable existence on shellfish and sea-eggs.

'These poor wretches were stunted in their growth, their hideous faces bedaubed with white paint, their skins filthy and greasy, their hair entangled, their voices discordant, and their gestures violent.' They had no religion, no belief in a future life and during periods of famine rather than sacrifice their dogs – useful for catching otters – they killed and ate the old women of the tribe. So much for Rousseau's noble savage.

Darwin visited the Galapagos, where he noted that the birds and tortoises of each island were different. He spent a short time in Australia, where he encountered aborigines, reminiscent in their backwardness of the Fuegians. He also landed at Cape Town, and met the astronomer Herschel, whose book he had read at Cambridge. Then, five years after leaving, he returned home. He was now a 'rather tall and rather broad-shouldered man, with a slight stoop, an agreeable and animated expression when talking, beetle brows, and a hollow but mellow voice'. His father found the shape of his head 'quite altered' but his dog recognized him, just as in Homer Odysseus's dog recognized his master when he returned from the Trojan War.

During the next two years Darwin sought to generalize from his experiences in six notebooks – a wonderful record of the psychological evolution of the theory of evolution. From what he had seen in the Galapagos Darwin felt sure that isolation causes varieties to become split off from species, and eventually to form species themselves, while old species become extinct, thereby increasing the separation between the surviving species. But crossing of species led to sterility, as in mules. 'One is tempted to exclaim that nature conscious of the principle of incessant change in her offspring has invented all kinds of plan to insure sterility, but isolate your species her plans are frustrated or rather a new principle is brought to bear.'

What drove one species to change into another? Darwin sought an answer fruitlessly until, in September 1838, he read Malthus's *Essay on Population* and saw that the answer lay in one word – food. The struggle for food, limited by the finite size of the Earth, weeded out the weak and obliged the strong to evolve in the direction of more efficient food-finding. So was born Darwin's theory of evolution by natural selection from accidental variations.

Meanwhile, in two other notebooks, entitled *Metaphysics on Morals* and *Speculations on Expression*, Darwin was applying himself to a second equally formidable problem, one which followed from this entry in the Transmutation notebooks: 'Having proved mens & brutes bodies on one type: almost superfluous to consider minds ... yet I will not shirk difficulty – I have felt some difficulty in conceiving how inhabitant of Tierra del Fuego is to be connected with civilized man.'

Darwin then asked such questions as Why do we blush? – what is its animal origin? 'Is frowning result of straining vision, as savages without hats put up their hands ...' 'In the drawings of Voltaire why is under lip curled over upper with mouth shut expressing *cool* irony, *not biting?*' Darwin in short was seeking to show that human expression is an evolution of animal expression – the sneer derives from the snarl – and moral reactions like shame are an evolved form of animal instincts, such as self-preservation or preservation of species. 'Hensleigh [Wedgwood],' wrote Darwin, 'says the love of the deity and thought of him or eternity only difference between the mind of man & animals – yet how faint in a Fuegian or Australian! Why not gradation?' Darwin came to the conclusion that man's mind, even his morality, had like his body developed 'by gradation' from animal intelligence.

Darwin's religious views, meanwhile, were undergoing change. As a boy he had had little religious grounding and, he later admitted, 'I do not think that the religious sentiment was ever strongly developed in me.' As he came to believe that species had arisen through accidental variation, he saw that Paley's argument from the providential design of animal parts collapsed. 'But I was very unwilling to give up my belief; – I feel sure of this for I can well remember often and often inventing day-dreams of old letters between distinguished Romans and manuscripts being discovered at Pompeii or elsewhere which confirmed in the most striking manner all that was written in the Gospels. But I found it more and more difficult, with free scope given to my imagination, to invent evidence which would suffice to convince me.'

Darwin considered miracles, and in the light of the fixed laws of Nature found them 'incredible', especially as 'the men at that time were ignorant and credulous to a degree almost incomprehensible by us.' Darwin seems to be bracketing the Apostles with his Fuegians.

Limitations as a historian were matched by self-imposed methodological limitations. Emma Wedgwood, a pious Christian who became Darwin's wife in 1839, put the point tactfully in a letter to him: 'May not the habit in scientific pursuit of believing nothing till it is proved, influence your mind too much in other things which cannot be proved in the same way, and which if true are likely to be above our comprehension.'

By the age of thirty Darwin had rejected Christianity for Deism. That his strong aversion to authority played a part in this development emerges from the *Autobiography* which Darwin wrote in old age. There he was to describe the Old Testament God as a 'revengeful tyrant' and to single out the New Testament teaching that unbelievers will be punished in Hell as a 'damnable doctrine'. He ended life as an agnostic.

In 1844 Darwin wrote down a sketch of his theory of natural selection for his friend the geologist Lyell. Then he left the subject for twelve years in order to study barnacles. Why? One of the reasons may be this. Darwin believed that he had found a 'single law of general application' such as had fired his imagination at Cambridge, but the trouble was, his law could not be measured quantitatively and because of the millions of years involved could never be put to the test in a laboratory. It was a new kind of scientific theory, not susceptible to the old kinds of proof.

In 1856, pressed by his friends, Darwin resumed work on his theory but made slow progress. Perhaps he would never have written it up but for the arrival in June 1858 of a letter from Alfred Russel Wallace. A Welsh botanist, Wallace was then in the Moluccas assembling evidence for a theory of evolution. He had been struck down with malaria and, during a break in his fever, he had recalled Malthus's *Essay on Population*: 'There suddenly flashed upon me the *idea* of the survival of the fittest.' With his letter Wallace enclosed a paper 'On the Tendencies of Varieties to depart indefinitely from the Original Type' which, said Darwin, 'contained exactly the same theory as mine'.

Darwin thought the honourable course would be to leave the field clear to Wallace, but on the advice of his friends Hooker and Lyell it was decided that Wallace's paper and an abstract of Darwin's writings on the subject, partly dating from 1844, should be read at a meeting of the Linnaean Society. Darwin then hastened to complete his book – thereby exemplifying the importance in human as well as animal affairs of competition. It was published in 1859 as *Of the Origin of Species by means of Natural Selection*.

Darwin's theory that all species, including man, have evolved by natural selection through accidental variations was subjected during his lifetime to searching criticisms. From these it soon grew apparent, as Darwin in old age came to see, that selection occurs *after* the useful change has come into being: therefore natural selection can cause nothing but the elimination of the unfit, not the production of the fit. Yet even in this restricted form, the immediate effect of Darwinism was, in Britain, widespread and profound. The old proof of God from design was wrecked, it seemed irretrievably. Instead of a providential God, the law of natural selection operated in a way that seemed cruel, wasteful and devoid of intelligence. Moreover, if modern man and modern apes did indeed share a common ancestor, at what point, and how, had man been endowed with a soul? What happened to the *Genesis* account of the Fall and Original Sin?

The Church of Rome, already at odds with science since Galileo, placed Darwin's books on the Index; in England, however, the Church had enjoyed good relations with science, too good really, for it had set such store by natural theology that it was ill-prepared to adapt to the contents of Darwin's books. As a result, it attacked natural selection and the view of man's descent from an ape-like ancestor.

At a meeting of the British Association in 1860 the Bishop of Oxford, near the end of a long speech denouncing Darwinian doctrines, turned to Darwin's champion, T. H. Huxley, and mockingly asked him whether he reckoned his descent from an ape on his grandfather's or on his grandmother's side? – to which Huxley retorted, 'If the question is put to me, would I rather have a miserable ape for a grandfather or a man highly endowed by nature and possessing great means and influence, and yet who employs those faculties and that influence for the mere purpose of introducing ridicule into a grave scientific discussion – I unhesitatingly affirm my preference for the ape.'

The audience laughed; the meeting ended; the lamps were dimmed. But for several decades afterwards there was little laughter among thinking Christians in England, for over their land fell the shadow of the ape.

*

When they turned from Darwin's tortoises and savages to the planets and stars, Englishmen had at their disposal two new findings. The Earth, said geologists like Lyell, was several hundred million years old; and, according to the second law of thermodynamics established by the German physicist Rudolf Clausius, the amount of energy available for conversion into work decreases constantly, therefore the sun is losing heat and in the distant future the Earth will cool.

Neither of these findings was very sensational; both could have been worked into Christian world-pictures. But Englishmen came to them in a new mood. Aware that they and their forebears had been guilty of resting their case for the existence of God on a too facile reading of the evidence of Nature, they now went to the other extreme. For them the milky moonlight of Palmer's sketch had curdled. They reinterpreted cosmological data, including the findings of Lyell and Clausius, in the light of what Darwin called 'the clumsy, wasteful, blundering, low and horribly cruel works of nature'. Rejecting their old cornerstone, a Creator whose providence was detectable in the heat of the sun and the light of the moon, they embraced Laplace's theory of a fluid nebula starkly spinning the solar system into being, and added that the whole was slowly losing heat.

Biologists came to be haunted by a cosmic spectre of darkness which they had helped to raise. In 1866, dining with Kate Amberley, soon to be the mother of Bertrand Russell, Herbert Spencer confided his belief that the sun would come to an end and was losing all its force, while Darwin, writing his *Autobiography* in 1876, embraced 'the view ... that the sun with all the planets will in time grow too cold for life, unless indeed some great body dashes into the sun and thus gives it fresh life.'

So in the second half of the nineteenth century it was not just Darwin but a Darwinian view of the cosmos operating according to laws that take no account of man which changed Englishmen's world-picture. Hence the force of T. H. Huxley's assertion in *Lay Sermons* (1883) that it was astronomy more than any other branch of science which had made it impossible for people to accept the beliefs of their fathers. The two strands, astronomy and biology, merge most movingly in Tennyson's *In Memoriam*:

> I found Him not in world or sun
> Or eagle's wing, or insect's eye.

*

The effect of Darwin on the rising generation is well seen in the early life of Thomas Hardy. A countryman deeply responsive to the detailed order in Nature, Hardy knew the Bible very well indeed, loved it and intended to become a clergyman. Then, at nineteen, he read the newly published *Origin of Species*. Slowly his view of Nature changed and with it his faith collapsed. By the age of thirty he was an agnostic, more and more attracted to the idea of an implacable Destiny driving human lives.

Hardy was musical, and like many music-lovers felt drawn to the night sky, images from which abound in his work. In one poem the 'tired stars' of a sky breaking from night into dawn 'thin together', and in *The Return of the Native* the assured Mrs Yeobright is one of those persons who 'carry, like planets, their atmospheres along with them in their orbits'. In 1882, a year in which he attended Darwin's funeral, Hardy published *Two on a Tower*, which is partly spiritual auto-biography. The hero, Swithin St Cleeve, is an astronomer, and his mastery of this science is one of his attractions for the lady of the manor, Viviette Constantine. She visits St Cleeve in his tower observatory. He asks her how many stars she thinks may be seen with a powerful tele-scope. 'I won't guess,' she answers.

'Twenty millions. So that, whatever the stars were made for, they were not made to please our eyes. It is just the same in everything; nothing is made for man.'

He shows Lady Constantine the planets, the nearest stars, then the farthest, then 'those pieces of darkness in the Milky Way'.

'There is a size at which dignity begins,' he explains, 'further on there is a size at which grandeur begins; further on there is a size at which solemnity begins; further on, a size at which awfulness begins; further on, a size at which ghastliness begins ... And to add a new weirdness to what the sky possesses in its size and formlessness, there is involved the quality of decay. For all the wonder of these everlasting stars, eternal spheres, and what not, they are not everlasting, they are not eternal; they burn out like candles ... Imagine them all extinguished, and your mind feeling its way through a heaven of total darkness, occasionally striking against the black, invisible cinders of those stars ...'

St Cleeve goes to South Africa. The apparent hopelessness of his love of Lady Constantine is mirrored in what he sees there. 'There were gloomy deserts in those southern skies such as the north shows scarcely an example of; sites set apart for the position of suns which for some unfathomable reason were left uncreated, their places remaining ever since conspicuous by their emptiness. The inspection of these chasms brought him a second pulsation of that old horror which he had used to describe to Viviette as produced in him by bottomlessness in the northern heaven. The ghostly finger of limitless vacancy touched him now on the other side.'

Though *Two on a Tower* is given a happy ending, the reader comes away sharing Hardy's conviction, expressed in a letter to a friend, that the more we pry into Nature's secrets the more we gain only sadness. That is the mood of *The Dynasts*, Hardy's summing-up on man and on the cosmos:

> ... the roars and plashings of the flames
> Of earth-invisible suns swell noisily,
> And onwards into ghastly gulfs of sky,
> Where hideous presences churn through the dark –
> Monsters of magnitude without a shape,
> Hanging amid deep wells of nothingness.

On his deathbed Hardy asked his wife to read him a quatrain of the *Rubáiyát of Omar Khayyam*. FitzGerald's translation from the Persian had been published in 1859 but only after Darwinism had made its mark did English people turn to the pessimistic fatalism of an outdated Ptolemaic astronomy, adopting as their own such lines as these:

> And that inverted Bowl they call the Sky,
> Whereunder crawling coop'd we live and die,
>   Lift not your hands to It for help – for It
> As impotently moves as you or I.

It is necessary to return to the first half of the nineteenth century in order to trace the development of a thinker who published his most important book ten years after Darwin's *Origin of Species*, and whose subsequent influence was to run concurrently with the Englishman's. Karl Marx was born in 1818 in Trier, a town on the Mosel then belonging to Prussia. His father was a respected lawyer of liberal views; in origin Jewish, he had adopted a nominal Christianity to escape the disabilities to which Jews were subject. He was an upright man of high moral standards, and a Deist. He and his son had a very close relationship.

Karl Marx attended the Gymnasium, where his schoolmates loved him because he was always up to tricks, and feared him because of the ease with which he lampooned his enemies. As a student first at Bonn University, then at Berlin, he was a pugnacious young man, whom Jenny von Westphalen, his future wife, addressed in one of her letters as 'My little wild boar'. His self-assertiveness at eighteen was already causing his father some worry: 'Since [your] heart,' he wrote to Karl, 'is obviously animated and governed by a demon not granted to all men, is that demon heavenly or Faustian?' The question arises in a more acute form when we look at Karl Marx's earliest expression of his views about man in relation to the cosmos, which he wrote in two collections of poems, one dedicated to his father, the other to Jenny, both composed before his nineteenth birthday.

In the poem 'Creation' young Karl Marx expresses the Judaeo-Deist world-picture in which he had been raised:

> Voids pulsate and Ages roll,
>   Deep in prayer before his Face;
> Spheres resound and Sea-Floods swell,
>   Golden Stars ride on apace.
> Fatherhead in blessing gives the sign,
> And the All is bathed in Light divine.

This is not far from the thirty-eighth chapter of the *Book of Job*, or indeed from Young's *Night Thoughts*.

Then Marx goes to university, where he would have encountered the world-picture prevailing in intellectual circles in Bonn: Laplace's atheistic nebular hypothesis. The shock on Marx is evident in 'Song to the Stars'. The serenity of 'Creation' is replaced by a mood of sombre anguish, as Marx described his troubled soul:

> It turns on you its look
>   Darkly, compellingly,
> From you, babe-like, would suck
>   Hope and Eternity.

> Alas, your light is never
>   More than aethereally rare.
> No divine being ever
>   Cast into you his fire.

> You are false images,
>   Faces of radiant flame;
> Heart's warmth and tenderness
>   And soul you cannot claim.

> A mockery is your shining
>   Of Action, Pain, Desire.
> On you is dashed all yearning
>   And the heart's song of fire.

Grieving, we must turn grey,
   End in despair and pain,
Then see the mockery
   That Earth and Heaven remain;

That, as we tremble even,
   And worlds within us drown,
No tree trunk's ever riven,
   No star goes plunging down.

Dead you'd be otherwise,
   Your grave the ocean blue,
All gone, the shining rays,
   And all fire spent in you.

Truth you'd speak silently,
Not dazzle with dead light,
Nor shine in clarity;
   And all round would be Night.

Laplace's world-picture had left Marx grieving, grey, sunk in despair and pain. Worse, it had made a mockery of his earlier joy and sense of communion with a spiritual cosmos. The result was a deep bitterness. Marx rounded on those who worshipped, as he had done, a Creator now found to be non-existent. We find this bitterness in Chapter 16 of a fragmentary humoristic novel, *Scorpion and Felix*, which Marx included with the poems dedicated to his father. Marx quotes the opening lines of the Gospel according to St John, then comments: 'Innocent, beautiful thought! Yet these associations of ideas led Grethe onward to the thought that the Word dwells in the thighs ... she saw in the thighs [the Word's] symbolic expression, she beheld their glory and decided – to wash them.'

Behind the sarcasm we sense the pain. Yet for Marx the main legacy of his encounter with Laplace's purely materialistic cosmos was something more positive than pain: a sense of having been cheated, intense anger and aggression. These Marx describes in the most personal of his poems, 'Feelings':

Worlds I would destroy for ever,
Since I can create no world,
Since my call they notice never,
Coursing dumb in magic whirl.

Dead and dumb, they stare away
At our deeds with scorn up yonder;
We and all our works decay –
Heedless on their ways they wander ...

Swiftly fall and are destroyed
Halls and bastions in their turn;
As they fly into the Void,
Yet another Empire's born.

So it rolls from year to year,
From the Nothing to the All,
From the Cradle to the Bier,
Endless rise and endless Fall.

The importance of these poems, which have only recently been published in full, is unmistakable. They show that as early as the age of eighteen, before he adopted first the idealistic philosophy of Hegel, then the materialism of Feuerbach, Marx had already made up his mind about man's place in Nature. Since the stars are merely matter condensed by chance, neither they nor anything else in Nature has anything to tell us about our destiny. More than that, their 'dumbness' is a 'mockery', stirring man to action over against Nature. By the time he went to Berlin, where he wrote his doctoral thesis on a comparison of the two main materialist cosmologies of ancient Greece, those of Democritus and Epicurus, Marx had found two heroes: Epicurus, who attacked religion from the standpoint of atomic materialism, and Prometheus, 'the foremost saint and martyr in the philosopher's calendar'. By the time he left Berlin Marx was an atheist, a materialist and a positivist, convinced that it was Nature and institutions that had to be overcome, not any aspect of man himself.

Marx's youthful views developed with his growing experience of political economy, and they were intensified by later events in his life; like Dante and Einstein he was an exile – as Shelley was also in a sense. But a course had already been set and declared: 'Worlds I would destroy forever, / Since I can create no world.' The course was set towards political revolution, *The Communist Manifesto* and *Capital*.

Marx believed that a cosmology is to be explained by the social order and economy of the day. On that theory, incidentally, it is difficult to understand why the shepherd-lad James Ferguson should have arrived at the same cosmology as Herman Melville, writing *Pierre* in Pittsfield,

Mass. Applying his view to a Paleyish book by the poet Friedrich Daumer, Marx wrote in 1850:

> Nothing but Christian-Germanic-patriarchal drivel on nature ... We see that this cult of nature is limited to the Sunday walks of an inhabitant of a small provincial town who childishly wonders at the cuckoo laying its eggs in another bird's nest, at tears being designed to keep the surface of the eyes moist, and so on. There is no question, of course, of modern sciences, which, with modern industry, have revolutionized the whole of nature and put an end to man's childish attitude towards nature as well as to other forms of childishness ... For the rest, it would be desirable that Bavaria's sluggish peasant economy, the ground on which priests and Daumers likewise grow, should at last be ploughed up by modern cultivation and modern machines.

Later, in *Outlines of the Criticism of Political Economy*, Marx wrote: 'Nature becomes ... pure Object for man, a pure thing of utility; it ceases to be recognized as a power for itself; and the theoretical knowledge of its autonomous laws itself appears only as a stratagem for subjecting it to human needs, be it as object of consumption, or means of production.' Later still in *Capital*, Marx describes Nature as 'the original larder', 'the original tool house'. 'The external physical conditions fall into two great economic classes, natural wealth in the *means of subsistence*, i.e. a fruitful soil, waters teeming with fish, etc., and natural wealth in the *instruments of labour*, such as waterfalls, navigable rivers, wood, metal, coal, etc. At the dawn of civilization, it is the first class that turns the scale; at a higher stage of development, it is the second.'

Marx's view of science allows no place for such disinterested studies as astronomy and cosmology. They are 'idealistic', not useful, a relic of man's earlier stage of development.

It has recently been established that a letter to Marx from an English scientist declining the dedication of *Capital* is not, as was long believed, from Darwin. But Marx did undoubtedly esteem Darwin. He wrote to Engels in December 1860 of *The Origin*: 'Although it is developed in the crude English style, this is the book which contains the basis in natural history for our view.' And in 1861 to Lassalle: 'Darwin's book is very important and serves me as a basis in natural science for the class struggle in history.'

When Marx died in 1883, Engels said at his grave: 'Just as Darwin discovered the law of evolution in organic nature, so Marx discovered the law of evolution in human history.' This is an important statement of what Marxists then believed, and still believe, but it is in fact misleading and calls for comment. Darwin's natural selection is not a law but a hypothesis which evidence from many sources is rendering probable; Marx's dialectical materialism is not a law either but a hypothesis which the evidence of the last hundred years does not appear to support.

Perhaps a truer affinity between Darwin and Marx is that both published theories about man's relation to Nature which choose to work on the assumption that God is absent from the cosmos. The consequences of this were to prove even more serious than Galileo's trial. Science diverged from natural philosophy; scientists left to humane studies the investigation of man's spirit and of the question of a First Cause. The 'two cultures' were born.

A hundred years on, we look more searchingly at the basic assumption. We see the limitations, even the occasional illogic of those two geniuses. What were men about, expecting universality from a single individual? His very brilliance as an observer of structures perhaps unsuited Darwin for doing justice to man's spiritual side, and to the everyday evidence he had in his wife of the fact of religious experience. As for Marx, his choice of atheism, under the influence of the nebular hypothesis and of the Greek philosophers who happened to be the subject of his dissertation, was less a reasoned weighing of all the evidence than an impassioned commitment. Furthermore, Marx's central claim that progress is a universal law has no scientific basis and indeed contradicts his materialism; as Bertrand Russell pointed out, 'Marx professed himself an atheist, but retained a cosmic optimism which only theism could justify.'

France, meanwhile, had been developing a world-picture different from those in England and in Germany. France was little touched by Darwinism. The French saw – as did Darwin himself in later life – that natural selection might explain the extinction of uncompetitive species but not the origin of superior qualities, and they preferred to believe one of their own countrymen, the naturalist Jean Baptiste de Lamarck (1744–1829), who declared that movement up the scale of being comes from a drive (*élan vital*) within organisms. The French therefore retained what the English Darwinians had been obliged to abandon: the notion of design in Nature.

Secondly, atheistic materialism, as taught by Marx, at this stage made little impact in France. That kind of thinking had had its day during, and immediately after, the Revolution. Now the Catholic Church was in the ascendant: few of the leading thinkers and imaginative artists were *pratiquants*, but the more influential among them were *croyants*. Furthermore, at this period a concern for individual freedom ran strong in French breasts and, as Kant saw, the notion of freedom perhaps presupposes the existence of God.

Victor Hugo epitomizes what was happening in France. Leader of the Romantics after the youthful success of his swashbuckling play *Hernani*, Hugo entered public life as a *député* and after Louis Napoleon's *coup d'état* chose exile. For nineteen years as a voluntary prisoner in Guernsey he learned the truths that prisoners learn. In 1870 France suffered defeat by Germany, Napoleon III fell, the Commune seized power. Victor Hugo returned and composed a long poem, one of his best, about that cataract of events: *L'Année Terrible*.

Writing as a convinced European, Hugo deplores the fact that Germans and Gauls should have been massacring one another; he mingles his own personal tragedies with the wider disaster, concluding that only after death will he understand why there have been so many battles, tears, regrets: '*Et pourquoi Dieu voulut que je fusse un cyprès / Quand vous étiez des roses*' – 'And why God willed me to be a cypress / While you were rose trees.' Then, in the last stupendous pages, transcending personal, national and even European concerns, Victor Hugo stands under the night sky and makes an affirmation of hope, remarkable from a man aged seventy:

> Qu'importe le zénith sombre si nous voyons
> Des constellations se lever, des rayons
> Resplendir, des soleils faire un échange auguste,
> Là le vrai, là le beau, là le grand, là le juste,
> Partout la vie avec mille auréoles d'or!
> Vous, vous contemplez l'ombre, et l'ombre, et l'ombre encore,
> Soit. C'est bien. Vous voyez, pris sous de triples voiles,
> Les ténèbres, et nous, nous voyons les étoiles.
> Nous cherchons ce qui sert. Vous cherchez ce qui nuit.
> Chacun a sa façon de regarder la nuit.*

---

* What matters the darkness overhead, provided we see constellations rising, rays of light glowing, distant suns shining majestically one on another; *there* lies truth, beauty, grandeur, justice, ubiquitous life with a thousand golden haloes! You, on the other hand, fix your gaze on shadow, shadow and still more shadow. So be it: I have no quarrel. Under that triple veil you see darkness, while what we see is the stars. We are searching for what can help, you for what is hurtful. Each has his own way of watching the night sky.

Sixteen years after those lines were published, they were to have repercussions on the spirit of a Dutch painter working in France.

'*Admire* as much as you can, most people *do not admire enough*,' wrote Vincent Van Gogh aged twenty-one, and the next fourteen years may be described as an attempt to find an environment he could admire completely. Then, in February 1888, Van Gogh arrived in Arles, under snow, and as spring broke, found the light and colours he had been seeking. Everything appeared to him dazzling, as though newly created, and the simplest objects – a sunflower, a chair with a straw seat – became charged with a cosmic glory. Already he was far sunk in schizophrenia, but even his illness contributed to this euphoric vision: 'The more ugly, old, vicious, ill and poor I get, the more I want to take my revenge by painting in brilliant colours, well-arranged, resplendent.'

Ever since boyhood Van Gogh had read much and his reading influenced his painting. Dickens left a strong mark on his early sombre painting of the poor, and now Van Gogh was reading the Bible, Balzac and Hugo. In 1888 he wrote to his brother and beloved confidant, Théo: 'I have just read Victor Hugo's *L'Année Terrible*. There is hope there, but ... that hope is in the stars. I think it is true, and well told, and beautiful, and indeed I should be glad to believe it myself. But don't let's forget that this Earth is a planet too, and consequently a star, or celestial orb.'

Slowly Hugo's lines transformed Van Gogh's imagination. One night in June 1888, at Les Saintes-Maries, he went for a walk. 'In the blue depth the stars were sparkling greenish, yellow, white, pink, more brilliant, more sparkling gem-like than at home – even in Paris: opals you might call them, emeralds, lapis lazuli, rubies, sapphires.' He began to find night 'far more alive and richly coloured than day'. He had found artistic fulfilment in Provence, but he wrote to Théo, 'That does not prevent me having a terrible need of – shall I say the word? – of religion. Then I go out at night to paint the stars and I am always dreaming of a picture like this with a group of living figures of our comrades.'

So Hugo's vision of hope seen in the stars met Van Gogh's need for religion and comradeship. The evidence for design in the cosmos appeased the Dutchman's sense of failure and self-doubt. Here at last in its fullest form was something to admire, therefore to paint. So poor Van Gogh would settle himself on the bank of the Rhône, sticking candles on the brim of his hat to give light, and paint the stars. In

September 1888 he sent a batch of pictures for Théo to put on sale – as unsuccessfully as their many predecessors: 'I should not be surprised if you liked the *Starry Night* and the *Ploughed Fields*, there is a greater quiet about them than in the other canvases.'

'Quiet' may seem an odd word to apply to Van Gogh's night scenes, but in the light of Hugo's equation between the stars and hope, the noun becomes understandable. As for the spinning motion of the stars in *Starry Night*, Van Gogh may have seen magazine engravings of photographs taken in 1887 by the Englishman Isaac Roberts, which showed the Andromeda nebula to be spiral.

Personal quiet, however, was always to remain just out of Van Gogh's reach. Two days after Christmas 1888 he cut off the lobe of his ear and presented it to a prostitute. The gesture was probably an act of self-punishment for having failed to establish 'normal' relations with society.

Van Gogh was shut up in an asylum and in May 1889 moved to Saint-Rémy. There the quest for hope and quiet continued. In June he painted perhaps the most powerful of his night scenes, yet another *Starry Night* in which whirling stars spin above windswept trees. Hugo's words are met and matched in the Dutchman's paint, one of man's most profound visions of the cosmos. The painting now hangs in the Museum of Modern Art, in New York. In that city of artificial light the real stars are seldom seen; but Van Gogh's Provençal stars are there, a reminder to others, as Hugo's lines had been to him.

While Van Gogh probed space with his brush, further north a Frenchman was probing it with his pen. Jules Verne was born in Nantes in 1828, eldest child of Pierre Verne, a solicitor, and Sophie Allotte de la Fuye, descended from a Scot named Allott, who had been ennobled for services as a member of Louis XI's Scots Guard. Jules spent a happy, conventional childhood in a Catholic home. He had a passion for the sea and at eleven tried to ship as a cabin-boy on a three-master bound for the Indies.

After his schooling Verne went to Paris to study law. He was then a man of middle height and athletic build, with the head of a Greek hero, blue eyes with a glint of humour and an occasional far-away look. He was high-spirited, generous, got on well with people and was much liked, even if his atrocious puns were not. He became friendly with Alexandre Dumas *père* the novelist and began writing, at first with little success. Then he married a young widow from Amiens with two children and in order to support them became a stockbroker. But he continued to write and in 1860 he had a musical comedy put on. *Monsieur de Chimpanzé* tells the story of a monkey dressed up as a man who finds his way into high society and behaves a good deal better than most of those around him.

Verne had always been interested in scientific gadgets ever since as a boy he had asked to be given a model telegraph set. Now he became friendly with Nadar, a pioneer of photography and ballooning. This turned him to writing an adventure novel about a balloonist. A contract with an enterprising publisher, Jules Hetzel, followed and Verne was set on a course which was to lead him, imaginatively, to the centre of the Earth and to twenty thousand leagues under the sea.

When he was thirty-seven Verne wrote his first space novel, *From the Earth to the Moon*. At the end of America's Civil War veteran artillerymen form a Gun Club and conceive the idea of firing a man to the moon in a hollow shell. The idea is contested, then developed and finally the Columbiad, a giant gun, is mounted at Tampa Town, Florida, and fires into space an aluminium projectile carrying three men and two dogs. The journey is estimated to take 97 hours but the projectile is pulled off course by the gravitational attraction of a meteor and fails to reach its destination; the book ends with astronomers at Cambridge Observatory watching the projectile orbiting the moon; the Director predicts either that the attraction of the moon will eventually cause it to fall, or that it will continue to orbit forever.

A sequel, *Round the Moon*, describes events in the projectile. At the moment when the gravitational forces of Earth and moon cancel out, the cosmonauts hover above the floor, 'like figures in Raphael's *Assumption*': actually weightlessness would have persisted during the entire voyage. At one point the cosmonauts are on collision course with a meteor, but suddenly the meteor bursts 'like a bomb, but without making any noise in that void where sound, which is but the agitation of the layers of air, could not be generated'.

The cosmonauts notice that 'the constellations shone with a soft lustre; they did not *twinkle*, for there was no atmosphere to produce scintillation. These stars were soft eyes, looking out into the dark night, amidst the silence of absolute space.' They observed the surface of the moon and identified the biggest of its craters – 32,856 of them according to Johann Schmidt's recent map.

The Director of Cambridge Observatory's predictions turn out to be not the only options open to the cosmonauts, for they have rockets aboard. They fire the rockets, hoping to land their projectile on to the moon but something goes wrong and they find themselves travelling back to Earth. They splash down safely in the Pacific – only three miles from the point where, a century later, Apollo 13 was to splash down.

> And now will this attempt, unprecedented in the annals of travel, lead to any practical result? Will direct communication with the moon ever be established? Will they ever lay the foundation of a travelling service through the solar system? Will they go from one planet to another, from Jupiter to Mercury ... ?

To such questions, says Verne, no answer can be given; but he adds that, knowing the ingenuity of the Anglo-Saxon race, no one would be astonished if the Americans were to seek to make some use of this bold first attempt.

Jules Verne knew the second law of thermodynamics quite as well as his English contemporaries but he did not let it get him down. In *Round the Moon* the American, Barbicane, talks gravely about the cooling of the Earth: 'According to certain calculations the mean temperature will, after a period of 400,000 years, be brought down to zero!' 'Four hundred thousand years!' exclaims his French colleague, Michel Ardan. 'Ah! I breathe again. Really I was frightened to hear you; I imagined that we had not more than 50,000 years to live.'

Love of the stars often goes with a love of the sea, of music and of freedom; certainly they combined in the person of Jules Verne. He spent his spare time sailing a cabin cruiser, the *Saint Michel*, he composed songs, mainly nautical, and when he settled in Amiens he played an active part in local politics. He also carried his literary imagination into real life by giving fancy-dress balls. Everything went well for the successful author until the age of fifty-eight, when his teenage nephew ran berserk and shot him in the foot. The bullet could not be removed and Verne was left lame. He gave up sailing and wrapped himself in his work, exploring every aspect of planet Earth.

Verne believed in evolution, but disagreed with Darwin's theory. 'I am astonished,' he wrote in 1862, 'at the foolish satisfaction which many good people experience in giving the name of chance to the activity of a higher power which it would be just as simple, and more logical, to call Providence.' A staunch Catholic, he criticized Poe – from whom he had learned much – because the American 'is always an apostle of materialism'. In a late work, *The Eternal Adam*, Verne expressed a view opposed to Marx's: 'Man's true superiority is not to dominate and conquer Nature. In so far as he is a thinker, his superiority lies in understanding Nature, holding the immense universe in the microcosm of his brain; in so far as he is a man of action in keeping calm in face of recalcitrant matter, in saying: "Destroy me, yes, move me, never!" '

Verne did share with Marx an optimistic belief in the powers of technology, in miracle-machines. But Verne had a long first-hand experience of politics at local and national level which the German lacked and this convinced him that the machines which enabled man to master Nature, and brought about a restructuring of society, would never enable man to master himself. Indeed, in his later stories Verne concentrates on what would happen should scientific power fall into the hands of men evil or deranged.

In *The Purchase of the North Pole*, for instance, two members of the Baltimore Gun Club decide to exploit the mineral treasures buried beneath the Poles. They propose to tunnel into the side of a mountain, load a cannon with super-explosive and fire a shot with such momentum that its reaction will tilt the Earth's axis! The consequences may be harmful, the world is alarmed and the American Government vetoes the project, but since the scientists are in hiding what can be done? The gun is fired but owing to a miscalculation the explosion is harmless, leaving the Earth untilted. But the danger of scientific power on a cosmic scale in unscrupulous hands has been movingly conveyed.

When he died in Amiens in 1905, Jules Verne may be said to have domesticated the cosmos. He made man feel at home in all the continents, under the Earth and under the sea, at the tropics and at the Poles, amid meteors of outer space and in orbit round the moon. Often with extreme precision he foresaw the revolution in man's world-picture which would result from advanced technology. In a real sense he prepared the twentieth-century world-picture and was a pathfinder for the space journeys that were to follow. Admiral Byrd flying to the South Pole, and later the astronauts heading for the moon, could truthfully say, 'Jules Verne brought us here.'

CHAPTER ELEVEN

Mind over Matter

Henri Becquerel's discovery of radioactivity – dynamic matter – Edwin Hubble and the expanding universe – new powers claimed for the mind – G. B. Shaw's impatience with the body – Dunne's precognitive dream of Mont Pelée's eruption – Virginia Woolf watches an eclipse – Einstein's Special Theory of Relativity – D. H. Lawrence's cult of the sun – Einstein describes cosmic order as a 'miracle' – Schönberg, angels and prayer – the Futurists paint speed – Surrealism – Mondrian and Calder – sky-scrapers – the atomic bomb and the hydrogen bomb.

On 20 January 1896 a slightly built, neat-featured gentleman with a thick moustache and spade beard could have been seen leaving the house occupied by the Professor of Physics in the Museum of Natural History, Paris. Henri Becquerel, aged forty-three, was on his way to the usual Monday meeting of the Académie des Sciences. Crossing the Seine, he entered the sixteenth-century Institut de France and took his seat with other frock-coated members in the Physics section.

The meeting was concerned with the work of a German. Wilhelm Röntgen had lately been conducting experiments with electricity, known to be polarized energy but still mysterious. While sending an electric discharge through a vacuum tube, Röntgen noticed that a screen lying on a bench some distance from his tube emitted a bright glow. Nothing curious about that save that his vacuum tube was completely covered with a black card. Röntgen concluded – correctly – that the particles of negative electricity discharged from his tube's negative pole could produce rays which had penetrating properties. Three days before Christmas Röntgen passed some of these rays across his wife's outspread hand and photographed her hand. He found that the rays passed through her flesh but were obstructed by her bones. What he had was a photograph of part of his wife's skeleton, and it was a copy of this photograph which was shown to the physicists in Paris.

Becquerel's special field was the absorption of light in crystals, so he was much interested by Röntgen's photograph, and by the information imparted by his fellow Academician Poincaré, that Röntgen's mysterious rays – they were termed X-rays – arose from a greenish-yellow phosphorescent spot where a beam of particles struck the glass wall of his tube.

Becquerel knew that certain crystals, having absorbed sunlight, later emitted light as phosphorescence. Did they also emit X-rays? He turned to his laboratory in the Professor's House – almost a family laboratory, for his father and grandfather had both held his chair of physics – and began an experiment. He had in his possession a beautifully phosphorescent salt – the double sulphate of potassium and uranium. Uranium incidentally had been discovered in Saxony in 1789 and named in honour of the planet Uranus, which Herschel had first sighted eight years earlier.

Becquerel exposed some of his crystals to sunlight, in order to render them phosphorescent, then spread them on sheets of metal. He wrapped sheets and crystals in black paper and enclosed them in a box next to a photographic plate. He hoped that the new kind of ray would escape from the salts, traverse the black paper and expose the plate.

That is exactly what happened. On 24 February 1896 Becquerel informed the Académie that he had been able to produce rays with properties like X-rays, but he had done so without recourse to electricity, merely by allowing salts to trap and later emit sunlight.

On 26 and 27 February Becquerel prepared other boxes of the same salt for further experiments, but both those days were cloudy and he could not expose them to sunlight. He put the boxes away in a drawer which happened to contain some photographic plates. The next two days were also cloudy, but on 1 March the sun shone and Becquerel took out his boxes. It was then that he showed outstanding thoroughness. Although it would go clean against the theory he had announced to the Académie, he examined the photographic plates in the drawer, just in case they had been affected by the salts, even though the said salts had not been exposed to sunlight and so were not in a state of phosphorescence.

To his astonishment Becquerel found that the plates were indeed exposed. They bore not just a faint, but a very pronounced, image of the salts which had lain next to them in the dark drawer. Becquerel concluded that crystals of double sulphate of potassium and uranium, without requiring the agency of sunlight, emitted continually penetrating radiation of a kind hitherto unknown.

The very next day Becquerel informed the Académie of his discovery. Then he continued his experiments. First he showed that, like X-rays, the penetrating rays from his crystals could discharge electrified particles (in modern terminology, could ionize the air they traversed). He tried other phosphorescent crystals and found that only those containing uranium emitted the penetrating radiation. Finally he tried a disc of pure uranium metal and found that the disc produced radiation three to four times as intense as that he had first seen with his salts.

Becquerel's discovery that uranium is radioactive seemed at the time a modest one. Neither he nor his colleagues suspected that it marked a turning-point in human history. It immediately changed man's view of matter; before long it was to show him the way to immense sources of energy, to allow him accurately to say how old the Earth is, and to understand the composition of the stars.

Physicists were naturally excited to find that one kind of atom at least was not the stable, static thing they had long believed, but dynamic. They began an intensive study of the atom's structure. Since man is about as much larger than a hydrogen atom as the sun is larger than man, they could observe the atom only through its effects. A young New Zealander, Ernest Rutherford, then working at the University of Manchester, directed a beam of alpha particles (the name given to the particles emitted by uranium) at a sheet of gold foil and observed where and how much the direction of motion of the particles changed when penetrating the metal. These measurements indicated that each atom of the metal was composed of a very small, dense, positively charged core, surrounded by negatively charged electrons. Rutherford's Danish assistant, Niels Bohr, showed that the atom's electrical energy was emitted not continuously but in packets of discrete energy, comparable to bursts of machine-gun bullets, called quanta.

Meanwhile, one of Becquerel's friends, Marie Curie, had found a second radioactive element, radium. This emits considerable quantities of heat and is able to go on doing so for long periods. Pondering such radioactivity, Albert Einstein, German-born but working in a hot-bed of inventiveness, Bern's patent office, sought to relate mass to energy by a mathematical equation. How could large amounts of energy be produced by the loss of a minute amount of mass? Only by multiplying the mass by a very large figure indeed. Einstein found the wanted figure in the speed of light, and in 1905 published his famous equation: energy is equal to mass multiplied by the square of the speed of light; more technically, $e = mc^2$, where e is energy measured in ergs, m the mass of a body in grams and c the speed of light in centimetres per second. In this formula, soon conclusively shown to be correct, Einstein was stating in general terms what experimental physicists were discovering in the laboratory: that we dwell in a dynamic world, where all mass is congealed energy, all energy liberated matter.

Rutherford was continuing to learn more about the atom. If the nucleus of an atom be represented by a billiard ball, the nearest electron would be one kilometre distant. So matter was immensely porous: chunks of space traversed by bursts of electrical energy.

In 1911 Rutherford suggested that an atom is constructed like a miniature planetary system. Its central nucleus is one or more positively charged protons, round which an equal number of negatively charged electrons orbit like planets round the sun. But in 1926 Erwin Schrödinger showed that this was too simple a picture. Electrons behave more like waves lapping a round buoy. Each wave has its own pattern. In an atom possessing two protons and two electrons the second electron has a different wave pattern from the first. In atoms possessing more electrons still, successive electrons have more complex wave patterns. It was always the wave pattern of the outermost electron that determined the configuration and chemical properties of the atom.

In each atom the electrons can change from one pattern to another and in doing so they emit frequencies specific to that kind of atom. These frequencies correspond exactly with the quanta, or bursts of energy, distinctive of the atom.

Schrödinger concluded that the quality of any element can be expressed in terms of the number of protons in the nucleus of its constituent atoms, and the number of electrons, normally the same as the number of protons, this number in turn determining the electrons' wave patterns. Gold is gold because it has 79 protons and 79 electrons, lead is lead because it has 82. In the sense that all reality exhibits orderly structure Pythagoras's guess that matter resembles a gemstone, Kepler's that it resembles the hexagonal snowflake, had both been close to the truth.

*

Since Homer's day men had believed that space was filled with very fine matter called aether. In 1887 two Americans, Albert Michelson and Edward Morley, conducted an ingenious experiment with light beams which were thought to be 'ripples' in this aether. The Earth's motion should have made the measured speed of light different in different directions. To their surprise, they found no such effect. Gradually it became clear that aether was just a fiction, and that space was empty. Between Earth and stars, between star and star there was no weight of matter, only void.

As Van Gogh had noticed, some stars are coloured. Antares is red, Capella yellow, Vega bluish-white. Colour was found to reveal a star's temperature, size and even age. Furthermore, when viewed through a spectroscope, every star could be seen to emit more precise colours, and these revealed its constituent elements. For instance, if there were potassium in a star, it would show a red and violet band; if calcium, one orange, two yellow and one violet bands, and so forth. Analysed thus, stars were found to be composed of the 92 elements familiar to man on Earth; in other words, throughout the cosmos identical building blocks were in use.

Colour and periodic fluctuations in the light of some stars enabled astronomers to calculate the distances of the stars, which were now found to be much greater than had previously been thought. The nearest star to Earth, Proxima Centauri, was 40 million million kilometres away. To avoid using such big figures, the distance was stated as $4\frac{1}{4}$ light years, because light, which travels at 9·46 million million kilometres in a year, needs $4\frac{1}{4}$ years to travel from the star to us.

In 1927 a Dutch astronomer, Jan Hendrik Oort, found that the Milky Way, now usually known as the Galaxy, was rotating. Its motion was not like that of a rigid wheel but, in conformity with Kepler's laws, at a greater and greater speed the closer the stars were to the centre. In the solar system the speed approached 270 kilometres a second.

Our sun was relegated, in 1918, from its privileged central position to marginal status some 65,000 light years (now known to be 30,000 light years) from the centre of the Galaxy. It was no longer thought that the sun and its planets (to which, in 1930, was added Pluto) had originated in the spinning of a nebula, as Laplace had held, for such a hypothesis could not explain their angular momentum. One widely-held theory, proposed by the English astrophysicist James Jeans, held that a near-collision between our sun and another star had resulted in gaseous matter being pulled from both, and this had condensed into planets. Given the distance between stars, such an encounter would be exceedingly rare, and it was calculated that in the Galaxy there could be at most only ten other planetary systems.

Edwin Hubble, athlete, boxer and fisherman as well as astronomer, was using the new 100-inch telescope at Mt Wilson in California to observe nebulae, which smaller telescopes showed as mere blurs. Hubble concentrated first on the Andromeda nebula. He distinguished within it globular clusters and even individual stars, some of which periodically changed brightness. By measuring the time between their maximum brightness Hubble could deduce their absolute brightness, and estimate their distance by means of the rule that the brightness of a star falls off as the square of the distance. He concluded that these stars, and hence the Andromeda nebula itself, were 900,000 light years away.

Hubble next turned to the nebula in the Triangle. Here too he found stars whose light varied periodically and again he was able to estimate their distance: 700,000 light years. By 1929 Hubble had studied 18 nebulae, and got enormous distances for them all. The distances far exceeded the dimensions of the Galaxy, hence an inescapable conclusion: the nebulae were systems *outside* ours, and must be called other galaxies. As geocentrism had been proved illusory by Copernicus, galacto-centrism was proved illusory by Hubble, and man found himself dwelling merely in one galaxy among many.

Most astronomers would have been content with a catch like that. But Hubble was angling for something even bigger. He studied the spectra of the light emitted from stars in these new galaxies. When a star is moving away from us, the whole spectrum of its light exhibits a shift towards the red, or long-wave, end. Hubble found that the stars in his galaxies showed red shift, in other words the galaxies were moving away from us. They were also moving away from one another. And they were doing so not randomly, but with astonishing regularity. In a general way the velocity of recession was proportional to the distance; if one galaxy was twice as far off as another, it was racing away at twice the speed. This rule has become known as Hubble's law, and the number which relates distance to speed the Hubble Constant.

*

When we set these discoveries beside those of the atomic physicists, we see that in a most remarkable way they complement one another.

Movement was found to be integral to all matter, invisibly small and unimaginably vast. The static world-picture of the nineteenth century had yielded to ubiquitous dynamism. Even man, being composed of atoms, was dynamic, and he lived in a dynamic cosmos. Moreover, this dynamism gave every sign of following detectable rules: it appeared to be orderly.

Discoveries of such magnitude naturally produced a profound effect on men's views of the world and of themselves. These, we shall see, were many and varied. But one main feature is plain. There was a sense of deliverance from the inert matter which had weighed on many backs in the nineteenth century. The emptiness of space and the new spacious matter left room, it seemed, for mind and spirit. In a famous phrase T. H. Huxley had compared consciousness to 'the steam whistle which accompanies the working of a locomotive engine'; that view now lost credence, and an immediately noticeable feature of the first half of the twentieth century is the new claims made for mind.

In his influential *Man the Unknown* (1935) the French physiologist and Nobel Prizewinner, Alexis Carrel, wrote: 'Thought seems to be transmitted, like electromagnetic waves, from one region of space to another. We do not know its velocity. So far, it has not been possible to measure the speed of telepathic communications ... Telepathy, however, is a primary datum of observation.' That was a large claim, and Carrel went on to say that clairvoyance too was a fact. Both claims were soon to find strong support from Professor Rhine's card-reading tests at Duke University.

As the wireless, or radio, became a familiar piece of furniture – Marconi sent his first message in 1895 – so there grew the notion that space was a medium for messages, even from 'beyond'. In many a Bayswater drawing-room, as indeed in Boston and Berlin, grey-haired ladies sat in the dark waiting for the tapping and even the turning of tables. Sensitive souls claimed to feel auras, and strong personalities were described as 'magnetic'. Belief in ghosts grew and was fed by anthologies of ghost-stories.

George Bernard Shaw was one of the first to see that the energy and regularity lying in the atom upset the Victorian view that Nature is wasteful and wayward. He too believed in the new potential of mind, and combined both ideas in an original theory of evolution. *Man and Superman* (1903), like many Shavian plays, is really about Superwoman, in this case Ann Whitefield, a determined young lady who, under a gentle manner, unscrupulously tracks down her quarry, the political pamphleteer John Tanner, and despite his reluctance leads him to the altar. Ann embodies the Life Force, which is leading man willy-nilly upward to wider, deeper, intenser self-consciousness – indeed to pure thought. In *Back to Methuselah* (1921) Shaw took this theme a stage further by suggesting that man could dispense with a body.

THE HE-ANCIENT: Prehistoric men thought they could not live without tails. I can live without a tail. Why should I not live without a head?

THE NEWLY BORN: What is a tail?

THE HE-ANCIENT: A habit of which your ancestors managed to cure themselves.

THE SHE-ANCIENT: None of us now believe that all this machinery of flesh and blood is necessary. It dies.

THE HE-ANCIENT: It imprisons us on this petty planet and forbids us to range through the stars.

A somewhat similar form of evolution, but in a Christian context, was to be developed by the French palaeontologist, Pierre Teilhard de Chardin. Chardin claimed that Earth is surrounded by a huge mental aura, the 'noosphere'. As man becomes progressively more spiritual, so the noosphere will expand, and one day merge with God.

The energy it emitted suggested that matter shared characteristics with mind, especially when it was found that mind emits 'brain waves'. Two leading English mathematician-philosophers, Bertrand Russell and Alfred North Whitehead, went so far as to claim that the world is composed of a single neutral stuff, neither mind nor matter, which they termed *events*. In an attempt to solve the body-mind relationship, they held that the events that make a living brain are identical with those that make the corresponding mind, just arranged differently. For Russell events evinced no underlying order, but Whitehead disagreed: he held that the 'togetherness' of events is God.

James Jeans, champion of the 'near-collision' origin of Earth, went even further than Whitehead towards philosophic idealism. He wrote in *The Mysterious Universe:* 'The universe can be best pictured, although still very imperfectly and inadequately, as consisting of pure thought, the thought of what, for want of a wider word, we must describe as a mathematical thinker.'

Another who asserted the power of mind, though in a very curious way, was J. W. Dunne. A volunteer in the Boer War, in spring 1902 Dunne was encamped with the 6th Mounted Infantry near the ruins of Lindley, South Africa; he had just come off trek, and mails and newspapers arrived but rarely. One night he had an unusually vivid dream. He seemed to be standing on high ground, the upper slopes of some spur of a hill. Here and there were little fissures, from which jets of vapour were spouting. He recognized the place as an island he had occasionally dreamed of before. 'When I saw the vapour spouting from the ground,' Dunne recalled, 'I gasped, "It's the island! Good Lord, the whole thing is going to *blow up!*"' He was seized with a frantic desire to take off the unsuspecting inhabitants in ships. He tried to get the incredulous authorities – who spoke French – to despatch ships, but was sent from one unhelpful official to the other. The number of people in danger obsessed his mind. 'Listen,' he cried, '4000 people will be killed unless –' He woke to find himself clutching the neck of one of the regiment's horses, believing it to be pulling away the carriage of an official who would not heed his warning.

Weeks later the London *Daily Telegraph* arrived with headline news. At 8 o'clock on the morning of 25 April the volcano Mont Pelée, which had been quiescent for a century, erupted on the town of St Pierre in Martinique, killing 40,000. Dunne checked the date of his dream and found that it had occurred *before* the eruption. It had corresponded to reality in all particulars, except that he had omitted one nought in the casualty figures.

Dunne argued that we cannot be immediately aware of the future (as opposed to merely inferring it inductively) unless the future is already there for us to be immediately aware of. So the future does exist, quite as much as the past. To the growing list of man's mental powers Dunne now added precognition, and he claimed that in precognitive moments we entered a different time-band, comparable to switching from a 'live' radio programme to a pre-recorded programme that will go out on the air only in a week's time. Through Dunne's own books and three plays by J. B. Priestley 'serial time', as it is called, became widely known and quite a few people in the 'twenties and 'thirties adopted it as a working hypothesis in their daily lives.

*

The novels of Virginia Woolf provide a good example of mind much to the fore, shaping the material world. In *The Waves* (1931) she follows six people: as children spending a summer by the sea; then, in middle age, meeting at Hampton Court; and finally as memories in the mind of one of them, grown old. The image of her title suggests both the wave motion of electrons and the waves of the brain, for here people and objects merge, as the author dispenses with narrative in favour of interior monologues by the characters in turn. Waves of colour, waves of a woman's beautiful hair, waves of desire – they rise, undulate and finally fall under the waves of the stream of everyday life: 'We are swept on by the torrent of things grown so familiar that they cast no shadow. We float, we float ...'

In the many beauties of trees, sunshine and water that crowd the novel Wordsworth, even Tennyson, would have found cause for joy. But to the reader's surprise the author does not respond in that vein. 'The light of the stars falling, as it falls now, on my hand after travelling for millions upon millions of years' elicits not wonder but 'a cold shock'. At the Hampton Court meeting, 'We had our bottle of wine, and under that seduction lost our enmity, and stopped comparing. And, half-way through dinner, we felt enlarge itself round us the huge blackness of what is outside us, of what we are not. The wind, the rush of wheels became the roar of time, and we rushed – where? And who were we? We were extinguished for a moment, went out like sparks in burnt paper and the blackness roared.' By the end of the novel the waves have become the symbol of eternal recurrence, 'the incessant rise and fall and fall and rise again'.

Why this marked note of unease? An answer can begin to be given by recalling that Virginia Woolf belonged to the Bloomsbury Set, of which E. M. Forster was also a member, and as such she rejected metaphysical systems. She had no time either for optimists like Shaw or radical systematizers like Dunne. Yet she could not help imbibing from her epoch, and did indeed imbibe, a world-picture with strongly metaphysical implications. We get hints of it in her Diary.

On the night of 21 June 1927 Virginia Woolf travelled with her husband and friends from London to the Yorkshire moors to watch an eclipse of the sun. 'Rapidly, very very quickly,' she wrote in her diary, 'all the colours faded; it became darker and darker as at the beginning of a violent storm; the light sank and sank; we kept saying this is the shadow; and we thought now it is over – this is the shadow; when suddenly the light went out. We had fallen. It was extinct. There was no colour. The earth was dead. That was the astonishing moment;

and the next when as if a ball had rebounded the cloud took colour on itself again, only a sparky ethereal colour and so the light came back. I had very strongly the feeling as the light went out of some vast obeisance; something kneeling down and suddenly raised up when the colours came. They came back astonishingly lightly and quickly and beautifully in the valley and over the hills . . . It was like recovery. We had been much worse than we had expected. We had seen the world dead. This was within the power of nature.'

Thirty pages on in her Diary Virginia Woolf asks: 'Now is life very solid or very shifting? I am haunted by the two contradictions. This has gone on for ever; will last for ever; goes down to the bottom of the world – this moment I stand on. Also it is transitory, flying, diaphanous. I shall pass like a cloud on the waves.'

The two passages show that Virginia Woolf held a world-picture devised by the already mentioned James Jeans, and very popular at that time. According to Jeans, the cosmos had existed, and would exist, for ever. It was a place of eternal recurrence, now expanding, now contracting. Our part of the cosmos, the solar system, was losing energy: the sun one day would grow cold, and that would be the death of Earth. An eternal cosmos, however viewed, has no place for a Creator – Jeans did not believe in one, nor did Virginia Woolf – and it reduces man physically to almost zero. This, the chance birth of Earth from a near-collision and its coming death – help to explain the note of unease in *The Waves* and in Virginia Woolf's work generally.

The other most influential English astrophysicist, Arthur Eddington, disagreed with Jeans. Eddington believed that the cosmos had a beginning in time and was finite in size. Eddington was a Christian, but held that the cosmos yields no evidence useful to the metaphysician, though it does yield intimations of God to the mystic. So despite their euphoria about the new discoveries and agreement that they indicated a marvellous complexity, Englishmen were divided about the conclusions to be drawn from them.

The situation was complicated by Einstein's Special Theory of Relativity (1905). Physicists had found that the mass of any object increases slightly when it is moved at high speed, and that all phenomena which change with time change more slowly when moving than when at rest. Einstein drew the conclusion that there can be no absolute measurements of space and time: they are always 'relative' to some arbitrarily chosen frame of reference.

Einstein was not saying as Kant had said, that space and time are imposed by the individual observer's mind. Indeed he emphatically states that physical laws describe a reality in interconnected space-time that is independent of ourselves. Nevertheless many, including a great religious poet, T. S. Eliot, saw the Theory as somehow destructive of traditional certitudes.

Eliot used it as the starting point for *Ash Wednesday*, written in 1930:

> Because I know that time is always time
> And place is always and only place
> And what is actual is actual only for one time
> And only for one place ...
> Consequently I rejoice, having to construct something
> Upon which to rejoice.

The 'something' is faith based exclusively on Revelation.

In *East Coker* (1940) Eliot accepts the new view that because the galaxies are receding space is dark and indeed will become darker. He identifies space with death, and this darkness with the Christian mystic's night of the soul:

> O dark dark dark. They all go into the dark,
> The vacant interstellar spaces, the vacant into the vacant,
> The captains, merchant bankers, eminent men of letters,
> The generous patrons of art, the statesmen and the rulers ...
> I said to my soul, be still, and let the dark come upon you
> Which shall be the darkness of God.

Eliot's retreat was not of course the only possible Christian response to an expanding universe. Alice Meynell was one who saw in it habitable planets and marvellous further scope for Redemption. In 'Christ in the Universe' she wrote:

> Nor, in our little day,
> May His devices with the heavens be guessed,
> His pilgrimage to tread the Milky Way
> Or His bestowals there be manifest.

But in the eternities,
Doubtless we shall compare together, hear
A million alien gospels, in what guise
He trod the Pleiades, the Lyre, the Bear.

*

There was much in the new physics to inspire nature poets, and one scientist at least, Eddington, held out a friendly hand to them: 'I am not sure that the mathematician understands this world of ours better than the poet and the mystic. Perhaps it is only that he is better at sums.' D. H. Lawrence, poet of instincts, doubtless welcomed those words; certainly he believed that Einstein arrived at his equations not logically but creatively, like a poet penning an inspired line. In general he was more receptive to the new physics than Keats had been to Newton's explanation of the rainbow, though being Lawrence he gave the discoveries an unduly irrational twist:

I like relativity and quantum theories
Because I don't understand them
and they make me feel as if space shifted about like a swan
that can't settle,
refusing to sit still and be measured:
and as if the atom were an impulsive thing
always changing its mind.

Lawrence too was always changing his mind, so in this respect he could feel at home with quanta.

It so happened that one of the few cosmic objects still imperfectly understood was the sun. Most physicists held that its heat came from contraction, though far-sighted Eddington surmised that it came from some form of 'sub-atomic energy' pervading all matter. This mystery suited Lawrence. Since antiquity the sun has been male and the moon female: Lawrence amplified that view, hailing the sun as the prime source of vital energy, and as a symbol of sexual virility. When he went to live in New Mexico, and visited Oaxaca, Lawrence adopted a mysticism of the sun, entwined with snakes and phoenixes, and he identified it – erroneously – with a pre-Columbian religion. He even envisaged an order of women consecrated to his sun cult, but these women were to pay homage to the sun not directly but through sexual receptivity to their menfolk:

when a man comes looking down upon one
with sun in his face, so that a woman cannot but open
like a marigold to the sun,
and thrill with the glittering rays.

*

With James Joyce we come to a writer who runs counter to the general trend in his denial of order in the universe. As a young man Joyce turned his back on Ireland and on his boyhood Catholicism; he therefore felt antipathy for a world-picture suggestive of Mind; moreover as a humorist he felt the need for surroundings where, as in Dublin, it is the improbable that happens.

In the glorious early days of the Royal Society Samuel Butler had laughed at astronomers mistaking a fly in their telescope for an elephant in the moon; now in the 1920's Joyce made fun of the latest world-picture. The cosmic, he declared, is comic, the universe, 'whorled without aimed', is 'chaosmos'. Splendid word!

Yet even in his idiosyncratic denial of order, Joyce was a child of his age. In 1917 Rutherford became the first man to change one element to another when he bombarded nitrogen atoms with alpha particles and converted them to oxygen atoms; five years later Joyce set about a comparable kind of alchemy, when he coined neologisms for the early pages of *Finnegans Wake*. That book is a rag, tag and bobtail of neologisms which cohere – if they do cohere – through the electromagnetic force of punning.

In the sixth chapter of *Finnegans Wake* Joyce indulges in a joke against himself. (I follow W. Y. Tindall's interpretation.) The dime-cash or time-space problem is the subject of one Professor Jones, an eminent 'spatialist'. Jones attacks Bitchson (Bergson) and Winestain (Einstein), then disposes of the quantum theory and, refuting Lévy-Bruhl, another addict of time, he takes his stand with the postvortex school of Wyndham Lewis, author of *Spice and Westend Woman* – really *Time and Western Man*, in which Lewis had attacked Joyce for immersing himself in Time Past, and cutting himself off from space, that is from present-day reality.

The fun continues. Joyce had started *A Portrait of the Artist as a Young Man*: 'Once upon a time and a very good time it was'; now, with a wink at Einstein, Joyce rewrites it as: 'Eins within a space and a wearywide space it wast.' Again, the cathode tube which had produced

Röntgen's X-rays was now being developed for television by, among others, Ferranti, and this provided Joyce with an opportunity for a fine tilt at the expanding universe: 'I can easily believe heartily in my own most spacious immensity as my ownhouse and microbemost cosm when I am reassured by ratio that the cube of my volumes is to the surface of their subjects as the sphericity of these globes ... is to the feracity of Fairynelly's vacuum.'

*

On the Continent meanwhile a debate raged about the implications of the new cosmology. Three voices were of special importance. The Vienna School, continuing the traditions of nineteenth-century positivism, held that science was self-explanatory and that metaphysical language was meaningless (it influenced the Bloomsbury Set). The Copenhagen School, headed by Niels Bohr, argued that the behaviour of quanta was in principle unpredictable, denied the notion of cause in the usual sense and assigned to Nature the ability to choose from among the infinite number of possibilities, all equally probable, the one that appeared to produce a coherent set of events.

The third voice was that of Albert Einstein. He spoke up for mind, for a universe finite in time, and for cause in the usual sense, claiming that underneath quanta's apparently uncertain behaviour lay comprehensible order. 'You find it surprising,' he wrote to his friend Solovine, 'that I think of the comprehensibility of the world (in so far as we are entitled to speak of such world) as a miracle or an eternal mystery. But surely, *a priori*, one should expect the world to be chaotic, not to be grasped by thought in any way. One might (indeed one *should*) expect that the world evidenced itself as lawful only so far as we grasp it in an orderly fashion. This would be a sort of order like the alphabetical order of words. On the other hand, the kind of order created, for example, by Newton's gravitational theory is of a very different character. Even if the axioms of the theory are posited by man, the success of such a procedure supposes in the objective world a high degree of order, which we are in no way entitled to expect *a priori*. Therein lies the "miracle" which becomes more and more evident as our knowledge develops. And here is the weak point of positivists and of professional atheists, who feel happy because they think that they have preempted not only the world of the divine but also of the miraculous.'

That is a considered reflection from his mature years, but in his youth Einstein was saying as much, and winning many hearers. In Germany, explicitly or implicitly, the view that there is a marvellous harmony between mind and orderly matter became a basis for literary Expressionism, which sought not just to describe externals but to convey the inner meaning of objects, situations and life, and for the group of writers who called themselves the 'seekers after God'. In music too it won an important disciple.

Arnold Schönberg was born in Vienna in 1874 of a Hungarian father and a Czech mother. He began studying the violin at eight and was soon composing songs in the style of Brahms. He worked for four years in a small bank, then conducted a metal workers' union choir. In 1907 his First String Quartet was performed but its polyphony, complex thematic development and counterpoint proved too much for the Viennese audience, some of whom walked out of the hall, one wag using the emergency exit.

In 1912 Schönberg composed *Pierrot Lunaire*, a setting of poems by the Belgian Albert Giraud in which a crazed harlequin, absinthe and moonlight blend in a sickly *fin de siècle* kaleidoscope. But he quickly moved on. Of Jewish descent, he had been reared as a Catholic, then become a Protestant. He began to see the space of the new astronomy as profoundly orderly and as a symbol of eternity. It is peopled by angels, and they can help man to union with God through prayer.

Schönberg asked Richard Dehmel to write him a poem for an oratorio on the following subject: 'Modern man, having passed through materialism, socialism and anarchy and, despite having been an atheist, still having in him some residue of ancient faith (in the form of superstition), wrestles with God (see also Strindberg's *Jacob Wrestling*) and finally succeeds in finding God and becoming religious. Learning to pray!'

In the end Schönberg composed the poem himself and around it, between 1917 and 1922, composed the oratorio *Jacob's Ladder*. Three groups of men, the discontented, those in doubt and the jubilant, are questioned by the Archangel Gabriel, who explains the direction they must take if they are to attain purity of soul. In the last bars Gabriel tells them: 'Whoever prays, has become one with God.'

It so happened that in his fellow-Austrian, Rainer Maria Rilke, Schönberg found another who sensed that the newly revealed message-way of space was peopled by angels: indeed in Rilke's *Duino*

Elegies angels, pure and powerful spirits on whom man can call for help, assume a greater importance than at any time since Milton. Schönberg set to music three poems by Rilke: 'Premonition', 'Make me the guardian of Thine immensities' and 'All who seek Thee'.

In art, said Schönberg, there is but one true teacher: inclination, and Schönberg's lay in the direction of complexity. A chess-lover, he played the game with four extra pieces on each side, using an enlarged board of a hundred squares. As a composer he had already discarded traditional concepts of consonance and dissonance, and in 1921 he developed a complex style in which major and minor keys were replaced by twelve-tone rows. Schönberg declared all twelve notes in the chromatic scale to be of equal importance and regarded none as a tonal centre; no note can be repeated before the eleven others have appeared, in order not to polarize the melody. This method, according to Schönberg's biographer Willie Reich, can be compared with certain models in contemporary physics. 'The introduction of rows is analogous, so far as the "microstructure" of the musical material is concerned, to the approach to the microstructure of matter produced in physics by the quantum theory and modern atomic theory.' Pythagoras and Kepler would have approved.

In 1933 Schönberg adopted the Jewish faith of his forebears and emigrated to the United States. In 1950, using the twelve-tone method, he began work on a series of *Modern Psalms*, a continuation of those of the Old Testament. Number 6 was to be an attack on the philistinism of atheist scientists who show contempt for miracles because they 'reveal what is unprovable', while number 7 was to reject the assertion that God no longer performs miracles, and to offer a new definition of the term 'miracle'. Number 1, the only Psalm to be completed before his death, beautifully conveys the theme that had so long haunted Schönberg: union with God by prayer.

*

We move now to the visual arts, and return to the beginning of the century. There lived in Paris then a brilliant and excitable young Italian, Filippo Marinetti, who wrote in French. In 1902 he published a tirade as furious as Marx's against the stars, prime examples of crass matter imprisoning man's spirit. The winds and the waves, personified as horsemen carrying flaming lances, launch an attack on the stars, which in glorious exultation they tear from the sky:

Et rien ne subsista dans le grand large infini
qu'une éphémère poussière d'argent, et des morceaux
de limailles méprisables et des sables noircis.*

Three years later, when Einstein published his equation $e = mc^2$ and his Special Theory of Relativity, Marinetti found himself freed from that weight of matter which had occasioned his tirade and entered a world of energy where the only absolute was the speed of light. In 1909 Marinetti published the manifesto of Futurism, a joyful renovation of all artistic activity in keeping with the new physics. 'Time and Space died yesterday,' declared the manifesto. 'We already live in the absolute, because we have created eternal, omnipresent speed.'

Marinetti wrote poems about express trains and racing cars, while photographers tried to capture them on film and Futurist painters on canvas. It proved no easy task to depict apparently dematerialized whirling wheels. Some tried to superimpose one object on another, which is more or less the technique of animation in cartoon films; others painted vibrations. Having declared that a running horse has twenty legs, and their movements are triangular, Carlo Carra painted just that in *The Red Horseman*, a successful work when you overcome a first impression that the rider is on a rocking-horse.

Other Futurists sought to break up ordinary objects into dynamic states. According to the 1910 manifesto of five Italian Futurists, 'Our bodies penetrate the sofas upon which we sit, and the sofas penetrate our bodies ... We declare ... that movement and light destroy the materiality of bodies.'

In 1915 the Futurists' hero, Einstein, extended his theory of relativity to include gravitation, which, he declared, is not a force, as Newton thought, but a curving of space-time. The planets describe curved paths because near the sun, just as in the neighbourhood of every concentration of matter, the universe is curved. So mass, gravitation and space-time are all interdependent. This was found to solve a long-standing puzzle: why Mercury's point of nearest approach to the sun changed from one revolution to the next. Due allowance made for all perturbations, there was still a mysterious one-way shift of Mercury's perihelion by an amount equal to 43 seconds of arc per century.

---

* And nothing survived in the infinite deep but fleeting silvery dust, contemptible fragments of filings and grimy sand.

Einstein in his General Theory explained that Mercury, being so close to the sun, moved in a region of such strong gravitational force that Newton's approximate theory of gravity was not good enough. Einstein's theory of curved space-time, on the other hand, produced a figure for the sun's gravitational force slightly more than Newton's inverse-square, and this exactly tallied with the perihelion shift which is observed.

Giacomo Balla chose to honour the inspirer of Futurism by painting several versions of *Mercury Passing in Front of the Sun.* They are fascinating attempts to convey the curvature of space around the sun by means of a convergence of lines, circles and triangles. The same artist's statue, *Boccioni's Fist* – Boccioni was a fellow-Futurist – is a construction in iron which well conveys the rapier-fineness as well as the blunt thrust inherent in dynamic matter.

*

Early Cubism, though contemporary with Futurism, probably drew its scientific inspiration not from Einstein but from Rutherford's 1903 paper describing radioactivity in terms of 'a chemical change in which new kinds of matter are produced'. Two years later Braque and Picasso began to break down an object and to reconstitute it in accordance with a rhythm determined by the artist himself. Mind mingles with, and reshapes, the external world, as in the novels of Virginia Woolf; persons and objects are composed of the same elements.

According to the new physics the behaviour of matter can be calculated, but no longer visualized. This loss troubled some artists, others it stimulated. In 1920 the Dutch painter Piet Mondrian, then aged forty-eight, wrote: 'Why should art continue to follow nature, when every other field [in particular, the sciences] has left nature behind?' Renouncing his earlier copying of the visible, Mondrian tried to express in abstract, non-representational terms an underlying order, invisible but structuring Nature. He favoured black horizontal and vertical lines that cross to form a right angle, with places of pure colour in between. 'The two oppositions (vertical and horizontal) are in equi-valence, that is to say, of the same value: a prime necessity for equilibrium.' In looking at such works 'man is enabled by means of abstract-aesthetic contemplation to achieve conscious unity with the universal.' The best of

Mondrian's paintings have the power of a Buddhist mandala – a symbol of the universe intended to lead to religious meditation and spiritual calm.

In Vienna meanwhile Sigmund Freud was expounding the hidden energy of the subconscious, a theory attractive to those who knew of the hidden energy of electrons within the atom. Strongly influenced by Freud, Max Ernst in 1924 originated Surrealism, proclaiming the universality of the irrational, as revealed by the subconscious. But Ernst's paintings, to my eye at least, reveal rather the provincialism of an ethnic subconscious when blatantly asserted. However, Surrealism at its best did implement Breton's declaration in the first Surrealist manifesto, that 'Nothing but the marvellous is beautiful'. Salvador Dali overemphasized perspective to convey infinity, the feelings of being lost and deserted in dream or life; he painted strange objects in space, and conveyed the wonders of physical change in his famous limp watches. These may have influenced Virginia Woolf, who wrote in *The Waves:* 'Everything became softly amorphous, as if the china of the plate flowed and the steel of the knife were liquid.'

The artist who most responded to Hubble's discovery of a multitude of galaxies was an American, Alexander Calder (1898–1976). The son of a successful sculptor, Calder stated in his autobiography: 'The first inspiration I ever had was the cosmos, the planetary system. My mother used to say to me, "But you don't know anything about the stars." I'd say, "No, I don't, but you can have an idea what they're like without knowing all about them and shaking hands with them." '

After taking a degree in mechanical engineering and going to sea as a stoker, in 1926 Calder, a big playful man, settled in Paris, where he attracted notice with performances of a miniature circus, the artistes made of wire. In 1930 he visited Mondrian's studio and became converted to abstract art. But he felt 'that art was too static to reflect our world of movement.'

So he devised the mobile. Although the Russian, Naum Gabo, had introduced motion into abstract art with his Kinetic Sculpture of 1920, Calder was the first to devise an art of motion. 'I felt there was no better model for me to choose than the Universe . . . Spheres of different sizes, densities, colour and volumes, floating in space, traversing clouds, sprays of water, currents of air, viscosities and odors – of the greatest variety and disparity.' His first mobiles were driven with a crank or by small motor, but repetitiveness lost its appeal for Calder, who always valued the element of the unexpected, and his later mobiles usually hang free, to move as air currents may take them.

Among Calder's most successful mobiles are *A Universe* (1934), painted steel pipe, wood, wire, string and motor, now in the Museum of Modern Art, New York, and in the same gallery what Calder calls a stabile, *Morning Star* (1943), painted sheet metal, rod and wood, in which Venus is balanced in space like a poised ballerina. Very fine too is the *Acoustic Ceiling* Calder designed in 1962 for University City, Caracas, on which curved, apparently floating panels of painted plywood suggest receding galaxies.

Calder's mobiles do not explain, as did Ferguson's orrery; they open our eyes. An American poet, Elder Olson, has put it well:

> ... to prove once more, by a new universe, more clearly seen,
> That things, whatever they are, propose a principle
> Without which they could not be, or be as they are,
> And to say, thus, what ships, stars, leaves,
> Fish,
> And other things have been saying a long time
> Though until now, until now, we did not hear.

There is one other element in Calder's mobiles to be noticed. As Rembrandt in his self-portraits seems to be saying, 'How strange that this lined, puffy-eyed, swollen-nosed face reflected in the glass should belong to me, that through it, and it alone, I must be what I am to others,' so Calder seems to muse, 'How strange that we, intelligent men who much of the time like to be still and think, should dwell in a spin-ball contraption that is never still and which, look at it this way or that, is really rather funny.'

In 1914 Antonio Sant'Elia issued a *Manifesto of Futuristic Architecture*: 'I proclaim ... that oblique and elliptic lines are dynamic, and by their very nature possess an emotive power a thousand times stronger than perpendiculars and horizontals, and that no integral, dynamic architecture can exist that does not include them.' It is a curious example of the recurrence in history of basic themes that Sant'Elia's demands could have been met by Gothic vaulting and tracery. However that may be, Le Corbusier responded to the Futurist's call and used cantilevers as oblique lines in some of his best buildings. Generally, however, Sant'Elia's demands proved too difficult, technically, to answer.

What we do find is the use of glass walls, as in Walter Gropius's Berlin Bauhaus. Glass walls allowed the man in the street to see inside the building, thus satisfying a feeling that government and the deliberations of public bodies should not be behind closed doors – the Escorial makes an interesting contrast – itself possibly reflecting the physicist's success in probing the inside of the atom. The critic Gieidion may be right in claiming that Gropius also intended the Bauhaus to have a specific space-time quality: 'the extensive transparency permits interior and exterior to be seen simultaneously *en face* and *en profil* like Picasso's *L'Arlésienne* of 1911–12.'

The most important architecture of the period, the skyscrapers of New York, grew out of the limited land available in downtown Manhattan. But there was no intrinsic need for them to rise so high. The daring of hundred-storey towers expresses the mood of confidence that accompanied growing understanding, and mastery, of the natural world. Stripes of the national flag rising to meet the stars, they came to be seen as the distinctive architecture of the first half of the twentieth century, and to be imitated in every continent. More than any other artefact of their period these pinnacles that defy gravity express the power of mind over matter.

This brief survey has revealed much diversity: cults of the Life Force and of the sun, paintings of express trains and mobiles of galaxies, intimations of angels. Behind the experimenting and subjectivism, as I have tried to show, lay a certain choice of themes and unity of mood: a sense of release from matter, a new trust in the powers of mind and an awareness, behind much complexity, of order. But there were other elements in the picture that in various ways made for division. Dynamism at the heart of nature was taken by some political parties as a justification for the mailed fist, and it is no accident that Marinetti the Futurist ended as a prominent Fascist. The disappointed or frustrated put their trust in a new elemental trinity: energy, mass and speed, one of whose avatars was Struggle. When the Life Force was added to support the crazy notion of a Super-Race, natural philosophy could almost be said to sanction conflict. Under this and other pressures the world drifted a second time to war, and it was against a background of bloodshed that the final chapters resulting from Becquerel's discovery of radioactivity unfolded.

As they learned more about the atom and probed the properties of uranium, physicists understood that if the nucleus of a heavy atom could be split into two parts, the energy released by the action of splitting would mean that the mass of its two fragments would be less than that of the original nucleus; the difference would be minute, but the energy released would be this minute mass multiplied by the square of the speed of light.

Were man to succeed in splitting the atom, could he use the dynamism in matter to make an explosive weapon of war? Some, including Einstein and Rutherford, thought not. Others, notably Niels Bohr, maintained he could. When the Second World War broke out, atomic scientists speeded their research. After Pearl Harbor the United States set up a small organization to develop an atomic bomb, under the code-name Manhattan Project.

Its director, J. Robert Oppenheimer, had been born in 1904, son of a German immigrant who had built up a textile business. He took a degree at Harvard, then studied at Cambridge under Rutherford. At the time of his appointment he was not only America's most gifted atomic physicist, but a man of keen sensibility and wide culture, who read the Indian Sacred Scriptures in Sanskrit.

Oppenheimer guided his research team in the desert near Los Alamos, New Mexico, while in the laboratories a race went on to develop suitable fissionable material, either uranium of mass 235, a refined form of the metal which had yielded Becquerel his discovery, or a heavy man-made element, plutonium. Plutonium was ready first and with this as its component the bomb was made.

On 15 July 1945 the bomb stood ready on a high scaffolding in the desert known as Point Zero. Ten miles away at base camp the Italian physicist, Enrico Fermi, offered to take wagers from his fellow scientists on whether or not the bomb would ignite the atmosphere, and if so, whether it would destroy New Mexico only or the world. In the control dug-out, five miles from Point Zero, were Oppenheimer and General Farrell, waiting for the weather to clear.

At 5.30 on the morning of the 16th the bomb was exploded. 'The whole country,' wrote General Farrell, 'was lighted by a searing light with an intensity many times that of the midday sun ... Thirty seconds after the explosion came, first, the air blast pressing hard against the people and things, to be followed almost immediately by the strong, sustained awesome roar which warned of doomsday and made us feel that we puny things were blasphemous to dare tamper with the forces heretofore reserved for the Almighty.'

No one saw the first flash of the atomic fire. It was only possible to see its dazzling white reflection in the sky and on the hills. Then came a bright ball of flame, growing larger and larger, a mushroom cloud, and intense heat. Into Oppenheimer's mind, unbidden, came lines from the *Bhagavad Gita* describing the transfiguration of Krishna: 'If in heaven together should arise the shining brilliance of a thousand suns, then would that perhaps resemble the brilliance of that God so great of self', and Krishna's prophecy: 'Time am I, wreaker of the world's destruction, matured – resolved here to swallow up the worlds.'

The explosion had been equivalent to 20,000 tons of TNT, causing a light to be seen by dwellers 200 kilometres distant. Their anxious enquiries were met with the answer that an ammunition dump had blown up – without, it was added, causing casualties. Three weeks later, on 6 August, a similar bomb fell on Hiroshima.

'The world stands in the presence of a revolution in earthly affairs,' wrote the London *Times* the following morning. The new power set a choice before the conscience of humanity 'and in a terrible and most literal sense it is a choice of life or death.'

'The shining brilliance of a thousand suns' – that image was not misplaced. For now scientists were fairly sure that the source of the sun's heat came not from the contraction of its mass, but from nuclear power. Hydrogen was converted to helium, by fusion, and this involved an even larger loss of mass than when uranium or plutonium broke down to smaller atoms, and correspondingly richer release of energy.

If hydrogen atoms could be fused in a bomb, the destructive force would be many times greater than in a bomb where plutonium or uranium was split. America pressed ahead with the development of such a weapon, despite the opposition, for reasons of conscience, of Oppenheimer.

The first hydrogen bomb was tested at Eniwetok Atoll on 1 November 1952. Its explosion, equivalent to three million tons of TNT, obliterated a small island and left a wave-washed crater almost two

kilometres in diameter. The fireball extended, within four seconds, to
$5\frac{1}{2}$ kilometres in diameter. Ten minutes after the blast the mushroom cloud measured 40 kilometres high and 160 kilometres wide.

It was an awesome All Saints Day, fraught with paradoxes. Having
probed the constructive power at the heart of matter man turned it
to destruction, producing on Earth during a fraction of a second the
flame that burns in the sun. Just as the greatest power was proved to
lie in the smallest unit, so in microcosmic man there lay henceforth
a macrocosmic power. In obliterating a small part of his world he
became aware of his ability to obliterate very much larger parts. His
moral responsibility extended not only to his own life and his
neighbours' but to all life on his planet. Good and evil that day
assumed cosmic dimensions.

CHAPTER TWELVE

Journey to the Moon

*The Wright brothers' flight – Saint-Exupéry, pilot and poet – man most fully
a man when flying* – The Little Prince – *the U.S. Space Programme –
astronauts as seen by Oriana Fallaci – the Apollo 11 mission – the first men
on the moon – celebration of Holy Communion – impressions of life in another
world – reactions to the moon landing – what the rock samples revealed –
changed place of moon and Earth in our world-picture*

In an age which regarded flight as an infringement of God's will
for man, Leonardo da Vinci had been secretive about his flying
machines and, as far as we know, conducted his experiments single-
handed. That may help to explain their failure, for flight, when it
came in the liberal eighteenth century, was to be a two-man affair. On
5 June 1783 Joseph Montgolfier and his brother Etienne, paper-makers,
succeeded in making a hot air balloon ascend into the sky, and six months
later the first free aerial voyage was made by Pilâtre de Rozier and the
Marquis d'Arlandes.

Balloonists described their smooth, silent passage, in such contrast to
the clattering of a coach on a rutted highway, and how droll it was to
see tiny houses and cows in a toyland landscape. But they drifted like
thistledown in the breeze and so, although high up, they were not in
control. They had no sense of power or exaltation. Ladies decorated
dresses, fans and fireplace tiles with balloons, but balloon-flight brought
virtually no enlargement of consciousness.

'What use is it?' someone had asked, pointing to an early balloon, and
Benjamin Franklin had answered, 'What use is a new-born baby?' Four
generations later the Wright brothers provided a more tangible answer.
Orville and Wilbur Wright were bicycle manufacturers in Dayton, Ohio
– practical men obliged to run their small business at a profit. That is
one thing that distinguishes them from most of their European counterparts, who were dilettantes and theorists. The other difference is that there

were two of them, with beautifully complementary characters: Orville intuitive and brilliant, Wilbur steady and persevering; but both down-to-earth nuts-and-bolts men. Laconic Wilbur spoke for Orville too when he said, 'I only know of one bird, the parrot, that talks; and it can't fly very high.'

The Wright brothers built and flew gliders. By the time they were in their early thirties they had made over one thousand glider flights. The particular challenge that intrigued them was how to power a glider with the recently-invented internal combustion engine. All available motor-car engines were too heavy, so in 1903 the Wrights built a lighter engine. It had four horizontal cylinders of 4-inch bore and 4-inch stroke, and developed some 12 horse-power. They fitted it on a linen-covered wooden glider, so that it would turn two airscrews in opposite directions.

The Wrights decided to try out Flyer I beside the Atlantic in North Carolina on a remote sandhill well-named Kill Devil Hill, for the brothers had had to fight many devils of prejudice against aeronautics. America's most famous astronomer, Newcomb of Harvard, had declared powered flight to be scientific nonsense, and once he slackened speed an intending aviator would fall – 'a dead mass'.

Orville Wright sets the scene: 'During the night of December 16th [1903] a strong cold wind blew from the north. When we arose on the morning of the seventeenth, the puddles of water which had been standing about the camp since the recent rains, were covered with ice. The wind had a velocity of ten to twelve metres per second [36 to 43 kph]. We thought it would die down before long, and so remained indoors the early part of the morning. But when ten o'clock arrived, and the wind was as brisk as ever, we decided that we had better get the machine out and attempt a flight.'

The brothers took it in turn to 'operate' the machine – the term 'to pilot' lay in the future. That morning it was Orville's turn. He laid himself flat on his stomach beside the motor, so bringing the weight of the loaded machine to 340 kilograms. 'After running the motor a few minutes to heat it up, I released the wire that held the machine to the track, and the machine started forward into the wind. Wilbur ran at the side of the machine, holding the wing to balance it on the track ... The course of the flight up and down was exceedingly erratic, partly due to the irregularity of the air and partly to lack of experience in handling this machine ... A sudden dart when a little over a hundred feet from the end of the track, or a little over one hundred and twenty feet from the point at which it rose into the air, ended the flight. It lasted only twelve seconds.'

To the Wright brothers, their family and a few friends those twelve seconds marked a triumph. But the Press and the public did not want to know. Indoctrinated to the impossibility of powered flight, they held the brothers to be cranks, and their short airborne lifts mere stunts. That lasted almost five years. Then in 1908 Wilbur performed a successful flight before a large, appreciative crowd near Le Mans, France, while Orville scored a similar triumph in the United States.

The following year the *Daily Mail* sponsored an international contest to fly the English Channel, and for the first time the world at large took an interest in flight. A manufacturer of motor-car lamps, Louis Blériot, won first prize, crossing from France in just over half an hour. A customs officer puffed up Castle Hill, Dover, on a bicycle to complete the formalities with a log-book of categories quite inappropriate to Blériot's 'vessel'. He decided to register it as a yacht!

Blériot, like the Wrights, was laconic about his aerial experiences. Before man could be fully conscious of his initiation into space there had to arise a pilot who was also a poet.

Antoine de Saint-Exupéry was born in 1900 in Lyon of a Dordogne family with more quarterings than money. His father, who had a job in insurance, died when he was four and the focus of Antoine's life became his mother. A beautiful lady from Aix-en-Provence with long dark hair and soft dark eyes, she was a gifted painter and musician. Antoine as a child would drag around a tiny green-satin armchair, so that he could sit down by his mother's side the moment she found a seat, and plead for one of the stories she told so well. Sometimes it would be a Hans Andersen tale, sometimes the story of Joseph and his brothers or of Rebecca at the well, and afterwards Antoine would tell it in turn to his younger brother and his sisters.

Antoine grew into a tall strapping boy. He liked romping and scuffling and was very unruly. In the summer he organized snail races, built himself a tree house and acted charades. He also wrote poems, the first at the age of six.

When he was nine the family moved to Le Mans, where Antoine attended school. With his round face, turned-up nose and tendency to daydream, he was nicknamed Pic-la-Lune. In Le Mans Wilbur Wright had set up shop building aeroplanes and young Antoine became a devotee of the new cult. He made a would-be flying machine by stretching a pair of old sheets over a frame of bamboo struts, attached to the handlebars of his bicycle. At twelve he went up for his first flight in an aeroplane. He did his military service in the Air Force and at twenty-one qualified as a pilot.

His first job was flying mail from Toulouse across Southern Spain to Casablanca. This was the heroic age of flight. The pilot, wearing a leather suit, fur-lined boots, thick gloves and goggles, map-case slung from his shoulder, climbed into the open cockpit of the Breguet biplane, powered by a 300 h.p. engine. A hand signal and the chocks were pulled away, he taxied over bumpy grass and rose into the sky. His dials were few: compass, altimeter, air speed indicator, rev meter. He controlled the machine almost as a horseman controls his mount, shifting his weight to keep it level in a high wind, kicking a jammed aileron free with his heel.

The engine failed on average once every 20,000 kilometres and Saint Ex, as his friends called him, had his share of crash-landings. But the mail was a sacred trust, to be delivered regardless of storm or blizzard. 'I didn't believe an aeroplane could take such blows without breaking into pieces,' he wrote after a dizzying 1000-metre drop which had finally bounced him on a firm cushion of air a bare hundred feet above the ground.

Saint Ex fitted in well to Aéropostale. Generous, brave and with a gift for funny drawings and card tricks, he was popular with his comrades. In 1928 he was sent to Buenos Aires to fly the more difficult South American routes. He crashed in Guatemala where, as he put it, he 'learned about gravity and remained eight days in a coma'. He also met a petite vivacious widow, Consuelo Gomez Carrillo, born in San Salvador, sculptress, poet, bohemian, whom he married in 1931. In 1939 he published *Terre des Hommes*, in which he expressed his philosophy that man becomes most fully a man, in the air.

'With an aeroplane you do a man's work and experience a man's cares. You are in touch with the wind, the stars, the night, with the sand and the sea. You outwit the forces of Nature. You await the dawn as a gardener awaits spring. You await the next stop like a Promised Land, and in the stars you look for truth.'

We have had a false impression of Earth, says Saint Ex. Like Catherine of Russia, we have been taken in by Potemkin's hastily erected façade villages. From the air we see the truth: beds of rocks, sand and salt, where only occasionally 'like a scattering of moss in the cracks of a ruin, life here and there dares to flower.'

Night flights were the real test of character. 'We were on the point of despair when suddenly a bright dot appeared on the horizon in front of us, to the left. I felt a tumultuous joy; Néri [the Corsican wireless operator] leaned towards me, and he was singing! It must be our next scheduled stop for at night the Sahara goes completely black and is a great dead expanse. However, the light twinkled a little, then went out. We had been heading for a star on the point of setting, visible for just a few minutes on the horizon between a layer of mist and the clouds.

'Then other lights came into view and with heavy hope we headed for each of them in turn. When the light continued steady, we made the crucial test. "We've sighted your beacon," Néri radioed to Cisneros airport, "switch it off, then flash three times." Cisneros complied but the light we were watching did not go off and on; it was an unblinking star.

'Although our petrol was running out, we would bite, every time, at the golden bait. Every time it was definitely the light of a beacon, every time it was our aerodrome and safety. Then, once more, we had to change star.

'From then on we felt lost in interplanetary space, among a hundred inaccessible planets, looking for the one true planet, our own, the only one to possess familiar landscapes, friendly houses and faces dear to us.'

In 1935 Saint Ex set out from Paris in a Simoun in an attempt to win a money prize for a flight to Saigon in 70 hours. He crashed in the Libyan desert. He and his radio-operator were incapable of pinpointing their position within 800 kilometres, and had only a little white wine and coffee in flasks. They lit distress fires and made sorties into the surrounding desert. They collected dew from the plane's wings but it was full of foul-tasting salts and their stomachs would not keep it down. Tormented by thirst, they struck out north-east. On and on they marched. Their strength began to ebb, and they started seeing things: oases, camels, cities, three dogs chasing one another. How long that lasted Saint Ex never knew. At last, on the brink of death, they were seen by a camel-riding Bedouin. Using the feather of a bird, he unstuck the lips of the airmen and gently pushed a lentil mash into their parched mouths – to have given them water then would have split their mucous membranes and killed them. Only when they had swallowed a little of the lentil mash did the Bedouin allow them the life-giving liquid.

'You are the world's greatest wealth,' Saint Ex wrote gratefully, 'and you are also the most delicate, so pure within the womb of Earth. A man can die beside a spring of magnesian water. He can die two feet from a lake of salt water. He can die beside two litres of dew in which certain salts are suspended. You refuse to be mixed with anything alien, refuse to be altered; you are a jealous god ... But through us you spread an infinitely simple happiness.'

In such passages as these Saint Ex sketched a new world-picture for man in the age of flight. By virtue of biological needs, such as that for water, he is utterly dependent on the Earth. But by virtue of a deep urge always to drive himself harder and to seek what lies beyond, he transcends the Earth. He transcends it most when in a state of wonder, and for Saint Ex at the controls of his plane nothing excited wonder more than star-filled space.

Saint Ex wanted to have children, but Consuelo did not, and the only child the airman-poet ever had was the hero of his children's story, *The Little Prince*. This boy is sole inhabitant of an asteroid not much bigger than himself, where he tends three knee-high volcanoes and roots out incipient baobab trees. One day, propelled by migrating birds, he visits other planets hoping to make a friend. He meets a businessman who owns stars, as others own stock or real estate, but is too busy reckoning his paper wealth ever to glance up at a star; and a geographer too occupied writing text-books ever to visit any of the wonders he describes. Then, more satisfyingly, under an apple tree he meets a fox. The fox explains to the boy the way to make friends. The secret is just to be in the company of the other person, not doing anything special, 'wasting time' as humans would call it, until gradually the stranger emerges as different from anyone else, unique and so irreplaceable.

Finally the star-boy meets the author, marooned in the North African desert after a crash. They become friends. When at the end of the book the boy returns to his asteroid the author knows that in future for him one of the points of light in the night sky will be his friend's little planet, he won't know which one, and so all the points of light will be signs of potential friendships.

As a young man Saint Ex had lost his Christian faith. Later he came to see that what he had been seeking in star-filled space was God. But for all his searching he could not actually find him. What then was he left with? Humanism? Precisely because he prized human solidarity so much, Saint Ex believed humanism was inadequate to sustain it. Men could only be brothers in something higher than themselves. According to Saint Ex, the best they could do was continually to strive upward towards a goal they dimly surmised but would never in this life be allowed to see or know.

Saint Ex's life followed the pattern of his books. When the Second World War broke out, after a stay in the United States, he persuaded the Air Force authorities to let him fly on active service. He performed many high-altitude photographic missions in an unarmed single-seater Lightning. During one of these missions, between Corsica and the Alps, he disappeared into the sea, in silence and without witnesses, like Icarus in the painting by Brueghel. He was just forty-four.

*

Rockets developed in the war that cost Saint Ex his life made it possible for man at last to escape from the gravity of his planet. Sputnik I went up on 4 October 1957, to be followed by Explorer I. In April 1961 Yuri Gagarin became the first man to fly in space, circling the Earth in 108 minutes. Six weeks later President Kennedy told Congress; 'I believe the nation should commit itself to achieving the goal, before the decade is out, of landing a man on the moon and returning him safely to the Earth.'

What sort of man? Plainly someone physically tough, highly intelligent, balanced in judgment. Former combat pilots, most with university degrees, went through more voluntary ordeals than the Thebaid hermits. Deprived of sleep and food, they were torrified in an oven at 60°C, then pitched on to slabs of ice. They were placed on the centrifuge which turned them round faster and faster, and when the force of gravity became 16 times normal, when their capillaries were bursting and they felt as if their teeth were leaping out of their mouths, the wheel would be stopped in a few turns. They had to memorize the positions of every wire, fuse, valve and control on the most complicated machines ever built. They had to learn a new language – of figures, acronyms, abbreviations and technical jargon, in which 'go' means 'ready', 'read' means 'hear' and 'lox' means 'liquid oxygen'. In one respect particularly they differed from Saint Ex. Whereas the Frenchman had thought nothing of taking off in an unchecked, battered old plane, spacemen were trained to check every piece of equipment three times, and every key device was supported by a back-up system.

Those early Mercury flights in a capsule the size of a telephone booth had an endearingly human touch. John Glenn found weightlessness a pleasant, floating sensation; over the Indian Ocean he steered by Orion; and he found that, once outside the Earth's atmosphere, space is always black. Night, not day, provides the truth about the sky. Scott Carpenter lost precious fuel chasing sunsets and sunrises with his camera: he was fascinated by their glittering iridescent arcs stretching for hundreds of miles. Amid the exaltation, they made one humbling discovery. Of all man's many buildings the only one visible on orbit was the Great Wall of China. All the rest were as though they had never been.

From Italy there arrived an Italian journalist to watch the spacemen train. Oriana Fallaci was petite, and very pretty, with fun in her dark eyes: at sixteen she had been a Resistance fighter, now in her early thirties she hoped to find in the spacemen the qualities she admired: courage – she was pretty sure she could count on that – but also a sense of humour and of wonder, and awareness of the contradiction and paradox at the heart of human activity.

Oriana warmed to one of the spacemen: Pete Conrad, a small, boyish, laughing Navy pilot who liked drinking root beer and making paper aeroplanes. They invented and shared a joke that they would start a chain of drive-ins on the moon. 'Obviously as well as root beer we'll sell hot dogs, French fries, hot doughnuts. Nothing fancy of course: no knives, no forks.'

Oriana also warmed to Theodore Freeman because he rode a bicycle – almost unheard of in Houston – and wrote poetry. 'How a poet managed to wind up as an astronaut I don't understand ... A poet today is in every sense a danger. You send him to the moon to take rock samples, for example, and he stands lost in wonder in front of a ruby, using up his oxygen supplies. You send him to Mars to make a technical report and he comes back with a sonnet that goes: "Soft silver hills, how I recall the sight ... The bright air lighter than a bridal veil ..." For Christ's sake, what the hell does that mean, air that was lighter than a bridal veil! Can't you just tell us what's the percentage of hydrogen?'

The other spacemen were quite different. Alan Shepard, a hard-headed, thrusting Navy commander, said of the moon project, 'At bottom it's another commercial enterprise.' Later he was to become a millionaire. John Glenn, freckled and public-spirited, was liked by most but not by Shepard: 'John always acts as if he were being watched by an army of boy scouts or children, even when he's scratching his nose or peeing.' Walter Schirra Oriana found 'a potential Saint-Exupéry' but 'dammed by the dykes of his rationalism'. Apart from Conrad and Freeman, Oriana found the spacemen dull, overcontrolled. 'Most of them were bald and you'd have thought they were older than they were ... So when I ask myself what the new astronauts are like, I answer they're old. In a land where youth is a pagan and cruel cult, the representatives of youth are old.'

What was the point of sending such men to the moon? Lacking the imaginative gaiety which for Oriana was essential to being human – did she mean European? – they would bring back a misleadingly limited picture of Earth's companion. 'Wake up,' she wanted to shout at them, 'Stop being so rational, obedient, wrinkled! Stop losing hair, growing sad in your sameness! Tear up the carbon-copy. Laugh, cry, make mistakes, disobey! Break the nose of that Bureaucrat [monitoring the eleven-minute Press interview], break his stop-watch. I say this to you in all humility, with affection, because I admire you, because I see you as better than myself and wish you would be much better than myself, much: not so little. Or is it too late now? Or has the System already broken you, swallowed you up?'

A crew of three was chosen for an actual moon landing and, as we shall see, they were a good deal less lacking in a sense of cosmic wonder than Oriana Fallaci believed. Neil Armstrong looked to the Italian 'like John Glenn's young brother: the same freckles, the same fair colouring, the same ease, he had even been born in Ohio ... His mouth was ironical, but an irony full of caution. His voice was quiet, his movements economical.' In his personal kit for the moon flight he chose to place a piece of the wing fabric and propeller of the Wright brothers' Flyer I.

Edwin Aldrin was a major in the Army, who had graduated number 3 from West Point and been a test pilot; to some he seemed an obvious choice, since his mother's maiden name was Marian Moon. The third member of the crew, who would orbit the moon without actually landing, was Michael Collins, also a West Point graduate, former Air Force test pilot and on Gemini 10 a walker in space.

The three held Press conferences. 'Why are you going to the moon?' 'Oh, it's just a job, like any other.' Norman Mailer, writing a book on the moon flight, was as disturbed as Oriana by the controlled reasonableness, the *a priori* exclusion of mystery. Finally, pressed by the journalists, Armstrong admitted, 'We're required to do these things just as salmon swim upstream.' That was the moment when Mailer warmed to him.

To reach the moon before the Russians had been the original purpose of the Space Programme, but as the Cold War grew less cold, that was considered no longer sufficient. So the spacemen would be going also as scientific discoverers. Admittedly, the prime object of interest – life – was ruled out by the absence of an atmosphere, but they could learn much about the moon's surface, its dust and rocks, they could look close-to at its craters and learn from their shape why and when they were made; they would even be able to learn the moon's age and whether it was ever one with Earth.

On the morning of 16 July 1969 the three ate a light breakfast and received a last-minute briefing. No foreign currency, no passports to worry about: instead the regulating of valves on their heavy space suits. They were driven to the sleek white Saturn V rocket, taken up by lift to their spacecraft – locked into the rocket – and installed for take-off.

At seven minutes past seven a.m. local time the liquid oxygen and kerosene was ignited, and the 3000-ton rocket rose from Earth. Two hours and forty-five minutes later the final stage of the rocket fired, releasing Apollo 11 into the freedom of space.

The Earth receded: after four hours of flight Armstrong could see from one small window the whole of North America. The sun however was omnipresent and potentially dangerous. The crew had to control the spacecraft's temperature by a slow rotation not unlike that of a chicken on a barbecue spit. Weightlessness at first they found entertaining but, as Aldrin said, 'You get tired of rattling around and banging off the ceiling and the floor and the side.'

The crew's instrument readings, and descriptions of them eating and drinking could not indefinitely sustain the world's curiosity. So reporters cut to the space-wives: Jan Armstrong watering her Japanese garden, Pat Collins reading *Peanuts* and drinking a Coke, Joan Aldrin listening to old Duke Ellington records.

Three hundred and eighty thousand kilometres. On the fourth day out the moon loomed very large. Lacking an atmosphere, it does not get dimmer towards the edges, like the Earth, and was bright all over, 'like a reflectorized highway warning sign'. Apollo 11 went into orbit round it, becoming a moon of the moon. The crew examined its bashed, embattled surface, and curiously in this ultra-modern mission only the moon was out of date, its craters named after forgotten Greek astronomers, its plains of dust perpetuating in their names mistaken seventeenth-century beliefs about a watery surface: Ocean of Storms, Sea of Rains, Sea of Tranquillity.

Armstrong and Aldrin crawled into the lunar module Eagle, and during Apollo's thirteenth moon orbit parted from the command module. Eagle went into an orbit that brought it within sixteen kilometres of the moon. Armstrong fired the descent engine, and, legs foremost, Eagle made for the pre-selected landing spot in the wide plain known as the Sea of Tranquillity. But at 1000 metres Armstrong saw that a computer error had brought Eagle off course. It was heading for a crater filled with big boulders and surrounded by rocks. A landing there, and Eagle might easily crash and tip over, wrecking their chances of ever leaving. So Armstrong took over control from the automatic guidance system, and while Aldrin read off angle, height and speed from the instrument panel, he guided Eagle's horizontal movement with his left hand, and its direction with his right, while he searched for level ground. He had to be quick, for fuel was running out.

Several times Armstrong spotted what seemed a suitable place only to reject it on coming closer. Twenty-five metres up, with one minute's fuel remaining, Armstrong found a sure site and guided Eagle towards it. The module's exhaust stirred up moon dust, making distance-judging difficult. But at last, twelve and a half minutes after beginning the powered descent, the module's legs gently touched down. Armstrong reported to Houston: 'Tranquillity Base here. The Eagle has landed.'

They had arrived in another world. That was wonderful, but they had not time yet for wonder. Their immediate task was to test their instruments one by one and ensure all were intact. That done, they unscrewed and removed gloves and helmets. From the small windows they saw they were on a level plain, pitted by thousands of one- and two-foot craters, with a low hill in the distance. Even within Eagle, they were aware of a gravity one-sixth that of Earth. If they let go of a pencil it no longer floated but dropped slowly. Aldrin found it agreeable, 'less lonesome' than weightlessness.

For an hour they made measurements, took bearings on the stars and did the sums needed for eventual ascent. They ate a meal and relaxed. Then Aldrin came on the air and spoke words very different from the spaceman's usual computer language, words which disprove Oriana Fallaci's misgivings.

'This is the Lem pilot. I'd like to take this opportunity to ask every person listening in, whoever and wherever they may be, to pause for a moment and contemplate the events of the past few hours, and to give thanks in his or her way.'

Aldrin's way was to take from his personal kit consecrated bread and wine, transferred there from his local Presbyterian church, where he was an elder, and a chalice. He put them on a small table in front of him, read some passages from the Bible, and partook of the bread and wine.

Aldrin's action appeared to some idiosyncratic. I do not see it so. I see it rather as giving expression to his whole personality, man as well as moonman. It could even be called traditional: had not Columbus raised a large cross on Guanahani, and Edmund Hillary a small cross on top of Everest? What Aldrin was doing when he poured the slow-falling wine into the chalice was to give ritual recognition to what for him, and many of the astronauts, was the most important fact about the moon landing: that all the skill and techniques and wise decisions would have been insufficient had they not taken place within the framework of an orderly and intelligible cosmos.

Armstrong and Aldrin prepared to go outside. They bled the pressure out of the module and helped each other place on their backs the portable life-support systems, which supplied them with oxygen, removed carbon dioxide and water, and pumped a coolant through fine tubes in the inner layers of their suits. They opened the hatch. Armstrong wriggled on to the platform, edged his way down the ladder, and stepped on to the moon.

To compensate for the pull of his support system, he walked arms forward like an ape, but in the moon's low gravity he weighed as little as his twelve-year-old self on Earth, and so he walked bouncily, cockily. Because his boots had less friction than on Earth, he had to plan well ahead when he wanted to turn or stop. He took photographs and collected samples of soil and rock from beneath an all-encumbering dust.

He was joined by Aldrin. There is no air on the moon, therefore no sound. Even if they had been free to remove their helmets the astronauts could not have exchanged words. They spoke to each other, as they spoke to Houston, by radio. They could not light a fire, because a flame cannot burn without oxygen. Nor could they have fetched from their module a glass of water to quench their thirst, for in the vacuum on the moon any liquid would have immediately evaporated.

After Aldrin had tried running and making kangaroo hops, they set up an American flag. Again because the moon has no air, it has no breezes either. Left to itself the Stars and Stripes would have hung limp, so the flag was stiffened out with wire. Against the monochrome moon its red and blue showed well.

Armstrong and Aldrin took stock of their surroundings. In appearance the region was somewhat like a desert on Earth, but even more arid, without those drops of salty dew Saint Ex had collected in Libya. Under a layer of dust the soil was firm to their boots. But there looked to be precious little variety, no rubies such as Oriana Fallaci had envisaged. There was no colour, nothing to catch a mineral-collector's eye, for the absence of geological activity and an atmosphere had left virtually unchanged the primary minerals which in our world have been weathered to beautiful, complicated and varied forms.

Aldrin noticed that the horizon was definitely curved: moon-dwellers if there had been any could never have believed in a flat moon. The absence of trees, of any vertical lines at all, had an effect on balance. The spacemen found it difficult to know when they were leaning forward or backward and to what degree. 'It felt as if you could lean farther in any direction, without losing your balance, than on Earth.' It was odd, too, when they tossed aside a lanyard, say, or a retaining fastener for which they had no further use, to see that they 'would go away with a slow, lazy motion'.

The two men's impressions were mainly visual or related to gravity. From the 'feel' of the moon, they were excluded by their space-suits. Subconsciously every minute of their time on the moon was dominated by awareness of their space-suits, hence of the hostile environment: a temperature in sunlight of plus 100°C, and no oxygen. They were truly out of their element, survivors by grace of artificial umbilical cords.

So they could not look very lovingly at the moon. 'Magnificent desolation,' commented Aldrin, but it was the last word he stressed. Even on Earth we can rarely love virgin territory until we associate it with people dear to us or see it through the eyes of a poet who has detected latent beauty. Perhaps later cosmonauts would come to love moonscape: for the first visitors it was what the Alps had been to medieval man.

For two hours they walked on the moon, collecting samples and positioning scientific equipment. Then they re-entered the module, removed their helmets and discovered that their 20 kilograms of rocks gave off a pungent smell, like gunpowder or spent cap-pistol caps. So the moon which knows neither the resin of pines nor the scent of cowslips does have a mineral aroma.

The spacemen were told to sleep. But the sun had set and the extreme heat suddenly became extreme cold – almost that of Antarctica. According to the official report: 'they felt as if they were freezing to death.' And they got no sleep at all before leaving the moon, some twenty-two hours after touchdown.

*

When the moonmen landed I was in the remote French countryside. The house had no television, so I followed events on the radio. This had advantages. Hearing without seeing, I was conscious of the immense distance the American voices traversed, and there was a penumbra in which my imagination could work. Like most people, I felt prickles of suspense when Armstrong climbed down the ladder, I caught my breath as the button was about to be pressed to make Eagle's engine fire; I felt a glow of relief at the docking, admiration for the technology and teamwork, and at the end even an exultant sense that all mankind had grown an inch.

Then came the newspaper photographs, fuzzy and blurred, followed by the sharper, more detailed photographs in magazines, and they elicited a different reaction. For many, including me, moonlight had been the scene of some of the marking experiences of our lives. It had turned to billows of white a hawthorn hedge in May and, at the gate in the hedge, made an aureole of a girl's fair hair; there were verses somewhere to prove it had not been illusion. Most of all, full moon had meant wave upon wave of foaming light, lifting one to feelings more intense than any in daytime; it had revealed the possibility of so much more to life than anything we had managed until then, and set it before us in terms of a choice between darkness and this pure light; it had been a time when irrelevancies were dimmed, and the essentials came clear. Then we had made midnight resolutions; almost we charted our course, like shellfish, by that vivifying light of the moon.

And now the photographs showed us a world of dark grey basalt, a wilderness of cinders and ash. It was like an Ypres battlefield, but with more pervasive scarring. After the first shock I felt weighed down for a long time by the incongruity between effects and cause.

The house where I was living has murals of the Bayeux Tapestry, and my eye happened to fall on those six Englishmen pointing trembling at the comet of 1066. Today we know that a comet is just a pellet of tiny iced particles, its tail ultra-tenuous ionized gases, and that awe is the least appropriate reaction to it. The moon for me had become like that comet. I saw that it had lost most, perhaps all, of its otherworldliness. There had been a gain in knowledge, but loss in mystery.

True, the night sky held other mysteries, but the moon, so close, so influential, had loomed far larger than any of them in our lives. Its reduction to material terms was almost as though an anatomist, dissecting Romeo's body, should announce that everything about Romeo, including his love for Juliet, could be accounted for by the tissue and chemicals on the dissecting table.

A French ecumenicist, Jean Guitton, took a more buoyant view. 'Yesterday,' he noted in his diary, 'I saw the cosmonauts on the moon, skipping about like young children. Symbol of a kind of renewal of life after death in another world.'

That was the reaction of a believer. But Guitton noted too the quite different reactions of his Marxist acquaintances. For them the landing marked the beginning of a new two-fold freedom – from a superstition-ridden past, and from the limitations built into the Earth. Should the Earth one day lose its atmosphere, man would emigrate to other parts of the sky. With that and replacement of ageing organs he was on the way to a different kind of immortality from Guitton's.

Lastly, an American reaction. Norman Mailer, watching the launch, felt tempted to see it 'as a species of sublimation for the profoundly unmanageable violence of man, a meaningless journey to a dead arena in order that men could engage in the irrational activity of designing machines which would give birth to other machines which would travel to meaningless places.'

Later Mailer adopted a more considered view. He set the landing in the context of his belief that a Force of Good is warring against Forces of Evil. Man, wrote Mailer, was 'designed from the outset to labour as God's agent, to carry God's vision of existence across the stars ... to reveal His vision of existence *out there*, somewhere out there where His hegemony came to an end and other divine conceptions began to exist, or indeed were opposed to us.' On Mailer's view, 'Apollo mission' becomes indeed a mission in the religious sense.

*

Apollo 11's return journey went according to plan. The spacecraft splashed down in the Pacific 195 hours and 18 minutes after take-off. Armstrong, Aldrin and Collins, having wrapped themselves in plastic suits to protect Earthmen from possible lunar germs, were flown to USS *Hornet* and quarantine. When that was over, they stepped out to receive the world's applause.

They had done their job. They had delivered impressions, measurements, photographs and rocks. Now it was up to others to make a new picture of the moon for man. And first to the geologists.

One of the moon rocks was put on display, behind two plate-glass panels, where Norman Mailer saw it. 'She was not two feet away from him, this rock to which he instinctively gave gender as she – and *she* was gray, gray as everyone had said, gray as a dark cinder and not three inches across nor two inches for width, just a gray rock with craters the size of a pin and craters the size of a pencil point ... Was she very old, three billion years or more? Yet she was young, she had just been transported here, and there was something young about her, tender as the smell of the cleanest hay, it was like the subtle lift of love which comes up from the cradle of the newborn.'

More prosaically, geologists categorized the samples. Type A had mineral grains too fine to be seen with the unaided eye, they were like the volcanic rocks known as basalts which we find around the craters of Vesuvius and Etna. Type B also resembled basalts, but contained easily distinguishable grains. Type C were composed of small rock fragments and minerals welded in a matrix of finely pulverized material. Type D consisted of miscellaneous particles smaller than a centimetre in diameter, material sometimes called lunar soil.

These and later samples were analysed by experts able to date minerals. Uranium decays through a series of radioactive elements at an established rate, ending with the stable substance lead; the amount of such elements in a mineral containing them allows us to calculate its age, that is, the time when it crystallized. By this method, the Earth was known to be about 4600 million years old.

The moon, the geologists declared, had come into being about the same time as the Earth, but it had never been part of the Earth – the rock samples showed it was too differently constituted. It formed by accretion of cosmic matter some 4600 million years ago, when the rest of the solar system was formed – also perhaps by accretion, for the theory of a near-collision between the sun and a star has been abandoned.

As the crust cooled the moon was bombarded by large meteorites, the impact of which produced the plains we call Seas, and rock Types C and D. Radioactive elements under the crust forced lava to the surface, filling some low-lying features. Then, about 3100 million years ago the moon cooled, since when it has been modified only by the impact of cosmic particles and meteorites.

It was certainly worth discovering that Earth and moon had separate origins. But in science the solution of one problem almost invariably gives rise to another. In this instance, the far side of the moon was found to be much rougher than the near side. The difference of elevation between high ground and low is about twice as great. And there are no large dark plains. Have huge asteroids knocked the crust off our side of the moon? It sounds improbable, but no one has been able to think of a more likely explanation.

The moon landings have profoundly changed man's attitude to the moon, and to himself in relation thereto. Let us look first at the losses. First, though perhaps least often expressed, a lot of people had been hoping that the moon would yield surprises: not life, of course, but something positively wayward, different, even perhaps funny – if not *Finnegans Wake*, at least a scene therefrom. But the moon's only surprise was that it had no such surprises. It turns out to be a most negative body, symbolically an anti-Earth.

For us on Earth the moon has lost its virginal quality. It has been possessed. Since the Apollo landing ten other astronauts have trodden its cindery soil and left behind three lunar jeeps. Linked to this, even our way of looking at it has changed. Behind a crescent moon now we are aware of the full matronly orb.

To the mother telling her child a goodnight story the moon is no longer a subject for fantasy. The day man walked the moon, the man in the moon took flight. Our children will grow up with other fairy stories, but they will not be about the moon.

The moon has been variously seen by pre-Space Age poets: 'a consumptive lying on the black pillow of the sky', 'a good old woman, with cotton wool in her ears', 'a hole in the world's skull', 'white as a knuckle and terribly upset'. The last section of the latest edition of *The Oxford Book of Twentieth-Century English Verse* contains no poem about the moon, a body which has acquired a complex identity of its own and can no longer serve as a mirror for personal moods.

Having become familiar, and on the way to being ordinary, the moon has been emptied of otherworldliness, as I mentioned earlier, and emptied also of some of its power. Since antiquity sun and moon had been visualized as a pair, male and female opposing yet balancing each other. Now the moon has been dethroned, and in our visible cosmos the sun rules alone. This will doubtless cause commotion in the unconscious, though only a very brave man would speculate how.

There have been gains as well. The moon has become more accessible to artists, and has entered the world of one distinguished English painter. Eileen Agar was one of the early Surrealists, a friend of Paul Eluard and of Picasso. In the 1930s she painted much in the South of France, using a palette of vivid colours to express a mood of joy. She tells me that for most of her painting life she associated the sun with the visual arts, the moon with music and geometry. So she rarely depicted the moon, and then only as a distant crescent. But in the year of Apollo 11 she came to a different view. She saw a big full moon filling her television screen. The moon seemed to be right inside her room. It began to become familiar to her, and she felt the urge to paint it, not as a mysterious object, but as a now familiar shape. Her *Room with a View of the Moon* makes an interesting contrast with Samuel Palmer's *Cornfield by Moonlight*. The top left corner shows a cratered moon on a television screen. The sky is black, as the astronauts say it is. Near the moon are spacecraft and stars. To the right on a table stand a green apple, a pot and a gourd. Below are a tablecloth and a leaf, with, on the left, a triangular basket of moon rocks. The colours are black, brown, moonlight blue, and some orange to emphasize the warmth of the room.

For the astronauts themselves there has been gain. When Yuri Gagarin circumnavigated the Earth in space, Khruschev remarked, 'He didn't see any angels.' Perhaps not. Perhaps the Americans didn't see any either. Yet I believe it is a fact that they were deeply affected by their cosmic experience, and the new perspective into human nature which it afforded. As a result, many have since suffered physical illness, mental breakdown and religious conversion of various kinds.

Perhaps the greatest gain in man's world-picture effected by Apollo 11 is the one described by Armstrong. Man went to the moon but found the Earth, and the Earth in a way is a spacecraft.

It's an odd kind of spacecraft, since it carries its crew on the outside instead of inside . . . Hopefully, by getting a little farther away, both in the real sense and the figurative sense, we'll be able to make sure people step back and reconsider their mission in the universe, to think of themselves as a group of people who constitute the crew of a spaceship going through the universe.

Matching that text is a photograph of the blue, cloud-swirled Planet Earth taken from a spaceship, so familiar from television and magazines that it does not require to be reproduced as an illustration here. On that blue globe no countries are marked in colour or stippling, no frontiers drawn.

At school we used to write to one another notes addressed to So-and-so in Such-and-Such a town, in Sussex, England, Europe, the World, the Universe, and perhaps, without knowing it, we were on the right track. That photograph of Earth from space has become part of our new world-picture. It is on page one of the geography books. Children are learning that before being citizens of this or that town, they are dwellers on a single planet. The fundamental cosmic fact about man is that we all share a single habitat.

This has engendered a new sense of solidarity, such as Saint Ex glimpsed in Aéropostale; now it is part of all men's experience. However much climate, latitude and migration may have caused us to vary, we are kin by virtue of our habitat. Those who see it like that are trying to strengthen their solidarity, while retaining the differences that permit competition and rivalry, of the kind for example that sent men to the moon.

There is a warning though in another space photograph. In it the Great Wall of China can be seen, bearing witness to man's divisive tendencies. We notice too, near the Wall, wisps of cloud and remember that man has the means of obliterating all such wisps in a giant mushroom cloud. We know that the continuance of Earth's physical atmosphere depends on a less tangible, equally fragile moral atmosphere. We know in short how we must act, or at least what we must avoid, if our Earth is not one day to look like the moon.

CHAPTER THIRTEEN

Happenings in Space

UFO's – two sightings and the difficulties they pose – Jung's explanation – futuristic pictures of space living – life in an orbiting steel cylinder – two kinds of Utopia – The First Men in the Moon *– The Book of Revelation and* The War of the Worlds *– changing themes of science fiction – 2001: A Space Odyssey – the appeal of science fiction – Ray Bradbury's stories – good and bad extra-terrestrials – the effect of science fiction on young people.*

One of the conclusions evident from this history of cosmologies is that man's imagination abhors a vacuum, and the space his intelligence wins his imagination will colonize. This chapter is devoted to considering some of the ways in which twentieth-century man has colonized, or intends to colonize, space: first with (as I believe) imaginary objects, then with projects for space living, and lastly with fictional creatures.

In Tehama County, northern California, on the night of 13 August 1960 two state policemen were driving a patrol car on Hoag Road, East Corning, searching for a speeding motor–cyclist. Charles A. Carson had served four years in the US Air Force and was accustomed to seeing planes by night, weather balloons and the Northern Lights. His colleague, Stanley Scott, had served as a paratrooper during the Korean War and he too was aware of the tricks light can play on the eyes during darkness.

The sky that summer night was cloudless. Ten minutes before midnight Carson and Scott saw what looked like a huge airliner coming in very low directly in front of them. Stopping their car, the policemen hurried out to see what they believed was going to be a plane crash. But to their surprise the airliner made no noise. They assumed that its engines had failed or been cut off, and they continued to watch.

'The object was about one hundred or two hundred feet off the ground when suddenly,' according to the policemen's report, 'it reversed completely at high speed, and gained an altitude of approximately five hundred feet. There it stopped. By now the object was clearly visible to both men. It was shaped somewhat like a [rugby] football, about one hundred and fifty feet long and forty feet high, was solid rather than transparent, and emitted a white glow that surrounded the whole object. Red lights glowed from each end and at times about five white lights became visible between them.'

Carson and Scott continued to watch in astonishment. They saw the object move in a variety of directions, now slowly, now at high speed. Sometimes it just hovered motionless. After a while the policemen returned to their car and radioed the Sheriff's office, asking him to call the nearest radar base. Then they tried to approach the object, but each time they drove forward, it retreated. They stopped the car. Then on two separate occasions the object came towards them, turning as it did so and sweeping the area with a huge red light. Scott thereupon swung the red light on the patrol car towards the object. At once it retreated. The object's red beam came on six or seven more times, sweeping the ground and the sky.

The object eventually moved slowly away in an easterly direction. The policemen followed. Near Vina Plains Fire Station they saw a similar object approaching from the south. The two objects came close to each other and stopped, occasionally shining their beams of red light. Finally both receded below the eastern horizon.

Carson and Scott looked at their watches. They had been observing the strange object for two hours and fifteen minutes. On their return to the Sheriff's office they found that the object had been seen by two deputies and by the prison jailer, who had taken his prisoners on to the roof to observe it. Their description tallied with the policemen's. Furthermore, the radar base had reported an unaccounted-for object on their screen in the vicinity of the policemen.

The Air Force made a full-scale investigation. They attributed the phenomenon to temperature inversion, which can cause a bright planet or star low on the horizon to be projected upward, but on the night in question Mars did not rise until one hour after the object appeared, and the one bright star, Capella, low on the horizon at 11.50 p.m., would have been far up the sky when the object disappeared below the eastern horizon. Carson commented, 'We find it difficult to believe what we were watching, but no one will ever convince us that we were witnessing a refraction of light.'

A no less remarkable sighting had occurred on 14 July 1952. Again the sky was clear. Captain William B. Nash was piloting a Pan American DC-4 airliner towards Miami. At about 8.10 p.m. near Norfolk, Virginia, Nash and Third Officer William Fortenberry, who sat beside him, saw six fiery red discs approach in line at high speed, about a mile below the airliner. 'Their shape was clearly outlined and evidently circular,' said Nash. 'The edges were well defined, not phosphorescent or fuzzy in the least.' The diameter of each disc was about thirty metres. When the line of discs was almost beneath the airliner, they turned up on edge, and flipped over, changing position so that the line was now in reverse order and moving away. They speeded off in a direction at an acute angle to their first course, and were joined by two similar but brighter craft which darted out from beneath the airliner. All eight discs 'climbed in a graceful arc above the altitude of the airliner and then blinked out one by one, but not in sequence.'

Captain Nash radioed a report to be forwarded to the Air Force. Next day he and Fortenberry were questioned separately by Air Force officials and their descriptions found to tally. Seven people on the ground had also observed red discs travelling extremely fast and making sudden changes of direction.

Since June 1947, when Kenneth Arnold, an Idaho businessman piloting his plane above the Rocky Mountains, saw a fast-moving formation of nine objects like metal discs and a local journalist christened them 'flying saucers', there have been no less than 60,000 reports of objects similar to those in these two examples. They may be disc-shaped, round, elliptical or cigar-shaped, orangey-red, white, metal-coloured or changeable, and a quarter of them carry lights. They cause cars to stall and car lights to go out; they upset plane instruments; they produce TV and radio interference and power failures. People seeing them experience a burning sensation, pins and needles, conjunctivitis, even momentary paralysis.

Only four thousand of the reports claim that the object landed, and only half that number claim to have seen one or more crew. The crewman is short – about 3 foot 6 inches – with small limbs and a large head. He wears a space suit and sometimes a metal helmet. When humans approach, he quickly returns to his craft.

In the United States, Project Blue Book examined 10,147 such reports between 1947 and 1965 and was able to explain most of them by reference to artificial Earth satellites, weather balloons, temperature inversion, electrical discharge from high tension cables and similar known phenomena. But there remained a hard core of 646 which the commission could not explain.

Most of those cases involve witnesses of undeniable truthfulness. Some are high-ranking Air Force officers, some clergymen. In 1971 Jimmy Carter, then Governor of Georgia, filed a report on a UFO he had seen four years earlier in Leary. These are not people likely to be taken in by optical illusions or to suffer hallucinations.

Moreover, sightings of UFO's are worldwide. In France 7 per cent of the population claims to have seen a UFO, and a Sofrès poll published recently by *Le Pèlerin* showed that 25 per cent of the French people, and 40 per cent of those under twenty-five believed that UFO's come from another planet. In the United States the figures are double. In Britain, where there have been fewer convincing sightings, people are more sceptical and some treat the subject as a joke: for instance, in London in 1979 I noticed a rash of car stickers in the form of a red and white striped flag surmounted by UFO.

If indeed there is a hard core of sightings which cannot be explained by known phenomena, it follows that they must originate in what is as yet not understood. Four explanations suggest themselves: atmospheric phenomena of a kind hitherto unrecorded; paranormal emanations from the human brain; secret weapons, presumably Russian; and extraterrestrial objects. I personally favour the first explanation, but the last is far more widely held and it is the one I want to consider here.

Its proponents point to the absence of a fully satisfactory terrestrial explanation of UFO's, and to the probability that there is intelligent life elsewhere in space – a point we shall discuss in the next chapter. Now if there *is* intelligent life in space, what more likely than that it should be interested in us, as we are in it?

Opponents of the theory invoke two facts. First, there is no close-up photographic evidence. The best photographs are two of a shiny silver disc taken by Paul Trent, a farmer in McMinnville, Oregon, on 11 May 1950. Recognizable details of farm and landscape validate the photographs but the disc is so small that it could be a blurred plane or bird. Secondly, even if we accept that UFO's travel at high speed, let us say at 19,000 k.p.h., which was the estimate of Pan American pilot Captain Nash, it would take more than 200,000 years for an extraterrestrial to arrive from the nearest star. Most people find it inconceivable that any being could live so long – and within a metal disc.

For my part, I see three further difficulties. One is the fact that UFO sightings come mainly in waves. Vintage years were 1950, 1954, 1957, 1959, 1967, and 1974. Now anyone with experience of journalism will know that once a story has been prominently headlined it goes dead for a period. No one wants to hear about it. But after two or three years when memories of the story have blurred a similar story can again be newsworthy. Claims to have seen UFO's follow this pattern, save for the gap between 1967 and 1974, when the attention of likely UFO-viewers was engaged by the moon-landing and its aftermath.

A second difficulty is the difference in size between UFO's seen in America and those seen in Europe. The American UFO's are generally very large: that seen by the policemen was fifty metres long, and those seen by the Pan Am crew thirty metres in diameter and five metres thick. European UFO's are rarely more than ten metres in diameter. What can explain this discrepancy better than the well-known American tendency to 'see big'?

The third difficulty is one of motive. Why do UFO's often circle but rarely land, and why do their crews take flight when men approach? One answer was given in 1977 in the journal *Science* by T. B. H. Kuiper and M. Morris. All higher civilizations have an urge to colonize their galaxy. UFO's carry beings from a higher civilization. One day they will colonize the Earth. But for the moment they refrain from doing so, for by intervening prematurely in our natural development, in the long run they could devalue the one resource they hope for from our planet: the culture and knowledge of Earth-men.

It is in some ways an appealing picture. Prince Charming falls in love with a young blue-eyed girl but waits for her mature beauty to develop before proposing. But I do not find it convincing because I doubt its basic assumption. We on Earth pride ourselves, perhaps mistakenly, on having outgrown colonization; why should we attribute that same foible to intelligent beings more advanced than ourselves?

Given such a weight of negative argument, why do so many people believe that UFO's come from another planet? That is the most astonishing fact of all in the whole UFO file. In 1960 Carl Jung gave an answer. Jung believes that all men, of whatever civilization, share a collective unconscious, which features certain basic objectives such as the circle and triangle, father and mother, earth and water, as symbols of important relationships. Now a round shape is a well-established symbol of the universe, of totality, of God. In agricultural societies the round shape will be the sun or moon; in a technological society it will be a machine, such as the flying saucer. The more a society is fragmented and troubled, the more it will tend to seek wholeness and reassurance in the form of such an object.

Accepting Jung's view that belief in UFO's is a symptom of a troubled society, perhaps we can modify his application of it. Our age is one where religious belief is weak or almost nil, and in any such age the void created by absence of the genuinely supernatural tends to be filled by the preternatural. One obvious example is the craze for mesmerism in eighteenth-century France and Germany.

Now in a space age the preternatural will manifest itself in space. But space is still, vestigially, heaven, still charged with Christian imagery. Among the best known of those images are angels. We learn about them in childhood, more as visitors than bearers of grace, then put them out of our conscious minds. But at the subconscious level we perhaps still need them.

On this view UFO's express, in modern form, our ambivalent attitude to angels. We are glad to be reassured that we are not alone in the universe, but are not sufficiently clear in our views about the role of superior beings to take matters a stage further. So when humans approach, the crews of UFO's nearly always vanish, without imparting their know-how – the secular form of grace.

Another remarkable phenomenon of our time is the widespread belief that in the not too distant future man will colonize space. The Americans, with their strong tradition of new frontiers, believe that since the moon is available, it had better be settled. Oxygen can be extracted from lunar rocks, and liquid hydrogen flown in from Earth; initially 600 tons will provide water for a town of 1000 people. Four solar power stations, one of which will always receive sufficient sunlight, will heat liquid nitrogen to drive turbines, generating electricity. The lunar towns will produce vacuum-cast alloys and vacuum welds. On Earth these require expensive vacuum chambers, but on the moon's surface they could be made cheaply in the ever-present cost-free vacuum.

A second project calls for making the moon a hospital. This is specially weird, since the moon was for long associated with sickness, wasting diseases and lunacy. The plan presupposes water and energy as for an industrial township, but the moon dwellers now will be doctors and nurses. Patients with weak hearts, muscle disease, arthritis and rheumatism will be flown by shuttle service for skilled treatment in the gentle gravity of the moon.

Even the most starry-eyed planners have to admit that the moon is not an ideal habitat. In 1974 Gerard K. O'Neill, a Princeton physicist, put forward the idea – sponsored since by NASA – of building habitable cylinders in space. Strip-mining on the moon would produce iron and aluminium which, with glass and concrete made by lunar colonists, would be catapulted into space. The construction site would be at a point on the moon's orbital path at a distance from the moon equal to that between the moon and the Earth; here a stable orbit can be maintained without the need for expending energy. Spacemen on the construction site would collect the lunar materials as they arrive and construct a cylinder 25 kilometres long and 6 kilometres wide. Another similar cylinder would be constructed 80 kilometres away, and the two tethered together by tension cables. Each will rotate in a different direction so that they are always pointed endwise towards the sun. The speed of rotation will be once every two minutes, thus generating an Earth-like gravity on the inner surfaces.

On each cylinder three huge rectangular mirrors reflect the sun through windows into the interior, providing light and heat, while on top of the cylinder paraboloidal mirrors concentrate the sun's rays to drive steam-turbine electric generators. Each cylinder is divided into six alternate strips of blue-tinted window and 'land' – an area composed of lunar soil and chemical additives together with vegetation and trees from Earth, and water made from lunar oxides and liquid hydrogen flown from Earth. The land areas are laid out as hills and dells, streams and lakes, parkland and forest. An atmosphere like ours is created within the cylinder, complete with protective ozone clouds which send down rain. Transport is by bicycle and electric car. Food is grown in non-rotating pods near the top of the cylinder under controllable conditions.

Inhabitants of such a cylinder will enjoy an ideal climate; they can swim in the lakes, fish in the streams, climb the hills. But they will not be idle. One of their principal occupations, we are told, will be the manufacture of other similar habitats! This work will take place outside the cylinder, where the vacuum and weightlessness of outer space facilitate heavy industry, the sun here as within providing limitless energy.

Such a space habitat could support well over one million people. It might eventually declare its independence of Earth, according to one suggestion in 2076, three centuries after the United States broke with Britain. And we are not to worry about boredom and uniformity. A pioneer who wants greater adventure than the steel cylinder can provide will rocket his family and mining equipment to the new Yukon – some asteroid rich in nickel and iron.

Those are three examples of projects for space living. Sometimes by their inventors, more often by popularizers, they are justified by high-minded motives. If our sun dies, if we lose our atmosphere; if a massive asteroid should head for Earth on collision course, then, we are told, men owe it to Life to effect a removal. Ray Bradbury goes so far as to say that men have an obligation to adapt themselves to the extreme conditions of space, even if that means they have to become lichens, insects, balls of fire.

I believe that the strongest motive behind such projects, and the enthusiasm with which they are greeted is one seldom avowed. The population explosion which has increased our numbers from 1·4 billion in 1900 to nearly 5 billion today makes us desperately worried about diminishing Earth space, and correspondingly eager for space elsewhere.

These trapeze-act space-living projects remind me of the undersea-towns planned in the wake of Jacques Cousteau's deep-sea explorations in the 1950's. The fear then was that we should run out of food, and therefore we should farm the sea-bed. The Japanese ate one variety of seaweed; our children would eat fifty-seven. Living in an undersea-town, man, it was promised, would breathe air purer than he would find in smog-bound Los Angeles or Tokyo. The costs were reckoned and found to be outweighed by the tremendous gain in food supplies, and it was even argued that an undersea-town was evolutionarily obligatory: on a certain day millions of years ago a fish heaved itself out of the sea and became the first amphibian; now that amphibian's descendants must, again for survival's sake, return to the same sea.

Where are the undersea-towns? Thirty years after they were on the drawing-board one does not see even an underwater-village. For the excellent reason that planners ignored a basic fact of human nature: man is attached to Earth, and usually to a particular spot on Earth, by ties which elude the logical theorist but which are yet as necessary as the umbilical: ties which involve a sense of belonging, continuity, family and love, and which help to make up our very identity. Aldrin complained that weightlessness was 'lonesome'; what would he say about the isolation of an undersea-town?

The same objection could be levelled against space habitats, and another objection too of a different kind. The futurists confidently claim that history is on their side. The inexorable logic of progress makes it inevitable that man will press ever forward and spend, as the Americans did in the 'fifties and 'sixties, larger and larger sums on space. Technological progress, we are told, will continue at its present rate of increase, and recent trends are extrapolated to show how many billions of dollars will be available annually for space habitats a hundred years from now.

This, I suggest, is a scientist's mis-reading of history. Inlaid cabinets of the reign of Louis XV are more elaborate and sophisticated than those of the reign of Louis XIV, and are in turn surpassed by those of the reign of Louis XVI. But would one thereby be justified in supposing that succeeding cabinets were finer still?

During the Reign of Terror no cabinets at all were made; under Napoleon and Louis XVIII cabinets became inferior to those produced in the previous two centuries. Society had changed, people's values, the demands on their money; and artistic technology in that particular field had reached a point of excellence where it might have been sustained but certainly could not have been improved. Today you can find mass-produced cabinets far inferior to those of the age of Louis XVI, but all the oil in Arabia will not procure you a new one that is superior to them. History, even economic history, does not proceed like a line on a graph, and progress is never mathematical progression.

Any design for living in the future, whether realizable or not, is a form of Utopia. Now there are two categories of Utopia, and it is important to distinguish them. One is the Utopia which seeks ways to improve human nature spiritually as a means to general contentment and peace; its archetype is Thomas More's book of that name. The other is the Utopia which seeks to satisfy men's desires by ever-improving social structures and technology; its archetype is *The Year 2440* written in 1770 by a Parisian, Sébastien Mercier, which promised happiness for all on the basis of a canal system, street lighting and inoculation.

Now most of the designs for space-living fall into the second category.

But their inventors or popularizers do not always state this; confusion arises, and the Utopias are often assumed to belong to the first category. Futurists encourage the confusion by writing as though once he is off Earth man will be purified, so to speak, by vacuum and weightlessness. Living as pioneers on the moon or in an orbiting cylinder will of itself damp down aggressiveness and ensure harmony.

Put like that, the fallacy is obvious. But when packaged in jargon it can mislead – sometimes harmfully. For it may divert us from the more essential category of futurist thinking, the Utopia which considers not outer space but inner man.

*

From the would-be facts of the future we turn to fiction set in the future: specifically to tales about inter-planetary man. Jules Verne had set man in orbit round the moon, H. G. Wells actually landed him there. *The First Men in the Moon* (1901) remains an important book, even though it has been overtaken by events, because it became the prototype of a genre. It describes how a scientist, Cavor, and a playwright, Bedford, travel from Earth in a sphere coated with a special substance which has the power of screening gravity. Astronomers of Wells's day knew that there was little or no atmosphere or water on the moon's surface, but Wells cleverly imagined the moon to be hollowed out, with water *inside*, 200 miles down, on which cities are built, reached by balloon. Through cavernous tunnels rise highly oxygenated air and water vapour, which in the cold lunar night falls as deep snow. Gold is as plentiful as iron on Earth, and weird plants grow rapidly from spores.

The moon is inhabited by mooncalves, leathery giant slugs 200 feet long that bellow, and by Selenites, creatures barely five foot high with short flimsy legs and whip-like tentacles, in appearance more like ants on their hind legs than human beings. They wear leathery clothes and goggles of darkened glass, and they communicate by twittering.

Cavor and Bedford eat an intoxicating fungus, which dulls their attention, and they are made prisoner by the Selenites. Bedford escapes and believing his friend dead returns to Earth. Then comes a twist. A Dutchman picks up radio messages from the moon. They emanate from Cavor. He has found many different types of Selenite, each physically adapted to its function and to the more intelligent he has taught his language. Eventually he is led by them before the Grand Lunar, who rules from a palace. Here Wells oversteps the edge of credibility. Himself a physically small man who had made his way by virtue of a brilliant brain set in a large head, Wells endows his Grand Lunar with dwarf body, elfin eyes and a brain case 'many yards in diameter'. As he listens to Cavor's account of mankind's illogical activities, the Grand Lunar's brain grows alarmingly hot; shadowy attendants busy themselves 'spraying that great brain with a cooling spray and patting and sustaining it'! Cavor injudiciously reveals that men rejoice and glory in going to war (the novel was written during the Boer War). As the story ends we realize that the Grand Lunar is never going to let Cavor return, lest he rouse men to war against the Selenites.

The First Men in the Moon epitomizes what may be termed the optimistic view of space. The Selenites are benevolent and their motives for holding Cavor wise. Wells's novel was to have many descendants, but before that happened, quite a different tradition came to the fore.

The West had long possessed a Christian view of the future, embodied in St John's *Book of Revelation*. It describes cosmic disasters, such as a vial of God's anger being poured over the sun, whereupon the sun afflicts mankind with burning heat, and stars falling from the sky; a seven-horned beast oppresses the righteous, as do plague-breathing horses and man-killing locusts. These disasters lead to God's Judgment of mankind (Doomsday), then to the creation of a new Heaven and a new Earth. Christians well into the nineteenth century actually saw the future in those terms. Then St John's vision began to lose its force, and H. G. Wells filled the vacuum with *The War of the Worlds*. This novel embodies the pessimistic view of space with an apparent scientific precision that appealed to an age making advances in science. Grotesque monsters arrive from Mars who plan to conquer Earth. With heat-rays and mechanical Fighting Machines the monsters subdue man, only to be destroyed in their turn by putrefactive and disease bacteria.

Two points are noteworthy. Wells's novel ends without that note of unqualified hope distinctive of the *Book of Revelation*; and novels of destruction in the same genre came to be called Doomsday fiction, but that is a misnomer, for Doomsday means Judgment, by God or some superior power, and no Judgment takes place in these novels.

Wells's *The War of the Worlds* was to dominate science fiction from about 1923, when Hermann Oberth's scientific book, *The Rocket into*

Interplanetary Space, aroused popular interest in space travel, until 1939. Book after book described rockets manned by bug-eyed monsters, ovoid, with a single red, lustful eye, green slime dripping from their beaky noses and cavernous mouths. Side by side, and sometimes overlapping, were the comic-strip heroes: Flash Gordon, who led a trio of Earthmen in a fight for survival on the distant planet Mongo; Superman, endowed with supernatural strength, the power to fly and X-ray vision, his origin explained in episode 1: 'Just before the doomed planet, Krypton, exploded to fragments, a scientist placed his infant son within an experimental rocket-ship, launching it towards Earth.' Found and adopted by an elderly couple, the Kents, the boy is named Clark Kent, and swears to devote his existence to helping those in need. No less high-minded is Batman, created in 1939 by Bob Kane as a rival to Superman. Batman has no supernatural power; everything he does is humanly possible.

Until the Second World War heroes of this ilk stood master over bug-eyed monsters and evil generally. Then the mood changed. Many books took as their starting-point a disastrous atomic war. In Walter Miller's *A Canticle for Leibowitz* (1960) the survivors are convinced that the written word is the root of all evil; if writings are destroyed then all knowledge of how to make the bomb will perish. A group of 'bookleggers' copy manuscripts and slowly build up the resources of civilization. A new technological age emerges, atomic war breaks out, and the book ends with the sole representatives of the human race, a rocketload of children, being despatched to outer space.

Among more optimistic science fiction writers the most prolific has been Isaac Asimov, an American of Russian descent who in the 1950's taught biochemistry at Boston University. Asimov's *Foundation* (1951) is set in the days of the Julactic Empire – a society of a million worlds throughout the Milky Way. The Empire is in danger of crumbling into barbarism; salvation arrives in the person not of a scientist or a Superman but of an academic. Hans Seldon and a band of psychologists create a new force – the Foundation – dedicated to art, science and technology. Using the techniques of psycho-history, they arrange to bring about a series of crises that will rescue the galaxy from barbarism and set it along the route to a new Empire, of which the nucleus will be the Foundation. They encourage trade as the surest basis of friendships between worlds, and to a lesser degree religion – not for itself but as an aid to the Foundation's secular goals. Spacecraft criss-cross the galaxy and there are also lightweight atomic blasters, from which a man protects himself with a special shield, but the most futuristic feature of *Foundation* is the notion that clever men can delicately manipulate the future, initiating crisis and danger in order to induce a spirit of adventure and creativeness.

With the 'sixties and plans for a moon landing, a new kind of fiction emerged, epitomized in Arthur C. Clarke's *2001: A Space Odyssey*. Nameless superior intelligences have long interested themselves in man's fate. Three million years ago they coaxed man-apes to tie a knot and to throw a stone straight, and on the moon they buried a magnetic arch which, once men became efficiently evolved to colonize the moon and excavate the arch, would emit a predetermined signal into space. That moment has now arrived, in 2001, as David Bowman in the spaceship *Discovery* flies to Saturn and lands on one of its moons, Japetus. Increasingly he feels himself to be under the care of superior intelligences. Having reached Japetus, he finds himself in what he believes to be a bedroom of a Washington hotel, complete with a replica of Van Gogh's *Bridge of Arles* on one wall and a local telephone directory. Only when he lifts the phone and there is no dialling tone does he realize that the hotel bedroom is just a replica designed by the superior beings to make him feel at home. Bowman is sucked through a hole in space-time, acquires immortality and briefly sees, as in a vision, what he takes to be the intelligences moving like a procession of stars. At the end of the book he makes his way back towards Earth, bent on helping forward his fellow men.

What are the special attractions of this kind of fiction? There are, I believe, four. First, the gadgetry: the stewardess's Velcro-soled sandals, which allow her to walk on the space-ship's Velcro carpeting without floating: *Discovery*'s cooking and washing area, which spins like a carousel to ensure enough gravity to keep liquids in saucepan and washbasin; the talking computer Hal, helpful at first, who under stress goes berserk and endangers *Discovery*'s mission.

Then there are descriptions of space which by their beauty induce wonder, such as this: 'The golden crescent of Saturn rose in the sky ahead. In all history, he was the only man to have seen the sight. To all other eyes, Saturn had always shown its whole illuminated disc, turned full towards the sun. Now it was a delicate bow, with the rings forming a thin line across it – like an arrow about to be loosened, into the face of the sun itself.'

A third attraction lies in man's encounter with new dangers. Conan

Doyle thrilled readers with a story of pioneer aviators flying ever higher until they reach a point from which they find they cannot return; more modern perils include a space-walk at 20,000 m.p.h. to repair a mechanical space-ship failure and a session in the Hibernaculum, which will freeze your body for a set number of years, then release you unaged – if in the meantime it hasn't broken down.

The fourth attraction, and I believe the most important, lies in the character and psychology of the beings met in space. Clearly they must possess the chill of genuine strangeness yet also be credible. And to be credible, they must not be wholly strange. Usually they can be divided into those morally worse than man, and those morally better. Among the most convincing in the former category are the Martians in Ray Bradbury's stories *The Martian Chronicles*, entitled in Britain *The Silver Locusts*. Bradbury was born in Waukegan, Illinois, and Mars he conceives as a Midwestern prairie in a hot dusty summer, its rare habitations taut with suspicion. As might be expected from a writer whose re-creations are oil painting, ceramics and collecting primitive artefacts, colours are cleverly used to create mood, and his Martians are sometimes masked.

In one of Bradbury's stories, a spaceship lands on Mars. We follow events through a Martian farmer's wife. She tells her husband that she has met a spaceman who claims to come from Earth. Her husband says, 'The third planet is incapable of supporting life. Our scientists have said there's far too much oxygen in their atmosphere.' But his wife sticks to her story. The spaceman has sung her a song, 'Drink to me only with thine eyes'. Martians do not have songs, and she is haunted by its beauty. Her dour husband becomes jealous, angry and perhaps afraid. He goes out and shoots the spacemen.

In another story, again a rocket lands. The Martian authorities believe the spacemen are psychotics who have 'produced' their rocket from sick minds, for Martian psychotics are able to 'produce' little demons and oily snakes, hallucinations that can be seen and touched. The authorities lock up the spacemen in an asylum. But the rocket remains. One of the Martians, Mr Xxxx, puts the spacemen 'out of their misery' by shooting them, hoping now to see the hallucinatory rocket vanish. But it is still there. Now Mr Xxxx believes *he* has become insane. Perhaps he has, for he decides that to make the rocket vanish, he must kill himself. And this he does.

In a third story the crew of a spaceship land on Mars and are delighted to find an Earth-like Victorian township, peopled with men and women who claim to recognize the spacemen from some years before, and who entertain them warmly. It is all part of a clever plan whereby the Martians divide, conquer and finally kill the intruders.

With a virtuosity unequalled in this field, in the same book Bradbury also creates good Martians. 'The Off-Season' tells how Sam Parkhill, a colonist from Earth, builds a hot-dog stand on Mars. He is disturbed by a Martian, grows angry and, thinking the other is drawing a weapon, shoots him dead. Then many Martians arrive: 'Across the ancient sea floor a dozen tall, blue-sailed Martian sand-ships floated, like blue ghosts, like blue smoke.' Instead of avenging their compatriot, the Martians make Sam a gift of half their planet. Sam plans a big expansion of business, to cope with an expected five million immigrants and visitors from Earth. He and his wife Emma celebrate. Then they see that atomic war has broken out on Earth, part of which seems to fragment into a million pieces. Emma concludes, 'This looks like it's going to be an off-season.'

In 'The Fire Balloons' two Catholic priests plan to convert the Martians. In form the Martians are soft blue fiery globes. The priests follow them over rough country. An avalanche looks sure to kill the priests but in the last fraction of a second they are mysteriously saved, Fr Peregrine thinks through the agency of the blue globes. To test this belief, the priest points his pistol at his own wrist and fires. 'There was a shimmer of light, and before their eyes the bullet stood upon the air, poised an inch from his open palm. It hung for a moment, surrounded by a blue phosphorescence. Then it fell, hissing, into the dust.'

Fr Peregrine draws a round circle in the centre of a blackboard. 'This circle will be the Martian Christ. This is how we shall bring him to Mars.' The priests build a church, or rather, in an area cleared of rocks they erect an altar and place on it a fiery globe. They tap an iron bell – for their own comfort; they play music, but the Martians do not come. Then they pray, and the fiery globes begin to arrive. They explain that they have put away sin with their bodies and do not need religion. Fr Peregrine thinks as he leaves, 'We couldn't build a church for the likes of you. You're Beauty itself. What church could compete with the fireworks of the pure soul?'

These examples from Bradbury's work belong to the category of extra-terrestrials morally better than man, which includes Wells's peaceful

Selenites, Eigel and Schuster's Superman, and Arthur C. Clarke's speechless benevolent intelligences in *2001*. Man has always been fascinated by characters who act according to different values from his own, hence by the fairy world and folk heroes. What is really new is the space setting, and with it an otherworldly note.

In many science fiction books there is a clear-cut battle between good and evil of an oversimplified white and black kind, the latter usually polarized in an authority savouring more of the past than the future. Some of Asimov's books reproduce the conflict between Papacy and Holy Roman Empire, while in *Star Wars* a lowly farm lad, thanks to an intercepted message, is able to rescue a princess, no less, from the clutches of Darth Vader, who wears flowing black robes and a black metal mask – the theme is as old as the first fairy story. Old-fashioned too is the assumption by science fiction heroes that they have an unchallengeable right to colonize space.

The one undeniable fact about science fiction is that millions of people across the world enjoy reading it. It has joined the detective story and the spy thriller as one of the distinctive genres of our time. Why?

One reason, I think, is that we are now largely a world of city-dwellers. Living in small, low-ceilinged rooms, in crowded offices and the seats of motor-cars, people enjoy 'getting away' to space and its light-year horizons. Then again, we live in a world of specialization, which seriously circumscribes our freedom at work. This too the reader can compensate for by following the greatly extended powers of man in space. In doing so he can also harmlessly sweat out the toxins of aggression. The jingoism of Planet Z against Arcturus is unlikely to result in the extinction by heat rays of New York, London and Paris.

We live, moreover, in a world of individual fantasy. Take humour. It used to depend on the infringement of accepted manners; now it is largely of the Monty Python sort: one personal fantasy flares up for a few minutes, is obliterated by a huge foot descending, and at once replaced by another. Science fiction is popular because it meets this trend with an assortment of fantastic creatures. Robert A. Heinlein's man from Mars, for example, can get close to someone else only by sharing water with him or her – a glass of it, a swim, or a bath; while Frank Herbert has created Calibans, thinking stars who are 'jump-doors' to outer space, and never lie; not to mention bedogs and chairdogs, creatures of low-level intelligence on whom higher species sleep and sit.

Adults will react to such characters according to taste, perhaps with cool irony, perhaps with mild affection. But a young generation is growing up, formed on films and television programmes, comic-books and novels about man's adventures in space, and they are taking the genre more seriously. They have pin-ups in their bedrooms of *Star Trek*'s Captain Kirk; the sweets they munch are called Star Bar and Sky Dust; their Wild West is northward, up in another galaxy, approached by pop music and 'pot'. How will science fiction affect them?

They will be influenced by the near-absence of women. Even Dark Age sagas had wives tending the hearth, but for the hero of *2001* there is no Penelope waiting beside her knitting machine, since Bowman, like the rest of the crew, is unmarried: it would not be fair, he explains, to leave a wife for seven years. So physical love is excluded. This perhaps tells us something about our attitude to sex. In higher regions among higher beings sex does not exist! – a view curiously close to the Christian belief that in heaven there is neither marrying nor giving in marriage.

If women do appear in space, it is usually as action-persons, committed to men's values, not as wives and mothers. Against the heady excitement of intergalactic flight young readers may well conclude that bearing and raising children are tasks for 'creatures of low-level intelligence'.

Again, young people will find God conspicuously absent, yet the genre abounds with alternative religions: in Frank Herbert's *Dune* trilogy each of five sophisticated species has its own distinctive religion. Identified with Western civilization and hence with Earth, Christianity has to be absent, often too because the author feels that he has gone beyond Christianity. Ray Bradbury, for instance, confided to Oriana Fallaci, 'I say that *we* are God ... I see God as something that grows and expands through senses and thoughts, that wants to be mortal in order to die and be born again, and wants to move on, to push ahead with the human race, to spread and expand it throughout the whole cosmos!'

Yet the religions offered are curiously thin, and all too often pagan versions of some aspects of Christianity. Such are the intelligences in *2001*; and when we are told that Bowman has been 'reborn', yet will always be part of the entity that has taken him in charge 'for its unfathomable purposes', we have a sense of *déjà vu*; a river of stars has replaced the River Jordan.

For young readers the effect of this could be to call all religions in doubt. There is something more. In science fiction they read – and see on the screen – feats more wonderful than any in the New Testament. These modern miracles have the effect of devaluing the miracles of Christ. It is worth recalling that army chaplains in World War II became seriously concerned because many young American soldiers put their trust in Superman, a substitute-Christ who can throw off the disguise of a weakling bespectacled reporter to champion the oppressed. The soldiers behaved this way because the media allowed them easily to picture Superman, whereas Christ, dependent on the written and spoken word, remained shadowy. Some such substitution may be happening today. One recent poll in Britain's schools showed that more teenagers believe in extraterrestrial creatures than in Christ. The stable at Bethlehem for the young is make-believe, but frog-men zooming in from Venus are real.

Yet the effects of science fiction are by no means all adverse. Basic challenge-holding facts, such as the nature of man and his purposes in life, which may be taken for granted in ordinary fiction, are here, so to speak, photographed by infra-red light. 'Nothing but doth suffer a space-change into something rich and strange.' And wonder, as Aristotle remarked, is the beginning of philosophy. Again, one of the perennial characteristics of the genre is spacemen's longing – sometimes lust – for immortality. In so far as this expresses a genuine aspiration shared by young people, it is likely to start them thinking about an unfictitious goal where that desire may be realized.

Finally, if, as I believe, the essence of science fiction is an encounter with other intelligences and only man's intelligence has been able to conceive and create these other intelligences, however much the author may adopt a behaviouristic position, the genre itself assumes that man is more than the sum of his instincts, that he possesses indeed, besides a body and intelligence, a questing, soaring, decision-making spirit. Varied though settings and characters may be, one of science fiction's perennial themes is man's sense of being lonely and unfulfilled in himself, his search for beings or a Being who can help him to fulfilment. And that fiction happens to coincide with the truth.

I have criticized some aspects of science fiction as trivial or sham. But the genre as a whole undoubtedly represents a most important widening of our world-picture. Man is writing his soul large in the galaxies, tapping out his visions in the night sky as on a computer screen. Not since the Greeks began to tell stories about the constellations have we created so many myths involving sun, moon and stars. It has meant taking a hard look at accepted beliefs, but that is never a bad thing. We are gazing upward and outward, most actively, not only with our senses, instruments and intelligence, but with the full range of our imaginations. We want to feel ourselves belonging with our whole personality to this newly enlarged cosmos full of surprises.

CHAPTER FOURTEEN

A Violent Cosmos

Birth of a mountain-island – continental drift – our protective atmosphere – the planets found to be lifeless – man's disappointment – his concern with preserving his planet – the sun and solar wind – metamorphoses of stars – a supernova – the pieces of star in our bodies – radio astronomy opens a new window – pulsars and black holes – the galaxies – man's response to the new phenomena – our progenitor star now a black hole? – interstellar communication – the beings we hope to encounter.

On the sixth planet visited by Saint-Exupéry's space-travelling boy sits an old gentleman writing enormous books. He is a geographer and he asks his visitor to describe his home planet. I have three volcanoes, the boy replies, one of them extinct. And a flower.

'We don't put in flowers,' says the geographer.

'Why not? They're so pretty.'

'Because flowers are ephemeral.'

'What does "ephemeral" mean?'

'Geography text-books,' says the geographer, 'are the most precious of all books. They never get out of date. It very rarely happens that a mountain changes place. Or that an ocean dries up. We write about eternal things.'

'But extinct volcanoes can become active again,' the boy objects.

'Whether the volcanoes are extinct or active,' says the geographer, 'it is the same to us. What matters is that they're mountains. And mountains don't change.'

That is the background against which the following event took place.

On the morning of Thursday 14 November 1963 the motorboat *Isleifur II* was fishing for cod five miles west of Geirfugslasken, the most southerly point in Iceland. Here the depth of the sea is about 130 metres. There was little wind and the sea was calm. At daybreak, shortly after 7 a.m., the crew noticed an unpleasant smell of sulphur in the air, and then, away to the south-east, observed that something unusual was happening. Columns of jet-black explosive material were rising from the sea's surface sixty metres into the air.

The captain radioed to the mainland that what appeared to be a volcanic eruption had started on the ocean floor, and soon after ten o'clock aircraft flew to the scene. By then the fissure had become 500 metres long and was erupting in four places. An hour later the explosion columns were up to 150 metres high, lightning flashed out of them, while the eruption cloud attained a height of four kilometres, before a north wind began to disperse it. Black ash fell from the cloud, great circular waves, discoloured by the volcanic material, spread outwards, while in the lee of the eruption floated greyish aerated pieces of pumice.

Next morning observers glimpsed a dark patch beneath the cloud. As the wind blew the cloud away, it became clear that the north-western rim of the volcanic crater had emerged above the sea to a height of between eight and ten metres. Here a battle was taking place between fiery lava seeking to be born and the grey Atlantic, seeking to stifle it at birth.

On 4 April 1964 the volcano began to erupt fiery matter, above which whirled white clouds of steam. By summer fountains of red-hot lava were shooting skywards and rivers of molten lava pouring into the sea. This activity continued for three years. Gradually it built up an island one square mile in area and 180 metres above sea level, consisting to the north of stratified tuff, to the south of thin layers of lava.

The Icelandic Government named the volcano Surtur, after the Icelandic fire god, and the island Surtsey. Geographers added these names to their maps, geologists landed to tap and probe, a pair of kittiwakes nested on the bare rock and, in 1969, a pale mauve sea-rocket flowered.

The birth of Surtsey captured the world's imagination, for it showed how Earth evolved, some 4600 million years ago. Similar vulcanism released steam which condensed as water, not so much as to cover all land, yet plentiful enough to become the oceans and so produce a moist atmosphere. The volcanism also emitted carbon dioxide and other gases, that could sustain plants but would have been poisonous to animals. Eventually the carbon dioxide was changed by the action of sunlight on plant chlorophyll into the oxygen needed by higher life.

Surtsey also provided scientists with confirmatory evidence for a new and literally world-shaking view of Earth. Ships with echo-sounding equipment had recently discovered a volcanic ridge along the bottoms of the major ocean basins. Named the Mid-Ocean Ridge, it coils for 60,000 kilometres. It has been erupted like a chain of volcanoes by heat deep within the Earth, and the heat is still at work, causing the ridge to grow, and, as this happens, to push outwards the ocean floor on either side.

The ocean floors were found to be magnetically imprinted with a pattern of more or less regular bands, differing in intensity and sometimes in direction of magnetization. It was already known that in the course of the Earth's long history the direction of its magnetic field had changed several times, and in the year of Surtsey's birth two British geophysicists, Fred Vine and Drummond Matthews, suggested that the pattern of bands represented the imprinting of changes in the Earth's magnetic field on the ocean floors as the floors grew under pressure. If so, the bands of magnetic anomalies would record the development of the oceans as the rings on a stump reveal the process of growth in a tree.

This theory was put to the test. Results showed that the magnetic anomalies relating to Europe and North America could be reconciled on the assumption that 200 million years ago America and Europe were joined, and that since then America had slowly moved 30 degrees westward, impelled by growth of the ocean floor. Further evidence strongly suggested that all the continents once formed a single supercontinent, which geologists called Pangaea.

What split Pangaea apart? That same element which sent rays through Becquerel's closed box to expose his photographic plates. Uranium, and its close kin thorium, as they decay deep in the Earth, release vast quantities of heat. That heat caused the supercontinent to split into six plates of rock, each partly land, partly ocean. The plates are Eurasia, Africa, America, Antarctica, the Pacific, and India-Australia. Seventy kilometres thick, they ride on a layer of red-hot rocks 700 kilometres thick which collectively churn and twist and turn under the action of uranium and thorium. The plates are still moving apart and together – about three centimetres a year – and sometimes buckle in earthquakes and volcanoes. The most usual place for such buckling is along the Mid-Ocean Ridge, which corresponds to gaps between the plates. At one such point on the Ridge Surtsey was born.

The world-picture we learn at school doesn't usually strike us as wonderful, but a new one encountered when we are adults may have that effect. I for one find myself marvelling that even a continent can be 'ephemeral', that terra firma in the strict sense no longer exists, that we ride a series of rapids, like Canadian lumberjacks poised on six rafts. I wonder at the radioactive heat which propels those rafts, and equally that it causes little harm to man: in the four centuries between 1500 and 1900 volcanoes took only 190,000 lives.

The fire of Surtsey, the flux of continents make a suitable starting-point for an imaginary journey through the cosmos as astronomers now know it to be, for there too we shall meet fire and flux.

*

Our journey begins with Earth's atmosphere. Some four times as high as Everest, it not only allows us to breathe, but it keeps Earth's temperature equitable, by erosion turns rock to soil, and ensures a rhythm of sunshine and rain to make our crops grow. Its extremely complex long-term fluctuations depend in part on a magnetic field generated by the slowly turning iron core in the Earth's centre, and although it looks so tenuous its varied patterns moving over vale and hill are powerful enough to slow down or speed up the Earth's spin by a thousandth of a second a day.

Fifty kilometres up we meet a thin layer of ozone. Ozone is a very rare pungent-smelling gas formed by action of the sun's ultra-violet rays on oxygen and resembling oxygen except that it has three atoms per molecule instead of two. (It is different from the 'ozone' we claim to enjoy at the seaside; that is tangy salt-laden air, and happens to have the same name because 'ozone' in Greek means 'odorous'.) Ozone the gas is poisonous: formed lower down and more abundantly it would kill man; but just at this height and in the minute proportion of one part to four million it forms a screen against the sun's lethal ultra-violet rays. For the sun which beams us life also beams us death; it is that improbable ozone screen which allows life to pass and bars death.

Next we enter the ionosphere. This is another screen – of ions, atoms charged with electricity. It protects us against cosmic rays, the high-energy particles that circulate through space. To give an idea of their power, one such particle photographed in a cloud chamber in the high-lands of Bolivia smashed through five iron plates each half an inch thick before colliding with an iron nucleus in the sixth plate. Without his ionosphere man would long ago have been burned off the Earth or rendered sterile.

In the ionosphere the nearest objects are meteors, sometimes called shooting stars. Meteors are thought to be fragments of comets. As it swings in towards the sun, some of the comet's iced gas evaporates, shedding a little of its cluster of rock or iron; this is drawn Earthward by gravity, rendered incandescent by the friction of the ionosphere and usually volatized before it has time to reach the ground.

Above the shooting stars circle more than 2000 man-made satellites. Some, like crystal balls, furnish a picture of tomorrow's weather. Some, like the Persian King's 'eyes and ears', conduct espionage. Some carry instruments for enlarging our world-picture. One, it is hoped, will soon be equipped with a laser-reflector able to detect that estimated 3-centimetre annual displacement of a continent. And some relay the radio waves used in television and intercontinental telephone communication. By allowing peoples all round the globe to share simultaneously events of importance, they are man's corrective to the Tower of Babel.

One thousand kilometres up we enter space proper, pass by the moon and head for Mars. In the seven years between the moon landing and the first probe of that planet most scientists spoke confidently of finding life on Mars; the only question was, what form of life?

In July 1976 Viking I deposited on Martian soil a four-legged, one-armed 'lander', bristling with aerials, camera and electronic feelers. The lander showed the surface of Mars to be a reddish-orange boulder-strewn desert, the colour caused by a thin coating of iron oxide on rocks and particles. There are volcanic craters and if water ever issued from them there is none now. The so-called canals, which as late as the 1930s some astronomers believed had been cut by intelligent Martians, turned out to be optical illusions, while patches once thought to be vegetation were revealed as dark streaks of surface dust heaped up by strong winds. By and large Mars gives the appearance of a rusty moon.

Adding to its sinister appearance are two cratered bodies – the larger only 23 kilometres wide – that orbit Mars. Named Phobos and Deimos (Fear and Terror), once they were spheres, but collisions with inter-planetary debris have worn them to an unevenly ellipsoidal shape. They look like two small pitted omens of decay, and seeing photographs of them we begin to understand why we value spherical bodies, with their suggestion of perfection and permanence.

The lander's programmed arm stretched out, picked up a handful of Martian soil and analysed it three different ways for traces of life. Scientists waited expectantly, and when the results came through, their disappointment was something we all shared. There are no organic compounds in Martian soil. There is no vegetation, no life. Our nearest neighbour has neither sea-rocket nor eglantine, kitti-wakes or skylarks, only desert and dust, without even organ cactus to link a stark horizon to the pink Martian sky.

Mars's atmosphere proved just as disappointing. It is very thin, being 95 per cent carbon dioxide and contains less than half of one per cent oxygen, compared with Earth's twenty per cent. The thin atmosphere lets the sun's heat escape, so the average temperature is very low: $-25°$ C. Man might be able to land on this rusty Antarctica but he could not live there without taking with him all his air, food, water and heating.

＊

The next planet on our itinerary is Venus, that point in the morning and evening skies so brimming with light man has made it a symbol of love fulfilled. Again much was hoped for, as Venus is almost Earth's twin in size, hence could produce, by volcanism, an atmosphere, and retain it by gravitational pull.

The first thing noticed by spacecraft cameras were dense pale yellow clouds fifty kilometres thick: it is these that reflect sunlight and make Venus the brightest object in our night sky after the moon. Do these clouds, like Earth's, 'bear light shade for the leaves when laid in their noonday dreams'? Most emphatically no. They are largely droplets of sulphuric acid, rent by continuous thunder and lightning – up to 25 strokes a second – and weird chemical fires.

And the planet itself? Although two spaceprobes have landed and photographed the surface, their instruments failed after an hour in Venus's searing temperature of $475°$ C. But radar tells us that Venus is cratered, that it possesses a mountain higher than Everest and the largest canyon known. Its atmosphere is mainly unbreathable carbon dioxide.

Even could he protect himself from the heat, man would find on Venus no liquid water, probably no oxygen, and of course not even plant life. He would suffer agonies as sulphuric acid droplets burned his flesh, and he would be crushed by atmospheric pressure as intense as though he had emerged from an ocean diving-bell 1000 metres down.

What went wrong that Venus is so unlike Earth? Perhaps Venus's composition differs from ours – just as Venus's anti-clockwise rotation differs from the clockwise rotation of other planets – and contains more sulphur. Perhaps, being closer to the sun, any water vapour released by volcanism failed to condense into droplets of water, so there was nothing to stop the atmosphere of carbon dioxide growing and growing. Since Venus's sulphuric acid clouds are opaque, the little sunlight that filters through them is retained as heat, as happens in a greenhouse but here for no life-giving purpose. Decidedly the name-givers erred. Mars the 'fiery' turns out to be freezing, and the planet associated with a nude goddess rising from the Mediterranean is hidden in cloud, dry and blazingly hot.

Last of the inner planets, Mercury is mainly nickel-iron, battered by meteors but with no certain signs of volcanism, and possessing practically no atmosphere. While clouds hold in Venus's heat day and night, cloudless Mercury is mercurial: during its day, which is 175 Earth-days long, the temperature rises to 400°C, while at night it drops to −200°. Thus Mercury provides, in one circle, two kinds of hell.

We head outwards now and before arriving at Jupiter, we come to an area of space where small planets called asteroids orbit the sun. The largest, Ceres, has a diameter of 1000 kilometres, but most of the two thousand big enough to be catalogued are only a few kilometres across. Some have a surface rich in silicates, like Earth, but despite Saint-Exupéry's *Little Prince*, none is large enough to produce a volcano.

We approach Jupiter, which turns out to be aptly named, being more massive than all the other planets put together. Indeed it is near the upper limit of size for a planet: slightly larger and gravitational collapse would have caused it, in a way we shall see presently, to 'catch fire' – and be a star. As it is, Jupiter is mainly liquid hydrogen and helium, and has no solid surface. Its cloud cover is 100 kilometres thick, extremely turbulent and marked by white, reddish brown and red bands, plumes and streaks. Most prominent is the Great Red Spot, discovered by Jean Dominique Cassini in 1665 and long believed to be a vortex formed over a mountain or a deep depression. But Jupiter we now know has no land to form such features, and some believe the Red Spot to be a cyclonic eddy 30,000 kilometres long, which has lasted for centuries.

There is some water in Jupiter's clouds but they are so turbulent it is highly unlikely that life would have been able to form in them. Here, man would freeze in a temperature of − 145°C.

Jupiter is not only dead itself but radiates death in the form of cosmic rays, which the planet's huge magnetic field traps, then releases. These rays bombard Jupiter's four largest moons, so that man could not hope to land on any of them.

Yet the moons look interesting. Three have crusts of frozen water, while Io, about the size of our moon, has a sulphur and sulphur dioxide crust. Io is the only body in the solar system other than Earth to have active volcanoes. One erupts to the startling height of 280 kilometres. Volcanoes usually signify that an orb has the 'beating heart' of geological activity, but Io's are probably externally caused. As Io passes between its sister moons and Jupiter, gravity tugs at its surface, stretching and heating it to the point of eruption.

Beyond Jupiter are planets so far-off we know little about them, but that little is chilling. Saturn has a composition much like Jupiter's, its several hundred rings made from solid particles of ice and rock, one of the outermost rings being braided like a three-stranded rope in a way that puzzles scientists but would have delighted Leonardo, who had a passion for knots. Uranus and Neptune are composed of liquid gases poisonous to man. All are so cold as to preclude water-based life. Pluto, remotest and smallest, is also coldest of all, with an estimated temperature of − 230°C.

Before travelling further, let us pause to consider the effects of these new discoveries.

Since 1600, when Giordano Bruno dared to speak of inhabited planets, Western man had been counting on life somewhere in the solar system. Not necessarily human life: as Bertie Stanhope suggested in *Barchester Towers*: 'Why shouldn't there be a race of salamanders in Venus? And even if there is nothing but fish in Jupiter, why shouldn't the fish there be as wide awake as the men and women here?'

It has been quite a shock to discover that there is no life at all, that the bodies originally hailed as living gods and goddesses are pock-marked dummies. Of course we do not put a shock like that into words or images, at least not immediately. But it is there. And with it goes the realization that, within this solar system, man is alone.

Next, instead of sister or cousin lands, man has found alien orbs: lifeless, offering no chance of life to him, but death rather in a choice of styles: by freezing, by cyclone, by asphyxiation, by burning acid. However he wrap it in sugared jargon, the bitter truth remains: the planets are violent, hostile, thoroughly nasty places.

His fear of them has made him fear for himself. He has become preoccupied with that breathable mix other planets lack. He has recognized with alarm that if he goes on felling forests he will cut off his own oxygen supply, if he continues turning his city air into smog with pollutants, he will poison all life. If he goes on using aerosols at the present rate, by the end of the century chlorofluorcarbon will have destroyed an estimated 16 per cent of the Earth's ozone and perhaps put the race in danger of sterilization by ultra-violet radiation. Unless he does something about it, his world will end with a whimper.

So he has bestirred himself. As Saint Ex's small boy daily raked out the volcanoes on his asteroid, so we limit our use of aerosols, pollutants and defoliants. We purify the Mediterranean, rivers and lakes, we establish Plant-a-Tree Year. We are determined not to let our planet become as hideous and hostile as its neighbours.

One further reaction I have noticed: a swing towards 'natural' food or foods which advertisers claim embody Earth's 'natural goodness': whole grain bread, tomatoes grown without chemical fertilizers or apples from trees untouched by insecticides. There is a growing use too of homeopathic remedies and herbal cosmetics. The trend began some years ago but has gathered momentum since discoveries about the deadness of other planets have made us feel the need to draw close to mother Earth.

We reenter our imaginary spacecraft and travel to the star we call the sun, on whose mass everything we have so far seen depends. How did it come into being?

As clouds of gas and dust drift through the cold wastes of space, they sometimes meet. Then the clouds are flattened, and their gas forms clumps. Gravitation causes the clumps to contract and spin. Contraction continues, gravitational energy turns to heat, and when the temperature at the centre of the spinning mass reaches ten million degrees, nuclear reaction begins. Hydrogen nuclei fuse to make up helium nuclei, releasing energy, and it is this energy that powers the sun.

The heat of the gas at the centre increases the pressure it exerts on its surroundings; this pressure opposes the ever-present inward pull of gravitation and stops the sun shrinking. A perfect balance of thermonuclear heat and gravity has kept the sun shining steadily for five billion years, and will keep it thus for as long again. The bigger a star, the quicker it burns its fuel; twenty-five per cent bigger, and the sun would by now have stopped shining. So it is fortunate for us that the sun is just the size it is.

Despite the finely tuned balance within it, the sun is by no means the perfectly shaped, uniform body men so long imagined it to be. Its surface is torn by turbulent local magnetic fields which can shoot huge loops of gas a million kilometres high, while its general magnetic field reverses direction every eleven years. This initiates the formation of the cool depressions we call sunspots, which appear in largest number four to five years after the reversal and affect Earth's weather in a way not yet understood.

Besides heat and light the sun sends us the solar wind: a continuous breeze of highly charged particles that take five days to reach us. That old puzzle, why a comet's tail sometimes moves ahead of the comet, is now explained: the solar wind deflects it in a direction away from the sun. At it reaches Earth the solar wind compresses our ionosphere, again affecting our weather, and is drawn towards the North and South Poles by the Earth's magnetic field, thus producing the Northern Lights which terrified Gregory of Tours and strengthened Charlotte Brontë.

The Greeks boasted that man had stolen fire from the gods. Now we realize that the boast was hollow, that fire, or at least fire's basis, was a gift in the form of plant or fossil fuels – stored sunshine. The gift is running short and man has begun to worry. Having photographed the icy outer planets, he can easily envisage a world grown cold.

So, quite suddenly, he has become more aware of the sun than at any time since the Cheops pyramid. In Culgoora, Australia, 96 self-steering 'dishes', huge metal spider-web receivers 15 metres across arranged in a circle nine kilometres in circumference, scan the sun once every second for magnetic activity. We have begun to construct a new kind of house, on its attic roof indium oxide filters able to collect solar energy. We are planning to put a solar-cell power plant in a synchronous orbit to transmit sunpower earthward by microwave beam. We foresee huge mirrors spread across the world's deserts to capture, retain and impart to man the sun's warmth. And these doubtless are just a beginning.

*

We leave the solar system and head for the stars. Most of the constellations the Greeks named for us turn out to be unrelated stars: just chance arrangements along man's line of sight. But Orion is truly an association in one part of space.

The eleven prominent stars comprising Orion, his Belt and Sword were formed, in the manner of our sun, out of gas clouds: one such cloud, the Orion Nebula, is visible round the central star in the Sword. Orion's right foot, the star Rigel, is blue. This is because it is more massive than our sun, and massive stars, spending their nuclear fuel fast, produce great heat, visible to us as blue light.

The three stars of the Belt are brighter than the three of the Sword. This is because they are older and have converted all the hydrogen at their centres to helium. As this happens a star's core becomes choked with helium 'ash'; when the ash grows to one-tenth of the star's mass, gravitational pressure builds up, shrinking the core and creating an increase of heat, which shows as increased brightness.

But this hectic luminosity is, by cosmic time, brief. As gravity shrinks the core still further, the outer envelope of the star expands and cools, and the star becomes red. Betelgeuse, the star marking Orion's left shoulder, is at this stage, having expanded to become 400 times larger than our sun. It is categorized as a 'red supergiant'.

The state of redness too is ephemeral. As gravitation continues to press on the star, the heated core causes helium to produce carbon. The heat increases still further, and the outer layers of star expand, gradually puffing out into space as great shells of gas. Stripped finally of its outer covering, the star's central skeleton is small, very dense and white hot, hence its name of 'white dwarf'. It will cool over thousands of millions of years, becoming yellow, then red and finally fading to invisibility.

Such is the usual way a star 'dies'. But a massive star – ten times or more heavier than the sun and known as a 'supergiant' – suffers a more spectacular death. In old age, as its core contracts, its greater gravitational force heats up the carbon to silicon, then iron, which is too heavy to produce energy by fusion reactions. Now there is no outflow of energy from the core to balance the gravitational inpull. Gravitation makes the core increasingly hot until the iron nuclei disintegrate, and blast off the outer layers of the star. The star becomes a thousand million times brighter than the sun, its surface area increases a million-fold in a day, and it shoots out materials at speeds exceeding 5000 kilometres a second. Such an explosion occurs in our Galaxy about every thirty to fifty years; one was seen by Tycho Brahe. He called it a nova, which means 'new', but it is in fact a dying star. Today astronomers refer to it as a 'supernova', using nova for a much smaller partial stellar explosion.

A supernova is of the utmost importance to man, because it hurls into space elements such as carbon and silicon formed in its core, as well as complicated and heavy elements, such as gold, lead and uranium, produced in the intense nuclear collisions as it explodes. These mix with the hydrogen gas of space and in time condense to form a new star. This 'second-generation' star will, like the first, consist mainly of hydrogen, which will burn to become helium, but it will also contain heavier elements from the exploded first-generation star. We know that our sun contains elements such as iron and gold, and therefore must be a second-generation star.

The same material as formed the sun probably formed our planetary system too. Size conditioned composition. On small planets like the Earth, with a weak gravitational pull, light elements such as hydrogen escaped, and heavier elements – the ones formed in the first-generation star's death – remained. So says the astrophysicist.

It is, by any standard, an amazing story. Would any poet have dared claim that Earth is related to the stars, that it has been drawn not just from local space but from some perhaps extremely remote area of our Galaxy where a star shone dazzling bright and burst? What inspiration would have led him to picture Earth's almost inconceivably heavy core as a giant swarm of iron atoms that once flew unconfined and singly through space after their genesis in the decomposing hot body of a star? Pindar, who associated veins of gold with the sun, did not dream that the issuing mint was another sun, not ours.

The paradox does not end there. Our bodies contain three grams of iron, three grams of bright, silver-white magnesium, and smaller amounts of manganese and copper. Proportionate to size, they are among the weightiest atoms in our bodies, and they come from the same source, a long-ago star. There are pieces of star within us all.

I look at the night sky with new wonder. Particle-sized I may be in comparison with the orbs above, separated from the nearest by a desert of space so wide that the distances travelled by all the men who have ever lived, if strung together, would not bridge it, yet the very acts of looking and of wondering depend upon atoms within me which are also in many of them. I see the stars by courtesy of a star, the one whose death made possible my birth.

*

We turn now to celestial activity which we cannot see because it does not emit light, yet which betrays its presence in other ways. Turbulent gas speeding through a magnetic field emits radio waves, which since 1937 can be detected by radio astronomers, while gas at extremely high temperatures sends out X-rays, blocked by Earth's atmosphere but observable by satellite. These techniques have enlarged man's view of space much as did Röntgen's photograph of the bones of his wife's hand.

We saw that the bulk of a supergiant star is flung into space. But what happens to the dense central region? Physicists calculated that it would contract to the point where all its protons and electrons have combined into the uncharged subatomic particles called neutrons – so densely packed that a matchbox of them would weigh millions of tons.

In 1967 radio astronomers discovered such a star. It was rotating fast, its disturbed magnetic field emitting just over thirty bleeps of radiation every second. They named it 'pulsating radio star', pulsar for short. The following year they found a pulsar at the centre of the Crab Nebula, and identified it as the dense central region of the supernova whose explosion, recorded by the Chinese in 1054, brought the Nebula into being. It is a ball a mere 25 kilometres across yet – almost incredibly – it radiates more power than the sun.

Physicists further calculated that if a neutron star is massive – three times heavier than the sun or more – its neutrons cannot support themselves against the oppressive force of gravity. The star continues to collapse. As its core contracts, so does the gravitational field. Particles of light have more and more difficulty escaping. Eventually they cannot escape at all. In a fraction of a second the star goes dark: it has become a field of intense gravity known as a black hole. Black because invisible; a hole because it is round and empty, everything sucked into it being destroyed, or at least disappearing from our cosmos.

Though by definition black holes cannot be seen, in 1967 the Russian physicist Iosif Shklovsky showed how to find a black hole which was orbiting an ordinary star. When the star began to expand in the course of its evolution, it would dump gas from its atmosphere on to the black hole. As the gas orbited the black hole before falling in, it would be heated, compressed by the black hole's force of gravity. The temperature of the gas would reach millions of degrees and emit X-rays.

Such a source has been found by satellite in the constellation of the Swan. Known as Cygnus X-1, it emits both X-rays and radio waves, and is circled every $5\frac{1}{2}$ days by an optically visible supergiant star. Cygnus X-1 is at least six times more massive than the sun. Pulling in gas from its companion, it will continue its lethal existence for thousands of millions of years. In 1978 astronomers found another giant invisible object emitting X-rays. Known as V861 in Scorpius, this too is probably a black hole. There may be many others. According to Einsteinian theory, a black hole could exist massive enough to swallow stars, even a galaxy, each star blazing forth with unimaginable energy as it vanishes for ever down the gargantuan gullet.

Some people fancy black holes, as the Spanish Court fancied dwarfs. Some feel a stir of excitement, as blasé Romans felt at comets. Some see them as evil objects: Satanic anti-haloes, where darkness triumphs over light; while space travel sophisticates posit that any spaceship entering a black hole would be flung out in another part of space where the crew would travel faster than light, and so, as Plato had once imagined, move backwards in time.

Others are sceptical. They claim that a black hole is just a hole in our knowledge, like the medieval geographers' 'Here be dragons'. But the evidence for their existence is already strong and to me the black hole is the culmination of that mounting violence we have met throughout our journey, beginning with the lightning-torn sulphuric acid clouds of Venus, continuing through the sun's eruptive flares, supernova explosions and now the abomination of annihilation, gigantic spider-trap for firefly stars, crematorium of outer space.

We travel on, leave our Galaxy and look back at it from outside. It is spiral-shaped, measuring 100,000 light years across and containing perhaps many trillion more stars than we can see on a clear night. Orbiting it, as the moon orbits Earth, are two smaller galaxies, the Magellanic

Clouds, named for the circumnavigator. Galaxies tend to cluster, and there are about thirty in the group to which our Galaxy belongs.

Beyond lie very many other galaxies: according to Einsteinian physics a finite number in a universe that will not continue forever. Some galaxies are relatively quiet like ours; some emit very strong radio waves from their centres, evidence of such tremendous violence that astronomers cannot explain it. The most distant galaxies of all, the quasars – a name derived from quasi-stellar, signifying that quasars appear star-like on photographs because of their remoteness – produce as much light as a hundred galaxies from a region nearly a million times smaller than a galaxy. The power required for that is again something astronomers cannot explain.

As many as 3,000 galaxies may cluster into a super-galaxy, centred on massive elliptical galaxies, each a hundred times heavier than the Milky Way. These central galaxies have grown to such size by swallowing such other galaxies as may have happened to pass, adding the stars in them to their own populations. The turmoil within a super-galaxy is such as to heat vast gas clouds to 100 million°C – more than thirty times the temperature of the surface of the sun.

Galaxies can explode. At least one is known to have done so. The whole complex fabric of matter and space has somehow engineered its own destruction, shooting out millions of stars, and vast clouds of gas and dust, into space thousands of light-years across. One galaxy, NGC 5128, has exploded twice. Were these explosions caused by a black hole? Some astronomers think so. What is certain is that at the centre of NGC 5128 lies a source of radio and X-rays. Short flares in the X-ray brightness occur about once a month, and these may represent individual large stars or clusters of stars being swallowed by a black hole.

There are an uncounted number of other galaxies, some perhaps so distant and fast-moving their light or radiation can never reach man, but with these exploding galaxies our journey outwards comes to an end.

*

What are our reactions to these objects and events? First, and with good reason I think, we are likely to be frightened. It is no small matter to discover that the quasi-totality of the cosmos is composed of space so cold the thought of it makes us shiver, punctuated by gas and matter burning at temperatures higher than a hydrogen bomb, the whole constituting a jungle where inanimate predators batten on inanimate prey.

It is the extremes, the flux and the violence taken together that are frightening in themselves. But they frighten us also in respect of what man is. It has long been known that, by cosmic measurement, man is small; now we see he is fragile too. If he approached any of these cosmic objects, he would be frozen or vaporized, rendered sterile or worse by cosmic rays.

Next, I think, we have a feeling of repulsion. It is one thing to know we have atoms of star material in our bodies, quite another to find that our supernova progenitor may have become in senescence a black hole. It pains us to be linked so closely to cosmic flux, to the inexorable getting and spending of nuclear fuel. Whenever he has met it in Oriental thought, Western man has recoiled from the notion of total flux; for the personality he treasures requires a certain stillness around it, in order to attain steadiness and continuity within.

Also, I think, we feel a sense of loss. Though radio astronomy has opened a new window through which we descry marvellous objects, almost the whole of what has been discovered is in dimension, nature and behaviour so different from what we meet in our daily lives that it escapes us. A matchboxful of material from a neutron star that weighs ten million tons – a cosmos of this kind is no longer, qualitatively, on a human scale. The Greeks made of their night sky a friendly place, setting heroes there to live on as stars; they could watch Orion at his hunting feats, and the moon, who loved him, unwittingly shoot her fatal arrow. But today's stars are too hot and too heavy for us to handle, even in stories. We number them in catalogues, but sadly we can no longer 'relate' to them.

At the aesthetic level, to some extent we can forget their lethal nature, just as we forget that the Earth circles the sun. The stars at a distance are still beautiful, and add beauty to a face: starlight is still for kissing. Yet the close affinity between the night sky and all that is soft and tender – that has gone. A betrayal has taken place. The intimacy is unlikely ever again to be quite the same.

Fifty years ago man could gain solace in the struggles of daily life by thinking of the serene majesty above. Silent stars, galaxies drifting like peaceful islands represented a background against which he could see his troubles 'in proportion'. Now, plainly, the magnetic field of celestial symbolism has changed direction. Steadfastness has yielded to flux, instability, violence.

For some the ubiquity of violence is not just inhuman but anti-human. Thereby it appears to jeopardize man's traditional notions of stability and order. And because some of the sources of violence for the moment defy explanation, they also seem to squeeze out mind, so that the boast, 'Mind over matter', now seems black humour. We shall look more closely at this attitude when we consider, in the next chapter, the origin of the cosmos.

*

Our brief survey ends and we return to blue-and-white Earth. What are our reactions to it now? First, perhaps a welcome sense of tranquillity. In our part of space at least there is an absence of violence and an equitable temperature. There intrude no giant bubbles of gas at 2 million°C, no detectable black holes.

Arising from that we recognize, with a pang, our extreme limitations. Although we boast of living in the Space Age, we do not really belong in space. We cannot go to the moon even, without taking with us our atmosphere, food and water, without being as careful of our space-suits as a haemophiliac of his skin, while even a few weeks in space cause serious calcium loss from our bones. We see we are fooling ourselves when we boast of freezing our bodies to a 200-year standstill and travelling at the speed of light.

We realize that we are indeed creatures of Earth. This is where we belong. It is a different Earth from the one known to previous generations, for we know that it is distinguished by its atmosphere, if nothing else, from anything so far observed in the cosmos, but it is Earth all the same. There may be Surtseys, the continents may drift, but at least such disorder is marginal. Earth may not be the best of all possible worlds, but at least it has the primary virtue of allowing us to exist.

Yet human nature being what it is we feel withal a sense of frustration. Just when we learn that we are formed of star material, just when we extend our knowledge almost to the edge of the universe, we find that we are excluded from it. A hell it may be, but exclusion even from hell can be galling.

These considerations, coupled with those remembered glimpses of fearful violence, have led to another feeling: loneliness. Our forebears, when they tried to count the stars, felt loneliness too, but given the scale of our knowledge, their loneliness was paltry compared to ours. It is a loneliness seemingly built in to the nature of things.

Given that feeling, it was understandable for man to wish to seek the assurance of other minds. In the 1960's, particularly in America, interest focused on those astronomers who argued that since our sun is a commonplace star, other stars have planets, some or all of them likely to be inhabited.

The topic is important, so it is as well to place it in perspective. No one knows how planets originate. According to the most authoritative historian of planetary theories, the physicist Stanley L Jaki, 'In spite of countless and at times ingenious efforts making use of every tour de force of mathematical physics and observational lore, no theoretician has yet even remotely succeeded in turning our planetary system into a self-explaining unit, let alone into a very typical phenomenon.'

It is obviously hazardous in the extreme to argue from the existence of planets we cannot explain to the existence of planets elsewhere. This however was done, notably by Stephen H. Dole and Isaac Asimov in *Planets for Man*. Making assumptions that would not today be accorded – for instance, that a planet can exist in a binary system (a pair of stars orbiting each other), Dole and Asimov concluded that one in 166 stars in the Galaxy has a habitable planet.

Biologist-astronomer Carl Sagan and radio astronomer Frank Drake then estimated that one in seven habitable planets will have an advanced civilization. If they are distributed randomly, the distance between us and the nearest should be about 300 light years. The most efficient method of communicating with it is by radio waves, but on which frequency would intelligent beings be likely to listen for, or send, communications? Evidently frequencies with a universal significance.

Now the more abundant molecules drifting through space sometimes collide. This alters the movement of their electrons, causing each atom to emit radiation on a distinctive frequency. In the case of an atom of hydrogen, which has only one electron, the electron flips over so that its direction of spin is opposite to that of the proton comprising the nucleus. In so doing it emits radiation on a wavelength of 21 centimetres at a frequency of 1,420 millions of cycles per second (Hertz). Since hydrogen atoms comprise about three-quarters of the atoms in space, this is the frequency which would first spring to a radio astronomer's mind.

In 1960 Frank Drake began the first search for extraterrestrial intelligence. Project Ozma, named after the ruler of Oz in L. Frank Baum's children's stories, consisted of listening, at a frequency of 1,420 megahertz, with a 26-metre radio telescope in West Virginia, to two nearby stars, Epsilon Eridani and Tau Ceti, both about 12 light years away. Drake listened for four weeks, but heard no sounds suggestive of intelligence. Since then six similar projects have been unsuccessful, but the number of stars examined is less than 0·1 per cent of the number that would have to be investigated if there were to be a reasonable statistical chance of discovering one extraterrestrial civilization.

An alternative approach would be to eavesdrop on relatively weak signals such as television transmissions occurring in other worlds. For this a listening system would be required hundreds of times the area of any existing telescope. Project Cyclops envisages 1500 antennae each 100 metres in diameter, all electronically connected to one another and to a computer system capable of analysing and assessing the flood of radio sound which would come in. The array would cover about 75 square kilometres and cost up to 10 billion dollars, today a prohibitive sum.

Now for the sending of messages from Earth. What medium should we use? It is quite possible that intelligent non-terrestrials have as their dominant sense smell or taste, or both, and communicate therewith, in which case our contacting them would be ruled out from the start. Scientists are obliged to be anthropomorphic here and to suppose that extra-terrestrials will have ears and eyes. An audible message in whatever Earth language would present them with grave problems of translation, so the medium chosen was a picture.

In 1973 spacecraft Pioneer 10 and Pioneer 11 were due to leave Earth to fly past Jupiter and then escape from the solar system to begin a journey across our Galaxy. Perhaps they would be sighted by extra-terrestrial beings and recovered. Sagan, Drake and Linda Salzman Sagan decided to design a picture message to be engraved on gold-anodized aluminium, postcard-sized, attached to each spacecraft. What should go into the message?

I find myself stirred and moved by any such attempt to define one civilization to another. I recall during the Cold War a torpedo-shaped cupaloy container being lowered deep into an excavation site in New York. Constructed to last until 6965, it holds microfilm of the *Encyclopaedia Britannica*, of the Bible, the Koran and writings of Confucius, as well as of many scientific textbooks. There are photographs of works by Giotto, Leonardo, Raphael and Michelangelo. But also there are articles in daily use, such as money, cigarettes, chewing gum, an alarm clock, a child's doll and a woman's hat. Suppose, five thousand years hence, the key to our languages were lost, what would seventh-millennium man make of the civilization summed up in those objects and in 75 samples of metals, fabrics and plastics buried alongside? In the early nineteenth century scholars tried their hand at deciphering Egyptian hieroglyphics, most assuming that the bird, a basic unit in the hieroglyphics, played a prominent role in ancient Egypt. But when the Rosetta Stone allowed us to read the language correctly we realized that the ancient Egyptians had not been specially interested in birds, as birds! From the chance frequency of one visual symbol scholars had drawn the wrong conclusion. They may do so again, either on this or another planet.

Clearly on the small size of a spacecraft plaque something immensely less ambitious than the New York package could be attempted. At the bottom of the plaque is a diagram of the solar system. The spacecraft's trajectory is indicated by an arrowed line as it leaves the third planet, Earth, and swings by the fifth planet, Jupiter. To the right, standing against a spacecraft to show their height, are an unclothed man and woman. An attempt was made, says Carl Sagan, to give them 'panracial characteristics'.

This visual telegram is short but pithy. It is a statement about man in the cosmos: what he looks like, his size and his position in space. It includes one detail indicative of his spirit. The man raises his right hand, palm forward, in the gesture we on Earth associate with greeting. We are informing space – let us hope we are telling the truth – that we are friendly beings.

When it came to devising a picture message to be beamed to outer space by radio Frank Drake decided to cast it in binary code. This uses two digits, zero and one. Zero represents a white square, one a black square. Anyone receiving such a message must divide it into lines, each with the same number of digits, then translate the digits into black or white squares. The completed picture looks like a patterned handwoven textile.

The site chosen for transmission was in dramatic contrast to this futuristic technology. In Arecibo, Puerto Rico, there is an extinct volcano, formed thousands of years before civilization came to the Americas. In it a 300-metre antenna has been constructed, and from it in 1974 the first message from Earth was transmitted towards a group of stars called the Great Cluster in Hercules.

The message consisted of 1679 binary digits – 'bits' – which, to make sense, had to be broken up into 73 consecutive groups of 23 bits each and arranged in sequence one under the other, reading right to left and then top to bottom. The message conveyed the following information: numbers 1 to 10, atomic numbers for hydrogen, carbon, nitrogen, oxygen and phosphorus, formulas for sugars and bases in nucleotides of DNA (the molecules in a cell's nucleus that store genetic information), the number of nucleotides in DNA; then come visual representations of the double helix of DNA, of a human being, of the solar system (Earth being displaced towards the human being) and of the Arecibo telescope.

The message is, in effect, a miniature world-picture: man as he sees himself in relation to the cosmos, in particular to his solar system and to the elements composing it and himself. Because binary code yields only a crude picture, man has the rectilinear shape of a robot, while the sun is square.

The message, travelling at the speed of light, will take 24,000 years to reach its destination. Will it be understood? When Drake circulated a much shorter message to members of a Commission on Intelligent Extra-terrestrial Life, only the part of it relating to his own special field was decoded by this or that expert. Moreover, there is a strong probability that across such a distance reception would be imperfect. So Sagan prepared a message identical to Drake's shorter message but with one of the zeros missing, and submitted it to a distinguished group of physicists, chemists and biologists in Cambridge, Mass. They spent several hours trying to decode it – without success. Presumably the extra-terrestrial beings will have to be more intelligent than us if we are to get a reply 48,000 years from now. And intelligent in a specific direction, for even an Archimedes or an Aristotle would probably have failed to decode the Arecibo message.

What strikes me most about the Arecibo message is not its content or its chances of successfully reaching Herculean eyes, but the fact that man chose to send a message at all. Even now, when telescopes are so sophisticated that an astronomer can spend a lifetime scanning one small part of the sky for one particular kind of object, as Byzantine grammarians used to spend a lifetime analysing the use of those inconsequential Greek words called particles, even now the thing we most hope for is to get away from matter, be it supergiants or sub-atomic particles, and encounter mind or minds. In the oceans of space the most worthwhile prize, the one catch which, of we are honest, really interests us, really would justify all the labour, is intelligence. From the depths of an extinct volcano man has uttered his call to the stars, and it is a call for intelligence.

There are already several billion intelligent beings on Earth, in fact the Arecibo message gives their number. Why are we not content to communicate with them? Because I think we know already the kind of answers we should receive. We are seeking another style of civilization, with unfamiliar values, a different set of presuppositions. But that too we can find in this world, in the civilization of the Maya, say, or of ancient Egypt, or among the more than one hundred still unexplored tribes of New Guinea. It seems then that we hope for a more 'advanced' intelligence. Do we mean beings simply with quicker and more efficient spacecraft than ours, or ways of prolonging their life-span? I doubt whether most people would be satisfied with such technical know-how, especially second-hand, for the paradox still holds that man values only what he himself has won by long hard struggle.

Whom then would we be glad to meet? Two kinds of beings, I suggest. Either those morally much worse or morally much better than ourselves. Either evil beings bent on our destruction who would evoke from us unsuspected reserves of courage and ingenuity, and so help us to grow in stature, or beings who might provide insights, clues, even perhaps service manuals as to how we might improve as persons. These are the beings an earlier age called devils and angels, and our own age embodies in science fiction. A deep innate longing for heroism, be it in the Greek or Christian mode, persists into the interstellar age.

CHAPTER FIFTEEN

Chance or Design?

The big bang theory – formulated without reference to a Maker – reasons for its popularity as a world-picture – effect on men's beliefs – Camus, Beckett, Ionesco – the note of anguish – an absurd cosmos jars with the evidence of purpose in daily life – widening scientific explanation to include design – statistical evidence for design in the solar system – other 'knife-edge' factors prerequisite for life built in to the cosmos – tension between our glimpse of order and the scale of the phenomena – a historical constant: the penumbra of mystery.

'Go, and catch a falling star,' urged John Donne; modern astrophysicists have done that and more: in a network of graphs they have caught the receding galaxies and, heirs of the Greeks in their search for an *arche*, drawn them into a theory of how the cosmos began. They base their theory on two measurements. The first is the increase in the speed at which more distant galaxies are receding: 15 kilometres per second for every million light years' distance; from which figure – Hubble's Constant – galaxies can be followed backward in time to the moment when the matter composing them was joined.

The second measurement was discovered by two American engineers at a New Jersey satellite-tracking station. It takes the form of radio waves coming evenly from all directions. It sounds like a background of static on a radio, and it is audible evidence that everywhere in the cosmos there is a background temperature of $-270°C$: extremely cold by human standards but in a cosmic context 3 degrees warmer than the absolute zero to be expected. This slight heat can be explained by assuming it to be the remnant – an outer ripple – of an almost unimaginably intense heat aeons ago. Again working backward, theorists can plot the primal violence such as would have produced exactly that temperature today. Their theory runs like this.

In the beginning, some 15 billion years ago, was a ball of radiant energy. Nothing else existed, not even space beyond; the ball of energy was both picture and frame, both text and margin. Within it was all the matter that comprises today's cosmos, packed many million million times more densely than in a neutron star (where one matchboxful weighs 10 million tons), and the temperature in degrees C may have been as high as 1 followed by 32 noughts.

On Earth atoms of matter can be broken down into heavy particles (protons and neutrons), into smaller particles such as electrons, and into particles that have no mass or electrical charge whatsoever, such as neutrinos, which are invisible, and photons, which radiate light. Intensify conditions, as in a high-speed accelerator, and energy in the form of photons will produce small particles in pairs – such as one negatively charged electron and one positively charged positron. These immediately annihilate each other, producing more photons.

In the uniquely extreme conditions of the primitive ball scientists believe that conjectural, elusive particles even smaller than electrons and positrons – what they term 'exotic' particles – interacted to produce short-lived photons, and photons interacted to produce exotic particles. Everything in the ball was effectively at the same temperature, and the whole blazed with light.

This concentrated ball of energy exploded with tremendous force. One may speak of a sudden hot big bang. At about one hundredth of a second after the explosion, the earliest time scientists can describe with confidence, the temperature had dropped to six thousand times the centre of the sun, while matter had expanded fast, though still packed millions of times more densely than in a neutron star, and consisted mainly of photons, neutrinos and an equal number of electron-positron pairs. There was also a tiny number of heavier particles, roughly in the proportion of one proton and one neutron for every billion other particles, but of great future importance.

About 14 seconds after the explosion the temperature fell to 3 billion degrees, cool enough for electrons and positrons to begin to annihilate each other faster than they could be created out of photons. The energy so released temporarily slowed the rate of cooling.

Three minutes after the explosion the temperature had fallen to 1 billion degrees. A very small number of the protons and neutrons began to combine as nuclei of simple atoms, mainly hydrogen (one proton and one neutron) and helium (two protons and two neutrons). These, the remaining protons and neutrons, plus a small number of electrons that had escaped annihilation by positrons, made up an ionized gas in thermal equilibrium with the photons.

The gas continued to expand. All this while the universe was radiant with light; as yet darkness did not exist, nor night. At last, after 700,000 years, a critical point was reached: the density of the gas became greater than the mass density of radiant energy; in other words, matter got the upper hand of light. Chemistry could now take place, electrons combining with nuclei to form the first stable atoms.

The gas was opaque and not entirely uniform, 73 per cent of it being hydrogen, 27 per cent helium. Some parts were heavier than others, and now gravitation began its long career as the chief formative cosmic force. Under its pull those patches denser than others attracted gas from less dense regions. Each became the seed of a growing concentration and thus, one million years after the big bang, there came into being the first galaxies.

Within each galaxy, again under the influence of gravitation, hydrogen and helium atoms combined to form stars, and light reentered the cosmos. Some of the stars, as we saw earlier, would end their several billion-year lives by exploding as supernovae, so initiating a new kind of matter: heavy atoms.

Four and a half billion years ago, in our Galaxy, clouds of gas and dust, mainly hydrogen, nitrogen and carbon but containing some heavier atoms also, condensed into the second-generation star we call the sun. Planetary bodies formed, orbiting it, and on one of these, as it cooled, conditions became such that life arose, and eventually man, who with sophisticated instruments is now able to detect lingering traces of that moment of supreme heat when the fireball exploded, and to observe galaxies still receding under the momentum of the explosion.

Such, briefly, is the theory known as the hot big bang. Until recently it had a rival in the steady-state theory, which says the universe has always existed and has always been expanding, with hydrogen forming continuously and spontaneously to replace receding galaxies, but the steady-state theory cannot satisfactorily explain cosmic background radiation, and today the hot big bang commands wide scientific assent.

The big bang would allow a place for a Creator who 'programmed' and detonated the fireball – indeed the theory originated with a Belgian mathematician who was a Catholic priest, Abbé Lemaître, but equally it can be formulated without reference to a Creator. Now scientists, committed to finding physical causes, are understandably reluctant to speak of Divine intervention. In regard to a comet's appearance or the explosion of a supernova such silence is in place, but the big bang differs in kind from other explanations in that it purports to describe the origin of everything, so that when a majority of cosmologists from the late 1940's onward chose to describe the big bang without mentioning a Creator, their silence had philosophic and religious implications. It looked like an endorsement that the universe as we now know it arose from a chance explosion.

This view gained weight when cosmologists joined hands with certain biologists who claimed that life too had a chance origin. The cosmologists then extended their scenario – a favourite word, less modest than Laplace's 'hypothesis' – to run like this. A spontaneous hot big bang gave rise to the cosmos, to the solar system and to Earth, in whose primitive gaseous atmosphere, as laboratory experiments show, lightning flashes and the sun's ultra-violet rays could have produced organic molecules, notably amino acids, which are the building-blocks of proteins, themselves an essential constituent of the living cell. From such random processes life was sooner or later bound to arise. (Recently amino acids have been detected in a meteorite found in Antarctica, though of a different kind from amino acids in the known simplest forms of terrestrial life, so it is possible that life originated extra-terrestrially, but flourished, at least in our solar system, only on the watery planet, Earth.) Simple forms of life led inevitably, through struggle and through random mutation of genes, to the higher animals and man. It looked very neat: a fortuitous explosion at the beginning of time entailing, after millions of years, the fortuitous explosion of life.

Such is the theory that has been most popular in the West since 1950. We who have followed them down the ages know that a theory becomes a dominant world-picture only if it ties in with other salient experiences of the day. The big bang ties in with at least three. There is the violent cosmos man now knows surrounds him, and there is the atomic bomb, lurking at the back of his mind like a flicker of fear and a cloud of despair.

His third experience is of a somewhat different kind. Population increase has diminished not only each man's space, but also his time. For example: time restrictions on parking his car, the high cost of being anything but brief on the telephone, our frequent experience that a staple of friendship – small talk – has been rendered almost as obsolete as the pony-trap by the lack of time we all lament, which is really our haste to survive in a crowded world. Especially our time to think about a God has shrunk – though we delude ourselves that it is he who has receded.

Such is the background which has made it easy for the man in the street to accept the world-picture described above: the cosmos originating in a self-igniting big bang, which eventually produced the random group-ings of atoms we call the galaxies, and the equally random groupings of atoms we call life.

Upon that picture many distinguished scientists have set a seal of approval. Two astronomers already mentioned, Iosif Shklovsky and Carl Sagan, made it the basis of an influential book, *Intelligent Life in the Universe*. The elementary particle physicist Steven Weinberg wrote in *The First Three Minutes*: 'The more the universe seems comprehensible, the more it also seems pointless, though the effort to understand it is one of the very few things that lifts human life a little above the level of farce, and gives it some of the grace of tragedy' – tragedy because what is fortuitous is plainly pointless. The polymath biologist Jacques Monod wrote in *Chance and Necessity*: 'Man at last knows that he is alone in the unfeeling immensity of the universe, out of which he emerged only by chance.' To which Jean Guitton has added, 'Without kin, without neighbour, even without enemy.'

In late antiquity, during a period of turmoil much like ours, lying in prison under sentence of death, Boethius wrote his *Consolation of Philosophy*, which says in effect: look up at the stars, see the order there, that is evidence for a beneficent God; despite appearances, he is con-cerned about your ultimate welfare. But since 1950 or so the book placed in our hands has been entitled *Despair of Science* or words to that effect, and, as we have seen, takes precisely the opposite line.

Such a world-picture could be expected, like Darwin's, to make a shattering impact on men's values and sensibilities, and so it has proved. There is ample evidence of this in first-person writings and sociological surveys but, having begun this book with a quotation from Sophocles, now near the end I prefer to turn to drama, believing that it provides a reliable blood sample of public concerns. Albert Camus was born in Algeria in 1913 and spent most of his life in France, where he worked in the Resistance. Chief spokesman of his generation, he wrote half a dozen novels and four plays. He lived through the discoveries of violence in the cosmos and the elaboration of the big bang theory, though not the space probes, for he was killed in a car accident in 1960. Since then his name has been given to more than one lycée in France.

In his philosophic essay, *The Myth of Sisyphus*, Camus asks an imaginary scientist to explain the universe. The scientist speaks of atoms, electrons and models of matter, much as a big-bang theorist would. Camus is disappointed. This is not the level of explanation he wants. 'You are explaining this world to me with an image. Then I realize that you are reduced to poetry: I shall never know. Before I have time to wax indignant you have changed theory. In this way science, which ought to teach me everything, ends with a hypothesis, and lucidity founders in metaphor.'

The universe, Camus concludes, is evidently not accessible to reason yet man's deepest need is to apprehend the universe. This is the basis for Camus's famous contention: 'Human life is absurd.'

To deal with absurdity three courses are open: faith in God, which Camus rejects because an indecipherable universe offers him no evidence on which to base faith; suicide, which he rejects as an irrational escape; and protest, which Camus adopts. Man can fulfil himself only to the extent of protesting against his nonsensical fate. But protest will not change it. Like Sisyphus, he will continue to push his boulder up the slope, knowing that it must always roll back before reaching the top.

In his play *Caligula* Camus depicts the young Roman emperor as unbalanced and debauched. Night-time for him has meant incest with his sister Drusilla, whose death before the play begins induces thoughts of mortality. Searching for a meaning to life, Caligula becomes convinced that he will find one by sleeping with the moon. If he knows the moon carnally he will know and understand the cosmos.

While pursuing this goal Caligula acts on the assumption that life is absurd. He commits deeds of gratuitous cruelty; he identifies himself with Venus, 'goddess of hate', and obliges the patricians to chant to him: 'Instruct us in the truth of this world, namely the absence of truth.'

One day, while painting his toe-nails scarlet, Caligula confides to a courtier that he has at last slept with the moon – several times. 'She crossed the threshold of my room, and in her slow steady way came to my bed. There she unfolded herself, sweeping me up like a wave in her smiles and brilliance.' But this delusion of communion causes Caligula to behave more irrationally than ever. We the audience realize that even possession of the cosmos would not entail its comprehensibility.

Samuel Beckett, born in Ireland in 1906, resident most of his life in Paris, where he was a disciple of James Joyce, began as a novelist, then moved to drama. *Waiting for Godot* was performed in Paris in 1953, and in English the following year. At one level his play depicts two tramps talking, fooling and waiting for someone who at the final curtain still has not arrived; at another level it creates a whole world of hopelessness, expressed in Lucky's speech near the middle of the play. Voiced as by a defective record, it is a chilling implicit indictment of current cosmological speculation for promising much yet failing to provide man's longed-for link with the cosmos:

> in the light of the labours lost of Steinweg and Peterman it appears what is more much more grave that in the light the light the light of the labours lost of Steinweg and Peterman that in the plains in the mountains by the seas and by the rivers running water running fire the air is the same and then the earth namely the air and then the earth in the great cold the great dark the air and the earth abode of stones in the great cold alas alas ...

The last image, 'the great cold alas alas', is the key to the play's hopelessness. Comments by Beckett and related texts make clear that Lucky is alluding to man's spiritual alienation from the warming sun as an emblem of a possible Creator.

In *Endgame* (1958) Beckett is still haunted by that theme. One of the characters, Clov, clambers up a ladder and observes with a telescope.

CLOV: Let's see ... Zero ... zero ... and zero.

HAMM: Nothing stirs. All is –

CLOV: Zer –

HAMM (*violently*): Wait till you're spoken to. All is ... all is ... all is ... all is what? (*Violently*) All is what?

CLOV: What all is? In a word? Is that what you want to know? Just a moment. (*He turns the telescope on the without, looks, lowers the telescope, turns towards Hamm.*) Corpsed.

Clov continues to scan with the telescope.

HAMM: What's happening?

CLOV: Something is taking its course.

Pause

HAMM: Clov!

CLOV (*impatiently*): What is it?

HAMM: We're not beginning to ... to ... mean something?

CLOV: Mean something! You and I, mean something! (*Brief laugh*) Ah that's a good one!

Lack of meaning is carried further by the third playwright, Eugène Ionesco. Born near Bucharest in 1912 of a Rumanian father and French mother, Ionesco spent his childhood with his mother in France, returning at thirteen to Rumania, where he made his literary reputation. In 1938 he fled to Paris. He holds political views quite different from those of Camus and Beckett: evidence that the importance attached by all three writers to the absurd has a deeper origin than response to political turmoil. For the past twenty years one or more of Ionesco's plays has been on in Paris, a remarkable sign that he has touched a deep chord.

We enter Ionesco's world near the beginning of his first play, *The Bald Prima Donna* (1950):

MR SMITH   (*looking at his paper*): There's a thing I can never understand. In the births, deaths and marriages column why the devil do they always give the age of the deceased persons and never tell you how old the babies are. It doesn't make sense.

MRS SMITH: It's never struck me before.
                (*The clock strikes seven. There is a silence, then the clock strikes three. There is a pause.*)

And later:

MRS SMITH: We can never tell the time here at home.
THE FIREMAN: But the clock?
MR SMITH: It doesn't work properly. It's of a contrary turn of mind. It always strikes the contrary to the right time.

'I'm all for contradiction,' observes a character called the Author in another Ionesco play, *L'Impromptu de l'Alma*, 'everything is nothing but contradiction.' Not only do facts contradict other facts, but characters speak at cross-purpose. Many are senile, devoid of faith and hope, lacking sensuality and sentiment, now following one memory, now another that contradicts it. Ionesco's characters are more symbols than persons, and their irrationality obtrudes beyond mere words: one man may have green hair; a whole group may grow hides and horns and change into rhinoceroses.

Critics argued about the metaphysical grounds for this curious world until 1967, when Ionesco made them clear in *Journal en Miettes*:

An individual life is nothing unless it reflects universal life, unless it is both itself and something else; separate and it is nothing ... The laws I can know, but not the reason for the laws ... What is beyond, the unknown Person or Thing, that alone is worthy of our interest. We want to know that Unknown more intimately than we can through science, and we find it impossible. But I cannot resign myself to knowing only the prison walls.

As with Camus and Beckett, the gulf between rational mind and fortuitous, hence irrational, cosmos engenders anguish. In common with many young men of the late 'fifties and early 'sixties Ionesco felt driven to the one religion which accepts that all perceptual and conceptual phenomena are meaningless – Buddhism. Yet he could not go all the way. He insisted that the individual retain his individuality even in the very heart of Nirvana, and this thirst for an Absolute formed the main subject of *La Soif et la Faim* (1964).

The notes of anguish, loneliness and absurdity are not limited to the theatre, though there is space here to take only one another example. By dispensing with a narrator, one able to link his experience with the world at large, the anti-novel limits itself to descriptions of externals – a cross-section of the alien cosmos whose movements can be observed but whose inner springs are hidden. The anti-novel rejects as obsolete the view of D. H. Lawrence that the gulf between man and nature could be bridged by a love-affair with a farm lass. If, as one biologist claims, 'man is a gene machine, blindly programmed to preserve its selfish genes,' what remains of love? And if the sky is fierce with thermonuclear burning, how can we watery beings commune with it by romping in the hay?

I have lingered on the anguish caused by this world-picture because it was as deep as any in history and because my generation lived through it. For many of us it became a personal dilemma. Repeatedly we felt the arrangement of facts demanded by the outer picture clash with facts of personal experience. I am thinking of the conviction, day after day, that we were making free choices, and that sometimes they were not absurd. One had the experience of setting oneself a goal and pursuing it purposefully – perhaps without success but not therefore wholly in vain. There were experiences, too, of order not of our making. One encountered a person whose goodness was of such a kind it plainly originated elsewhere than in subliminal drives or enlightened self-interest. One had the experience, in a personal relationship, of incidents improbably falling into place, forming a pattern that one could only call providential, until acquaintanceship ripened into friendship. And for those who happened to have a religion there was a glimmer, occasionally, of whence that order might come.

But this private experience jarred with the world-picture. At Oxford, where I was an undergraduate in the years after Hiroshima, the rage was logical positivism, a philosophy born of materialist science which labels words like 'God' and 'soul' meaningless. Any dialogue with it was by its own definitions ruled out. There were times, at philosophy lectures, when I felt like that visitor to Leningrad who from a vantage point sees three prominent churches with his own eyes, yet can find none of them on his map; puzzled, he makes enquiries and learns that 'living churches' are not put on the map, only one 'dead church', the museum of atheism.

I recall turning to a scientist friend. If man, I said, can think his way from his immediate surroundings to the cosmos – surely remarkable enough – why shouldn't he go on from cosmos to God? But the scientist answered that since this second stage was not an experience common to all men, by the terms of science it could not be allowed. 'Read off the figures from Jodrell Bank,' he advised. 'There you have an unseen world that really exists.'

Such is the power of a dominant world-picture, it ends by asserting itself as surely as a sky-writer's advertisement. Then, just to keep going in day-to-day activities, one gradually has recourse to double-think. With me the cosmic and the personal, the without and within, divided. And for a long time, weakly, perhaps in shame, I averted my eyes from the stars.

Non-scientists, we hoped that a scientific voice might suggest a way out. Albert Einstein declared that fortuitousness has no place in a universe expressing at every so far observable point the wonderfully high order and rationality he termed God, but after Einstein's death in 1955 we listened in vain for others.

Then, beginning in the late 'sixties, a few scientists began to take a cold, hard look at this fortuitous cosmos that had won so many adherents. Some recalled a criticism Eddington had made in the days when the big bang theory was being sketched. We scientists, said Eddington, sweep design (which he preferred, noncommittally, to call 'anti-chance') out of the laws of physics. 'By sweeping it far enough away from our current physical problems, we fancy we have got rid of it. It is only when some of us ... try to get back billions of years into the past that we find the sweepings all piled up like a high wall and forming a boundary – a beginning of time – which we cannot climb over.'

This line of thought was taken a stage further by two astrophysicists, John Barrow and Joseph Silk. They started from the remarkable fact that the structure of the cosmos today is more than 99·9 per cent regular: that is, its large-scale features would appear the same to an observer in any galaxy no matter in which direction he looked. For this to be so, they calculated that the cosmos must have been regular as early as the first millionth of a second after the big bang: their precise figure is 10^{-35} second. In other words, homogeneity in the universe is not explained by the big bang, it is just placed so far back that no possible Earth-based experiment could prove or disprove it.

A different criticism came from Von Braun, who developed the moon rocket: 'The more [man] learns about natural science, the more he sees that the words that sound deep are poorly contrived disguises for ignorance. Energy? Matter? We use them but we don't really know what they are.' The same could be said of Self-igniting Fireball and Exotic Particles.

A more basic criticism of the fortuitous cosmos was voiced by the radio astronomer Bernard Lovell. In his Reith Lectures Lovell pointed to the narrowness of its frame of reference. There is, he said, a built-in limitation to all scientific explanation 'when faced with the great problem of creation', and, Goethe-like, he compared the limitation to that of 'the spectroscope in describing the radiance of a sunset or the theory of counterpoint in describing the beauty of a fugue.'

This view, that the scientist is listening in to reality on one frequency only and has an obligation to heed those listening on other frequencies was endorsed by non-scientists also: for example, by L. C. Knights of Cambridge, who wrote that 'reason in the last three centuries has worked within a field which is not the whole of experience, that it has mistaken the part for the whole, and imposed arbitrary limits on its own working.'

How does a scientist regain 'the whole of experience'? One way is through dialogue: it was by corresponding with Churchmen friends, notably Richard Bentley, that Newton came to see the metaphysical implications of his *Principia*. Another way is by observing how the non-scientist actually reasons. When shown an unfamiliar device or object, his first question is likely to be, 'What is it for?' When he enters a planetarium he may be given an explanation of the machinery in terms of molecules of steel and glass, and of light particles, but only when he has learned what the planetarium is for will he be satisfied. Yet the explanation in terms of function does not supersede or compete with the physical explanation; it complements and enriches it. For reason in the enlarged sense demanded by L. C. Knights – enlarged, that is, by reference to the last three centuries – both are necessary.

When I encountered these criticisms and the suggestions for looking afresh at the cosmos in terms of function and design, I became interested. At times I felt a stirring of hope. Was it not common sense, after all, to approach so great a matter as the cosmos with all the faculties at man's disposal? And as for considering Nature in terms of function, historically that had an impeccable ancestry running from Socrates through Aquinas to Einstein.

Here one foresees an objection. When Darwin showed that the giraffe had not been endowed with a long neck by a Designer but had evolved one in order to reach high foliage, did he not make such terms as 'design' taboo? The answer is that he may have done in biology, as studied in the nineteenth century, but not elsewhere, for when we move outside biology, living, struggling, adaptive processes plainly cannot be at work, therefore design is not ruled out *a priori*.

With an attitude modified by these considerations, and ready to admit evidence for design should it appear, we turn back to the cosmos as found in the last chapter. We re-encounter the frightening violence and flux. But need they entail such conclusions as were dramatized by Camus, Beckett and Ionesco? Indeed, are theirs the most probable conclusions? Surely quite as remarkable as the violence itself is the contrast between the violence in other parts of the cosmos and the quietness in our part, between the myriads of gas-spheres and our blue-and-white orb. Far from blazing iron furnaces that sometimes blow up, far from a drifting lethal gas bubble 1200 light years across, we find a leafy avenue of suburbia. We encounter serenity and, for structures higher than hydrogen and helium, a haven.

But there is more to Earth's situation than tranquillity. The same matter as is sterile on Mars is fertile with us. Water, as we saw earlier, has been released from Earth's depths in just the finely balanced amount as to sustain, without drowning, land-based life. We are shielded from extinction by two narrowly poised screens: ionosphere and ozone layer. We have a climate whose life-giving properties are again finely balanced, and which depends in part on Earth's magnetic field (which not all planets possess) and on the close orbiting of a large moon. There are many factors like these, adding up to the possibility of peace and enduringness. It looks indeed as though modern science has been holding back a revelation quite as momentous as Copernicus's: it is Earth, not sky, that is heavenly.

These factors, each looking highly improbable even within the context of the solar system, have had to intermesh in space and time in order to provide those narrow margins within which a living cell can exist. Indeed the situation in which we find ourselves is so remarkable that it prompts the question, What are the odds against one of the bodies orbiting the sun being habitable?

A young astrophysicist, Nigel Henbest, has made a special study of the subject, and his findings appear as an Appendix (p. 331). He holds that the following conditions are necessary if a planet is to be habitable.

1. Suitability of the sun. In order to have planets a sunlike star must be of a certain size and composition. About 1 in 10 sunlike stars satisfy this requirement.

2. Planet's mass and distance from the sun. Only bodies of a certain mass can retain the light gases which make a breathable atmosphere. This mass is known to be between half and twice the Earth's mass. Another condition of habitability is a temperature above the freezing point of water and below the boiling point. This habitable zone – an ecosphere – is narrow. If the Earth-sun distance is 1·00, the inner edge lies at 0·95, the outer edge at 1·004. The only planet in our solar system to satisfy this condition is Earth. The chances of a planet within the required mass limits lying within the ecosphere is 1 in 50.

3. Suitable amount of water. A comparative study of the land surfaces of Earth, Mars and Venus suggests that the chances of a planet having an amount of free water suitable for intelligent life is 1 in 2.

4. Absence of nearby cosmic catastrophe. The two most probable catastrophes would be a nearby supernova or collision with an asteroid. A nearby supernova (closer than 30 light years) should occur every 100 million years or so. It would destroy the ozone layer, possibly allowing ultra-violet through to damage and perhaps destroy intelligent life. An asteroid impact could be more serious; it has recently been invoked to explain the extinction of the dinosaurs. A conservative figure for the chances of such an event making the planet uninhabitable is 1 in 2.

Final probability. Multiplying together the above figures, we find that the chances of one of the planets of our solar system being habitable are 1 in 2000.

Such long odds do not of course disprove a claim that the conditions necessary for life within the solar system are fortuitous, but they certainly oblige us to treat such a claim – in the 'sixties and 'seventies it was often made glibly – with great caution. The figures show furthermore that we can no longer entertain the long-current assumption that matter merely had to thread its way through the eye of one needle – amino acids into self-replicating cell – in order to emerge as intelligent life. It first had to thread its way through the eyes of several other needles.

There is another aspect of those odds of two thousand to one. They do provide firm, quantitative evidence for asking a new question, Whether purpose may not be evident also in the very structure of the universe?

There are many factors built in to Nature without which life could not exist. They are not quantifiable in the same way as the chance of a habitable planet existing is quantifiable, because we have no other 'Nature' against which to measure them, but they are none the less remarkable, so much so that some scientists call them 'coincidences'.

One example may be given. As Freeman J. Dyson of Princeton showed in an epoch-making article in 1971, if the nuclear force binding two protons together happened to be only a few per cent stronger than it actually is, a nucleus of helium-2 could exist. (Helium-2 would have two protons and no neutron, in contrast to the two existing forms of helium, helium-3, which has 2 protons and 1 neutron, and helium-4, which has 2 protons and 2 neutrons.) If a helium-2 nucleus could exist, the proton-proton reaction in the sun would yield a helium-2 nucleus plus a photon, and the helium-2 nucleus would in turn spontaneously decay into a deuteron, a positron and a neutrino: in other words, our sun would have burned up all its hydrogen like a bomb billions of years ago. Only because the nuclear force is such that a helium-2 nucleus cannot exist does the sun burn gently, fusing its hydrogen into helium-4, and emitting the steady heat needed for life on Earth.

There are a number of such finely balanced margins in the structure of Nature. Those most relevant to cosmology have been described by Nigel Henbest in the second part of his Appendix (p. 337). They show that the cosmos rests on a narrow ledge of order between precipices of chaos.

Factors favourable to life built into the structure of the cosmos, as well as the five local factors quantifiable in terms of statistical improbability, oblige the scientist to ponder. Though he may have been trained in a college over whose gate is written, 'Abandon the word God all who enter here,' he has to re-think such terms as 'singularity' and 'intrinsic queerness', he has to ask how many 'coincidences' must pass, like black cats, before he queries the notion of chance. As a whole man in the sense defined I think he may well be led to postulate a creative Mind, acting with design. But then he may chide himself: 'I am indulging in the pathetic fallacy!' Surely not, since here no less than in his laboratory he would be following the logic of verified facts. More open to the charge of pathetic fallacy is the atheistic form of the big bang theory, since it argues from life as we happen to know it on Earth to cosmic conditions habitually 'seeding' life – against the evidence, accumulating year by year, of lethal extremes.

If our scientist looks outside his special field, he will notice an important change of opinion at the other end of the cosmos from the galaxies. Reductionism – the belief that it would be possible to explain a living organism exclusively in terms of chemical and physical processes – has lost the favour it enjoyed in the 'fifties and 'sixties. Joint work by the brain scientist John Eccles and the philosopher Karl Popper, the Serbelloni symposium organized by the biologist C. H. Waddington and Arthur Koestler's Aspbach symposium of neuro-physiologists, geneticists and animal biologists, have shown the inadequacies of Reductionism, the latter concluding that 'Reductionism requires at least as great a faith, if not much greater faith, than the organismic and hierarchic approach,' i.e. than theories which postulate a directing force.

Critics of Reductionism make three main points. First, the amino acids and nucleotides necessary for life have been found to be unimaginably more complicated than at first believed, so that the loose term 'building-blocks of life' is hardly more appropriate than were the molecules of glass and steel in our example of the planetarium. Second, if mere survival is the test, an amoeba is extremely fit to survive: it is easy to imagine that the first animal species promptly consumed all the available food and then became extinct; moreover, why should some animals, such as certain species of spiders when trapping prey, have evolved methods vastly more sophisticated than mere survival requires? Third, some neurologists now doubt whether today's human brain, in its newly revealed complexity, could have been evolved by random mutation in the mere one and a half million years since smaller-brained *homo erectus* emerged.

I mention these matters because it is interesting to find that the life sciences may soon, like cosmology, return to some notion of design.

If the foregoing evidence is found to be strong, what conclusion does it entail? It does not oblige us to reject the big bang theory, only to enlarge it. In the beginning was God, and God said, 'Let there be light' – the superdense fireball. It was God who caused it to expand in a way that suggests to us an underlying order, and perhaps a purpose. Part of that purpose, so it seems, was to make possible the existence of other minds – minds able, like that of Socrates, to reason from Nature to the existence of a Maker, and to deduce that man, endowed with mind and free will, is in a somehow special relationship with him.

That at least is the conclusion I was able to draw. And with it came a sense of relief. For the first time since boyhood I could look at the night sky without shuffling my feet. The purpose I had glimpsed in day-to-day life and that of my friends was not delusion, it tallied with that immense order in space. What was a black hole but a local quicksand in that wider bay which gave one footing? So the old dichotomy began to heal, the double-think, the anguish of absurdity.

The megalocosmos and the microcosm of the spirit rejoined, as in certain favoured periods of history, I became aware of a new consistency of behaviour. Rash the phrase might sound, but I could think of no other to express the experience: I felt at home in the universe.

Those who believe in Christianity find the purposeful world-picture specially welcome because it tallies with the Gospels. The star over Bethlehem, the darkening of the sky above Calvary, the earthquake at the third hour: why did the personal drama coincide with a cosmic drama if not because Christ was fulfilling an order already begun, drawing attention to those other favourable conditions as a way of showing what he was and is: the ultimate and completely favourable condition for fullness of life?

The intelligence some are seeking in interstellar space was, on the Christian view, here on Earth, treading our soil, only the other day. The God of the cosmos became the God of the Gospels to show us that He is a Person, and that we matter more to him than the galaxies. 'Why should Christ come to this pinpoint corner of the cosmos?' But that is the kind of thinking which prompted a question already satisfactorily answered, 'Can Christ come from Galilee?'

His claim that we matter would then explain those favourable factors in the universe, otherwise so puzzling. He has revealed that we possess something more precious and lasting than star material. Yet not everything has been revealed. We have not been told fully the why of suffering, or the why of evil. We still grope our way in the dimness of starlight.

*

In this newly serene frame of mind I decided to pay another visit to the Royal Greenwich Observatory in Herstmonceux, where I had done research for my book, in order to catch up on some of the latest work there. The ultra-modern equipment is housed in a moated brick medieval castle, on a site which has yielded Roman denarii.

It was a November afternoon. I arrived early and strolled between clipped yews beside a croquet lawn. I thought of the legionaries who had passed this way, worshippers of a Persian Sun God, of Harold's troops, uneasy about the comet, marching against Duke William, of Chaucer clip-clopping his way to Canterbury, perhaps pausing to measure with a pocket astrolabe the height of Venus and of Mars, of the speech Shakespeare wrote for Lorenzo: 'There's not the smallest orb which thou behold'st / But in his motion like an angel sings.' I pictured John Wilkins and Robert Hooke sketching a machine to go to the moon, their friend Christopher Wren designing the first Royal

Observatory, at Greenwich, and Thomas Hardy visiting that observatory before writing *Two on a Tower*.

I passed in review some of the many world-pictures I had encountered: Pliny's stars that perched on ships' masts, and Isidore's ashen stars without light, the Alexandrians' cringing view of the moon as Big Sister, and St Francis's view of it as a dear sister, St Gregory's Northern Lights portending disaster, Charlotte Brontë's offering hope.

I had arrived in good spirits, but as the sun set and the outlines of the castle grew dim, I thought with a touch of regret: the world-picture we hold today, that will grow dim also and become one line in a future history book. True, I know more about the cosmos than the legionary who passed this way, but am I thereby better off? He believed that what he did was contingent upon his birth-sign: I know that my knowledge is contingent upon my birth-date. Contingency, contingency ... As the castle faded, I felt less sure of being 'at home in the cosmos'.

In the azure of the eastern sky appeared the first stars. Not quiet points of light, still less 'patines of bright gold' but red giants and white dwarfs burning at many million degrees; and in interstellar space not Scipio, captains, merchant bankers but pulsars and quasars, heard but unseen violence, and black holes.

The evening was clear and cold, with a hint of frost, the full array of stars startlingly bright. Earlier ages had been able to see steadfastness there and a promise of peace. Mine could not. Even as I gazed and reminded myself of purpose, I could be sure that in one of the galaxies, visible perhaps only as a dot, some star was collapsing or exploding with all that that implied of apocalypse. Not the silence of infinite space but its roar of destruction – that was what terrified.

I lowered my eyes to the brow of the hill, seeking in the flint the reassurance Wordsworth, by Ullswater, had found in granite Helvellyn. But what was that hill but stage decor, papier-mâché absent in Act I and destined to be absent in Act III? There was no longer stability in sky or land. In a second form I encountered contingency, contingency ... and very dim now was my feeling of being at home.

Eight o'clock came and I drove, with sidelights only, for headlamps could efface some thin stream of light which had struggled for a million years to reach us, to the viewing domes, giant toadstools on the ephemeral hill. In the car park I was met and escorted to an administrative building, dark save for ankle-high shaded lights in the corridor. A glass door was opened. By flashlight we stealthily crossed a lawn to the stars, like burglars after diamonds.

The viewing dome, of steel with wooden panelling, sounded like a frog pool in summer, as bleeps signalled the passing seconds. My astronomer friend welcomed me. He is in his forties, tall, wears a short beard along the line of his jaw and has dark eyes that flash with *feu sacré*. I remarked on his Russian-style fur hat and he explained with a smile that he wore it not only against the cold but to protect his head when he knocked it, sometimes, against the many protuberances of his instruments.

He was photographing a star of the seventh magnitude. We looked at it, fuzzy, through the small guide telescope attached to the main instrument. In a few days he would compare his photographs with others, and be able to calculate the star's distance: one more number for the computers. Those were his orders for the day: in this advance post to pin down one particular star; but he spoke too of work nearer his heart. Two quasars had been detected very close together and with almost identical spectra. It would be an amazing coincidence if two such immense and unusual bodies with identical evolutionary histories could occur as neighbours; but some astronomers, including my friend, believed that there was only one quasar and that some object – probably a galaxy – near the quasar's line of sight was bending the light by means of its gravitational field in such a way as to give the appearance of two. It remained to gather evidence, and that my friend was doing.

Stirred by his account, I voiced the thought I had on coming to Herstmonceux. My friend listened but said nothing. 'This business of design,' I went on. 'Newton's disciple Bentley did the sums that could be done in those days. He calculated that if one crewless galleon sailed from the North Pole and a second from the South there was more chance of their colliding than there was of atoms conjoining to form the Earth ... Now you have far more facts at your fingertips than Bentley. How is it that you and your colleagues don't stand up and speak, as he did, of order and design?'

My friend considered this in a silence punctuated by the bleeps. 'For two reasons, I think. The first is, we're so very busy with the tasks we're paid to do. It takes us all our time simply getting our equations to come out right.' (Again, shortage of time in this crowded universe.) 'And the second reason ... We're aware of the great minds that have set themselves to understanding the cosmos, and I think we feel – well, humble.'

My heart fell and I swallowed hard. If he, the expert, felt humble . . .

We talked further. Regarding the facts he was hoping to establish that evening my friend said, 'Perhaps they'll solve one question, but in so doing they'll raise a second question, even two. And so it goes on.' He cited example after example, and I saw that the penumbra of mystery seemed never to lessen.

Contingency, contingency. Half an hour later I left the viewing dome in disarray. I had come to Herstmonceux with a neatly packaged view. In three separate ways it had been battered; now I held in my hands crumpled paper and broken string.

I walked across the lawn. The slice of sky visible from the viewing dome was now replaced by the whole cloudless vault. It had been, I realized, a form of hubris to believe myself at home there. However sure one might feel that the cosmos was underpinned by design, there would always be a nearly equal sense of contingency, an awareness that any theory was surpassed by the reality.

My eye settled on the twinkling Pleiades. I thought of how Homer had seen them, and Sappho, and I began to grope for a foothold. The Greeks had felt, as I felt, both the explicable order, and the inexplicable mystery. This, it belatedly dawned on me, was one of the recurring patterns in history, the sense that man rode the waves less as Sophocles' mariner than as a surf-rider, poised for a split second of comprehension, then diving into the trough of darkness. John Donne had felt it, and Pascal, and Van Gogh, and Saint-Exupéry.

This antiphonal chanting in man's spirit between the choir of understanding and the choir of mystery a protagonist of the absurd would describe as yet another of the torments of Sisyphus. I did not see it like that at Herstmonceux, nor do I now. For is it not also a fact – a remarkable one – that without this tension and dialogue there would have been no advance in man's understanding of the cosmos? If he had been too sure of his early world-picture his wonder would have ebbed and with it the urge to enlarge it; too bewildered by what he saw and he would have hidden in the irrational. If there is design in the cosmos, perhaps it is part of that design that we should continue to fluctuate between despair at ever understanding the cosmos and confidence that it can be understood, the wonder thereby engendered being of great value to man as a reminder of his peculiar condition: a spiritual amphibian completely at home neither in nature nor in supernature. Perhaps it is our lot continually to discover with a surge of excitement newer and more complete world-pictures, only to notice in each a flaw, as a reminder that the perfect picture, which Dante called the beatific vision, does not belong here below.

References and Notes

Place of publication of books in English is London unless stated otherwise, and of books in French Paris.

CHAPTER 1   The Greeks

p.15   'Wonders are many on Earth ...' Sophocles, *Antigone*, 332–5, trans. E. F. Watling.

p.16   'Earth, sky and sea, the indefatigable sun ...' Homer, *Iliad*, xviii, 483–9, trans. E. V. Rieu.

p.16   'When first the Pleiades ...' Hesiod, *Works and Days*, 383–4, trans. T. F. Higham.

p.17   'The moon is gone ...' Attributed to Sappho. *Oxford Book of Greek Verse in Translation* (1938), p.211.

p.18   'Their souls were received ...' *Epigrammata Graeca*, ed. G. Kaibet (Berlin 1878), 21 b, 1.

p.20   Homer on comets: *Iliad*, iv, 75ff.

p.20   'Is it a signal of war ...' Pindar, *Paean* ix, 13–20.

p.20   'Most of the Athenians ...' Thucydides, *Peloponnesian War*, vii, 50.

p.21   'Just because the Greeks ...' C. M. Bowra, *The Greek Experience* (1957), p. 168.

p.22   Socrates and Anaxagoras: Plato, *Phaedo*; D. R. Dicks, *Early Greek Astronomy to Aristotle* (1970), pp.94–5.

p.23   'Thou, Earth's support ...' Euripides, *Troades*, 884–8.

p.24   Pythagoras's life and teaching: Diogenes Laertius, *Lives of the Philosophers*, viii, 1–50; W. K. C. Guthrie, *A History of Greek Philosophy* I (1962), ch.iv.

p.26   'There's not the smallest orb ...' Shakespeare, *The Merchant of Venice*, Act V, Sc. 1.

p.28   'The Pythagoreans lost confidence ...' K. Popper, *The Open Society and Its Enemies* (1966) I, p.319.

p.28   On the knees of Necessity ... *Republic*, Bk X, 617ff.

p.28   Time starts to run backwards ... *Statesman*, 270.

p.28   The soul ... is like a winged charioteer ... *Phaedrus*, 246ff.

p.29   'Thou wert the morning star ...' *Oxford Book of Greek Verse in Translation* (1938), p.517.

p.31   'It is better to think the first mover ...' *Physics*, 259a6.

CHAPTER 2   From Alexandria to Bethlehem

p.34   The ship *Woman of Alexandria*: Athenaeus, *Deipnosophistae*, v, 206–9.

p.35   Archimedes' achievements: E. J. Dijksterhuis, *Archimedes* (Copenhagen 1956); Archimedes in Spain: Leonardo da Vinci, *Notebooks*, trans. E. MacCurdy (1939), p.850.

p.38   'When the moon rises ...' Strabo, *Geography*, III, 5, 8.

p.39   Stoic cosmology: S. Sambursky, *Physics of the Stoics* (1959).

p.41   A friend 'asserted that he saw the tiles ...' Seneca, *Questions about Nature*, vi, 31.

p.41   Littleness and imperfections of the world: Seneca, *op. cit.*, i, Pref. 8–11; iv, 11, 3.

p.41   'Much in life is sad ...' Seneca, *Letters on Morality*, vi, 6–7.

p.44   'You have to do it ...' W. W. Tarn, *Hellenistic Civilization* (1952), p.352.

p.47   The star of Bethlehem. Another opinion favours a triple conjunction of Saturn and Jupiter (in Pisces) of 7 BC. D. Hughes, *The Star of Bethlehem Mystery* (1979).

p.49   'The ball-player cannot catch the ball ...' Clement of Alexandria, *Stromata*, 2, 6.

p.49   'Whoever reflects on four things ...' *Hagigah*, ii, 1, following the opinion of Rabbi Akiba.

p.49   'When the season of spring ...' *Opera* Amphilochii Iconiensis (Louvain 1978), pp.155–6. The text of this previously unpublished homily was found in a manuscript in Sinai.

p.50   'flesh, marrow, bones, nerves ...' Clement of Alexandria, *Protrepticus*, x, 98.

CHAPTER 3   The Propinquity of Heaven

p.52   'In the middle was a gleaming cloud ...' Gregory of Tours, *The History of the Franks*, trans. O. M. Dalton (1927), viii, 17.

p.52   'On a voyage I have seen stars alight on the yards ...' Pliny, *Natural History*, ii, 37.

p.54   'Immediately the fire stopped ...' Gregory of Tours, *In Honour of the Blessed Martyrs*, ch.84.

p.54   'Taking the reliquary from my neck ...' *Ibid.*

p.55   'Night comes about ...' Isidore, *Etymologies*, v, 31, 3.

p.55   'Night, we believe, is for rest ...' Isidore, *Treatise on Nature*, ii, 1.

p.55   'These stars, seven in number, and shining brightly ...' *Ibid*, xxvi, 7.

p.56   'The stars have no light of their own ...' Isidore, *Etymologies*, iii, 61.

p.57   'He saw some kind of dark valley ...' Bede, *Ecclesiastical History of England*, iii, 19.

p.58 'Pepin: What is the sun? ...' Alcuin, in Migne, *P.L.*, ci, p. 978.

p.58 'Your silence can mean only one thing ...' *Mon. Ger. Hist. Scriptores*, II, p.604ff.

p.59 'Then over all England ...' *Anglo-Saxon Chronicle*, year 1066.

p.61 Guibert of Nogent: Autobiography, trans. C. C. Swinton Bland (1925).

p.63 The Burgundian's meeting with 'St James': *Ibid*, ch.19.

p.63 'Small of body and family ...' Abbot Suger, *On the Abbey Church of St-Denis*, ed. E. Panofsky (Princeton 1946), p.33.

p.65 'Souls originate in the sky ...' Macrobius, *A Commentary on Scipio's Dream*, i, ix.

p.65 'It seems to me that I see myself dwelling ...' Abbot Suger, *op. cit.*, p.65.

p.67 'Be praised, my Lord, with all thy works ...' F. C. Burkitt, *The Song of Brother Sun in English Rime* (1926).

CHAPTER 4   The Circles of Heaven and Hell

p.68 Gerard of Cremona: *Dictionary of Scientific Biography* (New York 1970–8), xv, pp.173–92.

p.71 'if the Earth's radius is taken as 1 ...' Ptolemy, *The Almagest*, trans. R. C. Taliaferro (Chicago 1952), p.175.

p.72 'where the sameness, good order, proportion ...' *Ibid*, p.6.

p.72 'It would be a shame to live in a house ...' Ristoro of Arezzo, *Composition of the World* (1282).

p.75 'We are born larvae ...' *Purgatorio*, x, 124–5.

p.77 'They are independent, liberty-loving, fond of arms ...' Ptolemy, *Tetrabiblos*, ii, 3.

p.77 'The majority of men are governed by their passions ...' Aquinas, *Summa Theologica*, Pt 1, Q. 115, A. 4.

p.80 'For Venus sent me feeling from the stars ...' *The Canterbury Tales*, rendered into modern English by N. Coghill (1951), pp.298–9.

p.80 'Now up, now down, like buckets in a well ...' *Ibid*, p.66.

p.80 'Mine is the prisoner in the darkling pit ...' *Ibid*, p.91.

p.81 'So long as I wear certain things about me ...' T. Favent, *Historia ... mirabilis parliamenti*. Camden Third Series, vol. 37 (1926), pp.17–18.

p.81 'Infants bearing it shall be very apt to learn ...' *The Book of Secrets of Albertus Magnus* (1973), p.23.

p.82 'Gold for the sun and silver for the moon ...' *The Canterbury Tales*, *ed cit.*, p.481.

p.82 'As for proportions, why should I rattle on ...' *Ibid*, p.479.

p.83 'Behold we not / What different form the fern hath got ...' *The Romance of the Rose*, trans. F. S. Ellis (1937), 16859–65.

p.84   'We observe material things of very different types ...' F. C. Copleston, *Aquinas* (Harmondsworth 1955), pp.121–2.

CHAPTER 5   Wayward Explorers

p.87   'The spider, thinking to find repose ...' Leonardo da Vinci, *Notebooks*, trans. E. MacCurdy (1939), p.1093.

p.87   'The image of the moon in the east ...' *Ibid*, p.238.

p.88   'When the hand of the swimmer ...' *Ibid*, p.472.

p.88   'Flying creatures will support men with their feathers ...' *Ibid*, p.1105.

p.88   Leonardo's flying machines: *Ibid*, pp.493–502.

p.89   'The machine should be tried over a lake ...' *Ibid*, p.498.

p.89   'The Earth is a star almost like the moon ...' *Ibid*, p.281.

p.91   'the needles pointed north-west ...' Columbus, *Journal*, 17 Sept. 1492.

p.92   'I found that it was not round ...' *The Four Voyages of Columbus*, Hakluyt Society (1933), II, p.30.

p.95   'Fifty of us soldiers ...' Bernal Diaz, *True History of the Conquest of Mexico*, Pt I, ch.5.

p.96   'Among the extraordinary, though quite natural circumstances ...' Cardano, *The Book of my Life*, trans. J. Stoner (1931), p.189.

p.97   Cardano's dream: *Ibid*, p.158.

p.98   Astolfo's flight to the moon: Ariosto, *Orlando Furioso*, canto 34, 69–92.

p.98   Cardano's interview with Edward VI: H. Morley, *Jerome Cardan* (1854), II, pp.135–8.

p.100   'I often considered whether there could perhaps be found ...' Copernicus, *Commentariolus*, trans. E. Rosen (New York 1939), pp.57–8.

p.102   'shaken off the authority ...' A. Koestler, *The Sleepwalkers* (1959), p.199.

CHAPTER 6   The Blemished Sun

p.105   'Blow, winds, and crack your cheeks! ...' *King Lear*, Act III, Sc. 2.

p.106   'The skies are painted ...' *Julius Caesar*, Act III, Sc. 1.

p.106   'There's not the smallest orb ...' *The Merchant of Venice*, Act V, Sc. 1.

p.106   'Comets, importing change ...' *1 Henry VI*, Act I, Sc. 1.

p.107   'If then the World a Theater present ...' Thomas Heywood, 'The Author to his Booke'.

p.107   'The heavens themselves ...' *Troilus and Cressida*, Act I, Sc. 3.

p.110   'There are no solid spheres in the heavens ...' T. Brahe, *De mundi aetherei ... phaenomenis*, Bk II, ch.10, in *Opere* (ed. Dreyer) IV, p.222.

p.110 The relationship between Brahe and Kepler: A. Koestler, *The Sleep-walkers* (1959), Part IV.

p.112 'rough and uneven, covered everywhere ...' *Messenger of the Stars*, in *Discoveries and Opinions of Galileo* (New York 1957), p.28.

p.112 'Me thinkes my diligent Galileus ...' William Lower, 21 June 1610, in F. R. Johnson, *Astronomical Thought in Renaissance England* (Baltimore 1937), p.228.

p.112 'When god commanded Abraham ...' Thomas Browne, Notes from Commonplace Books, in *Works*, V (1931), p.246.

p.113 'The entire planetary system practically disappears ...' Kepler to Herwart, 16 Dec. 1598.

p.114 'We have added to the world Virginia ...' J. Donne, *The Satires, Epigrams and Verse Letters* (1967), p.97

p.114 'The sun is lost ...' *The Poetical Works of John Donne*, ed. Grierson (1929), p.208.

p.115 'Admiration, wonder, stands in the midst ...' J. Donne, *Eighty Sermons*, no. 20.

p.115 'I am up, and I seem to stand ...' J. Donne, *Devotions*, no. 21.

p.115 'You which beyond that heaven ...' *The Poetical Works of John Donne*, p.295.

p.116 'O rack me not ...' G. Herbert, 'The Temper', stanza 3.

p.116 'Why are some big, some little ...' R. Burton, *The Anatomy of Melancholy*, II, ii, Mem. 3.

p.116 'The element of fire is quite put out ...' W. Drummond, *A Cypress Grove* (1623), in *Works* (Edinburgh 1913), II, p.78.

p.119 'There are countless suns ...' G. Bruno, *The Infinite Universe and its Worlds* (1584), Third Dialogue.

p.120 'The Nolan has given freedom to the human spirit ...' *Ash Wednesday Supper*, in W. Boulting, *Giordano Bruno* (1916), p.125.

p.122 'The historians of Mexico ...' B. Pascal, *Pensées*, ed. G. Haldas (Geneva 1947), p.155.

p.123 'I see these frightening spaces in the universe ...' *Ibid*, p.69.

p.123 'The eternal silence of these infinite spaces ...' *Ibid*, p.113.

CHAPTER 7   The Apple and the Comet

p.125 John Wilkins described by Aubrey: *Brief Lives*, ed. Dick (1949), iv, p.320.

p.126 'Since the sun is in their zenith ...' J. Wilkins, *Discovery of a World in the Moon*, in *Works* (1802), i, p.101.

p.127 ''Tis the opinion of Kepler ...' *Ibid*, p.111.

p.127 'a flying chariot, in which a man may sit ...' *Ibid*, p.128.

p.129 'The meaner mind works with more nicetie ...' Henry More, 'Cupid's Conflict'.

p.130 'this heaven, this Earth, this universe ...' Sister Maria Celeste, *The Private Life of Galileo* (1870), p.283.

p.130 '"Milton", continues Voltaire ...' *Essai sur la poésie épique* (1732), ch.9.

p.131 'Round he surveys ...' *Paradise Lost*, iii, 555ff.

p.132 'Her spots thou seest ...' *Ibid*, viii, 145–8.

p.133 'mathematical demonstrations built upon ...' C. Wren, *Parentalia* (1750), p.200.

p.135 'every nebulous star ...' *Ibid*, p.205.

p.136 Descartes' cosmology: *Les Principes de la Philosophie* (1647), Part 3.

p.140 Newton and the apple: W. Stukeley, *Memoirs of Sir Isaac Newton's Life* (1936), p.20.

p.142 'Halley "at once indicated the object of his visit ..."' D. Brewster, *Life of Sir Isaac Newton* (Edinburgh 1855), I, p.297.

p.143 'at Plymouth and Chepstow Bridge ...' Newton, *Principia*, Book III, Prop. 37, Problem 17.

p.144 'When I wrote my treatise ...' Newton to Bentley, 10 Dec. 1692.

p.145 'Nature, and Nature's Laws lay hid in Night ...' A. Pope, 'Epitaph Intended for Sir Isaac Newton'. Byron's tribute in *Don Juan*, Canto X, is also worth citing: 'Man fell with apples, and with apples rose ...'

CHAPTER 8   Who Bowled these Flaming Globes?

p.149 'the empty space of our solar region ...' *A Confutation of Atheism from the Origin and Frame of the World* (1693), Seventh Sermon. On Bentley and his correspondence with Newton see S. L. Jaki, *Planets and Planetarians* (Edinburgh 1978), ch.3.

p.150 Addison's Ode: *The Spectator*, no. 465.

p.150 'Thy voice produc'd the seas ...' Isaac Watts, 'The Creator and Creatures', *Horæ Lyricæ* (1709), p.15, cited by A. J. Meadows, *The High Firmament* (Leicester 1969).

p.151 'There is but one objection against marriage ...' Edward Young, *Correspondence* (1971), p.133.

p.151 Edward Young, *The Complaint: or, Night-Thoughts on Life, Death, and Immortality*, published between 1742 and 1746. Passages quoted are: ix, 1274–80; ix, 769–77; ix, 698–719; iv, 704–16.

p.153 James Ferguson: *Life* by E. Henderson (1867), supplemented by personal communication from John R. Millburn.

p.154 Date of the crucifixion: *A Brief Description of the Solar System* (Norwich 1753), appendix.

p.155 'ten thousand ten thousand Worlds ...' *Ibid*, p.12.

p.156 'Of Man's miraculous Mistakes ...' *Night Thoughts*, i, 399ff.

p.157 Madison's 'The Symmetry of Nature': G. Hunt, *The Life of James Madison* (New York 1902), p.101.

p.158 Voltaire's distorted portrait of Newton: *Eléments de la Philosophie de Newton* (1738), Part I, chapters 1–5.

p.164 'Go down to the sea shore ...' *Napoléon Inconnu*, ed. Masson (1895), II, p.303.

p.165 'The intelligence watching over the movement of the stars ...' G. Gourgaud, *Journal de Sainte-Hélène* (1966), II, p.315.

CHAPTER 9　From Detachment to Involvement

p.167 'He was so occupied with his own thought ...' J. H. W. Stuckenberg, *The Life of Immanuel Kant* (1882), p.126.

p.168 'Air, water, heat ...' *General History of Nature*, Preface; my italics.

p.170 'Air, fire, earth and water ...' G. H. Lewes, *Life of Goethe* (1864), p.14.

p.171 'As if, forsooth! ...' *Conversations with Eckermann*, trans. Oxenford (1874), p.181.

p.171 'number and measurement in all their baldness ...' *Die Schriften zur Naturwissenschaft* (Weimar 1947), I, 9, p.367, in H. B. Nisbet, *Goethe and the Scientific Tradition* (1972), p.49.

p.172 'It would have been for [God] a poor occupation ...' *Conversations with Eckermann*, p.569.

p.173 'Quis coelum posset ...' *Astronomica* ii, 115–16.

p.173 'To see a World in a Grain of Sand ...' Blake, *Complete Writings*, ed. G. Keynes (1966), p.431.

p.174 'Mock on, mock on, Voltaire, Rousseau ...' *Ibid*, p.418.

p.174 'The Sky is an immortal Tent ...' *Ibid*, p.516.

p.174 'Do you not see a round disc ...' *Ibid*, p.617.

p.175 'who appear'd to me as Coming ...' *Ibid*, p.606.

p.177 'between those heights / And on the top ...' *The Excursion*, ii, 719–24.

p.178 'Oh! there is laughter at their work in heaven ...' *Ibid*, iv, 956–62.

p.178 'I have seen / A curious child ...' *Ibid*, iv, 1132–42.

p.179 The mending of Earth's axial obliquity: *Queen Mab*, vi, 45–6 and Shelley's Note thereon.

p.180 'Do not all charms fly ...' Keats, *Lamia*, 229ff.

p.181 Thomas Dick's *The Sidereal Heavens* was in the Keighley Mechanics' Institute, from where the Brontës borrowed books. The other astronomy book on the Institute shelves, Robert Mudie, *The Heavens* (1835), does not discuss other inhabited worlds.

p.181 'I'll think there's not one world above ...' *The Complete Poems of Emily Jane Brontë* (1941), p.184.

p.181 'No coward soul is mine ...' *Ibid*, p.191.

CHAPTER 10   The Shadow of the Ape

p.184 'Choose any well levelled field ...' J. Herschel, *Outlines of Astronomy* (1893), p.353.

p.186 Samuel Vince's arguments: *A Confutation of Atheism* (1807).

p.188 'These poor wretches were stunted in their growth ...' C. Darwin, *Journal of the Voyage of HMS Beagle* (1906), p.203.

p.189 'One is tempted to exclaim ...' Notebook C 53, in H. E. Gruber, *Darwin on Man* (1974), p.167.

p.189 'Having proved mens & brutes bodies on one type ...' Notebook E 47, *Ibid*, p.458.

p.190 'Is frowning result of straining vision ...' Notebook M 95, *Ibid*, p.283.

p.190 'Hensleigh says the love of the deity ...' Notebook C 244, *Ibid*, p.454.

p.190 'But I was very unwilling to give up my belief ...' *The Autobiography of Charles Darwin*, ed. Nora Barlow (1958), p.86.

p.190 'May not the habit in scientific pursuit ...' *Ibid*, Note 4, p.236.

p.195 'the roars and plashings of the flames ...' *The Dynasts*, Part III, After Scene.

p.195 'And that inverted Bowl they call the Sky ...' E. FitzGerald, *Rubáiyát of Omar Khayyam*, stanza 72. The dying Hardy asked his wife to read him the stanza beginning, 'Oh Thou, who Man of baser Earth didst make ...'

p.196 'Since [your] heart is obviously animated ...' Letter from his father to Karl Marx, 2 March 1837.

p.196 Marx's youthful poems: Karl Marx and Frederick Engels, *Collected Works* I (1975), pp.517–615.

p.197 *Scorpion and Felix*: *Ibid*, pp.616–32.

p.199 'Nothing but Christian-Germanic-patriarchal drivel on nature ...' A. Schmidt, *The Concept of Nature in Marx* (1971), p.131.

p.199 'Nature becomes ... pure Object ...' Karl Marx, *Grundrisse* (Berlin 1953), p.313.

p.199 'The external physical conditions ...' Karl Marx, *Capital* (Moscow 1965), I, p.512.

p.199 Marx's dedication of *Capital*: L. Feuer, 'The Case of the "Darwin-Marx" Letter', *Encounter*, October 1978.

p.199 'Although it is developed in the crude English style ...' Marx to Engels, 19 Dec. 1860.

p.199 'Darwin's book is very important ...' Marx to Lassalle, 16 Jan. 1861.

p.200　Similarity of world-pictures despite different social backgrounds: see for example Melville's *Pierre* (1852), Bk. III: 'That the starry vault shall surcharge the heart with all rapturous marvelings, is only because we ourselves are greater miracles, and superber trophies than all the stars in universal space.'

p.200　'Marx professed himself ...' B. Russell, *History of Western Philosophy* (1946), p.816.

p.201　'Qu'importe le zénith sombre ...' V. Hugo, *L'Année Terrible* (1872), p.408. Memorable too is Hugo's 'Abîme' in *Légende des Siècles* (1878), when a long dialogue between the Milky Way and the Nebulae is capped by God's one line: 'Je n'aurais qu'à souffler, et tout serait de l'ombre.'

p.202　'The more ugly, old, vicious, ill and poor I get ...' Van Gogh, *The Complete Letters* (1958), III, p.444.

p.202　'I have just read Victor Hugo's *L'Année Terrible* ...' *Ibid*, II, p.615.

p.202　'In the blue depth the stars were sparkling ...' *Ibid*, II, p.589.

p.203　'I should not be surprised ...' *Ibid*, III, p.57.

p.205　'According to certain calculations ...' J. Verne, *Round the Moon*, ch.18.

CHAPTER 11　Mind over Matter

p.207　Henri Becquerel: *Dictionary of Scientific Biography* (New York 1970–1980), i.

p.213　'Thought seems to be transmitted ...' A. Carrel, *L'Homme, Cet Inconnu* (1935), p.315.

p.214　'The He-Ancient ...' G. B. Shaw, *Back to Methuselah*, Part V.

p.214　'The universe can be best pictured ...' J. Jeans, *The Mysterious Universe* (Cambridge 1930), p.136.

p.215　'When I saw the vapour spouting ...' J. W. Dunne, *An Experiment with Time* (1927), p.34.

p.216　'We had our bottle of wine ...' *The Waves* (1946), p.197.

p.216　'Rapidly, very very quickly ...' *A Writer's Diary* (1953), pp.111–12.

p.218　'Nor, in our little day ...' *The Poems of Alice Meynell* (1940), pp.126–7.

p.219　'I am not sure that the mathematician understands ...' A. Eddington, *New Pathways in Science* (Cambridge 1935), p.324.

p.219　'I like relativity ...' D. H. Lawrence, *Complete Poems* (1964), p.524.

p.220　'when a man comes looking down upon one ...' *Ibid*, p.526.

p.221　'You find it surprising ...' A. Einstein, *Lettres à Maurice Solovine* (1956), p.115.

p.222　'Modern man, having passed through materialism ...' W. Reich, *Schoenberg* (1971), p.87.

p.223　'The introduction of rows ...' *Ibid*, p.134.

p.224 'Et rien ne subsista dans le grand large infini ...' F. T. Marinetti, 'La Conquête des Etoiles'.

p.224 'Time and Space died yesterday ...' *Futurist Manifestos*, ed. U. Apollonio (1973), p.22.

p.224 'Our bodies penetrate the sofas ...' *Ibid*, p.28.

p.225 'Why should art continue ...' F. Elgar, *Mondrian* (1968), p.110.

p.225 'The two oppositions (vertical and horizontal) ...' L. J. F. Wijsenbeek, *Piet Mondrian* (1969), p.94.

p.225 'man is enabled by means of abstract-æsthetic contemplation ...' *Ibid*, p.94.

p.226 'The first inspiration I ever had ...' J. Lipman, *Calder's Universe* (1977), p.17.

p.226 'I felt there was no better model ...' *Ibid*, p.18.

p.227 'to prove once more, by a new universe ...' 'Mobile by Calder', in Elder Olson, *Collected Poems* (Chicago 1963), p.143.

p.227 'I proclaim ... that oblique and elliptic lines ...' *Futurist Manifestos*, p.171.

p.228 'the extensive transparency ...' P. Collins, *Changing Ideas in Modern Architecture* (1965), p.290.

p.229 'The whole country was lighted ...' General T. F. Farrell's report to the War Department, in W. L. Laurence, *Dawn over Zero* (1947), pp.161–2.

CHAPTER 12   Journey to the Moon

p.233 'During the night of December 16th ...' F. C. Kelly, *The Wright Brothers* (1944), p.80.

p.233 'After running the motor a few minutes ...' *Ibid*, p.82.

p.237 'You are the world's greatest wealth ...' *Terre des Hommes* (1939), p.187.

p.238 Early space flights: John Glenn and others, *Into Orbit* (1962).

p.239 'Obviously as well as root beet ...' Oriana Fallaci, *If the Sun Dies* (1967), p.343.

p.239 'How a poet managed to wind up as an astronaut ...' *Ibid*, p.295.

p.240 'Most of them were bald ...' *Ibid*, p.303.

p.240 ' "Wake up," she wanted to shout at them ...' *Ibid*, p.319.

p.246 'Man, wrote Mailer, was "designed from the outset ..." ' N. Mailer, *A Fire on the Moon* (1970), p.122.

p.247 'She was not two feet away ...' *Ibid*, p.380.

p.250 'It's an odd kind of spacecraft ...' G. Farmer and D. J. Hamblin, *First on the Moon* (1970), pp.337–8.

CHAPTER 13   Happenings in Space

p.252 'The object was about 100 or 200 feet off the ground ...' M. Sachs and E. Jahn, *Celestial Passengers* (Harmondsworth 1977), pp.132–3.
p.254 'Their shape was clearly outlined and evidently circular ...' *Ibid*, p.123.
p.256 Jung's theory of UFO's: *Flying Saucers* (1959).
p.258 Colonization of space: G. O'Neill, 'The Colonization of Space', *Physics Today*, Sept. 1974; Adrian Berry, *The Next Ten Thousand Years* (1974), ch. 12.
p.268 'I say that *we* are God ...' O. Fallaci, *If the Sun Dies*, p.35.

CHAPTER 14   A Violent Cosmos

p.271 'We don't put in flowers ...' A. de Saint-Exupéry, *Le Petit Prince* (1946), p.56.
p.272 Birth of Surtsey: S. Thorarinsson, *Surtsey: the new island in the North Atlantic* (Reykjavik 1964).
p.278 'Why shouldn't there be a race of salamanders ...' A. Trollope, *Barchester Towers* (1857), ch.19.
p.281 Present-day view of stars and galaxies: N. Henbest, *The Exploding Universe* (1979); P. Murdin and D. Allen, *Catalogue of the Universe* (Cambridge 1979).
p.283 Black holes. The theory of the reversing of time: A. Berry, *The Iron Sun* (1977), ch.2. In 'The Quantum Mechanics of Black Holes', *Scientific American*, Jan. 1977, Stephen Hawking argues that small black holes will destroy themselves. The Gorgon's head will have seen its face in the mirror.
p.288 'In spite of countless ...' S. L. Jaki, *Planets and Planetarians* (Edinburgh 1978), pp.248–9.
p.288 Interstellar communication: C. Sagan and F. Drake, 'The Search for Extraterrestrial Intelligence', *Scientific American*, May 1975. But the problem of finding a language understood by both parties is not new. A hundred years ago, when astronomers were thinking of communicating with supposed Martians, they proposed to light chains of bonfires in the Sahara, to form a diagram illustrating the theorem of Pythagoras, that the squares on the two smaller sides of a right-angled triangle are together equal to the square on the greatest side.

CHAPTER 15   Chance or Design?

p.294 The big bang theory: S. Weinberg, *The First Three Minutes*: A Modern View of the Origin of the Universe (1977).

p.296   Chance origin of life. On p.6 of *The Mysterious Universe* (Cambridge 1930), James Jeans wrote: 'At intervals, a group of atoms might happen to arrange themselves in the way in which they are arranged in the living cell. Indeed, given sufficient time, they would be certain to do so, just as certain as ... six monkeys would be certain, given sufficient time, to type off a Shakespeare sonnet.' This was a bold claim for one not a biologist to make, and Jeans offered no supporting statistical evidence. However, such was his reputation as astronomer and philosopher, that the claim was repeated in books for several decades. It became almost a cliché of popular science and was not seriously challenged until in 1977 Professor W. R. Bennett, Jr., of Yale University, probed the words 'given sufficient time'. Bennett, conservatively, analysed a less improbable case than that of Jeans: he asked how long it would take one monkey to type the first nine relatively simple words of Hamlet's speech beginning 'To be or not to be ...' Using a computer, Bennett calculated the time that would be required and published his results in *American Scientist* for November–December 1977.

        Bennett decided to ignore the comma and give the monkey a typewriter equipped only with the 26 letters of the alphabet and a space key. He assumed that a good monkey typed steadily at 10 characters per second. He then found that if the monkey were to type correctly the first nine words of 'To be or not to be ...' it would take him about 3×10^{43} seconds (approximately 10^{36} years) on the average. That means it would take the monkey many times longer than the universe has been in existence!

p.297   'The more the universe seems comprehensible ...' *Ibid*, p.154.

p.297   'Man at last knows ...' J. Monod, *Chance and Necessity* (1972), p.167

p.300   'An individual life is nothing ...' E. Ionesco, *Journal en Miettes* (1967), pp.52–4.

p.301   'man is a gene machine ...' R. L. Travers in Forward to R. Dawkins, *The Selfish Gene* (1976).

p.302   'By sweeping it far enough away ...' A. Eddington, 'The End of the World: from the Standpoint of Mathematical Physics', *Nature*, 127 (1931), p.450, quoted in S. L. Jaki, *Science and Creation* (Edinburgh 1974), p.339.

p.303   J. D. Barrow and Joseph Silk, 'The Structure of the Early Universe', *Scientific American*, April 1980.

p.303   'The more [man] learns about natural science ...' D. Lang, *The Man in the Thick Lead Suit* (1954), pp.23–4.

p.303   Limitations of scientific explanation: B. Lovell, *The Individual and the Universe* (1959), p.110.

p.303 'reason in the last three centuries ...' L. C. Knights, 'Bacon and Dissociation of Sensibility', in *Explorations* (1946), p.111.

p.306 Freeman J. Dyson: 'Energy in the Universe', *Scientific American*, Sept. 1971; *Disturbing the Universe* (1979), p.251.

p.307 Criticism of Reductionism: K. R. Popper and J. C. Eccles, *The Self and its Brain* (Berlin 1977); C. H. Waddington, *Towards a Theoretical Biology* I and II (Edinburgh 1968 and 1969); A. Koestler and J. R. Smythies, *Beyond Reductionism* (1969). As regards random mutation, John Thoday has written: 'The prerequisite of continued survival is ability to adapt to an *unpredictable* future. The only way to do that is to generate variance at random ... The need for a random chance component in evolution is built in to the design of the Universe if indeed the Universe is designed.' 'Chance and Purpose', in *Theoria to Theory*, II (1967), p.37.

A Chronological Table

A CHRONOLOGICAL TABLE

DATE AD		CHAPTER
1871	Victor Hugo's *L'Année Terrible*	
1882	Thomas Hardy's *Two on a Tower*	
1887	Michelson-Morley experiment: there is no 'aether' in space	11
1889	Van Gogh's *Starry Night* (Museum of Modern Art, New York)	10
1895	Marconi pioneers wireless telegraphy	11
1896	Becquerel discovers that uranium is radioactive	
1901	H. G. Wells: *The First Men in the Moon*	13
1903	First powered flight by the Wright brothers	12
1905	Einstein's equation: $e = mc^2$, and Special Theory of Relativity	11
1909	Rutherford isolates particles in the atom	
1909	Marinetti's manifesto of Futurism	
1915	Einstein: General Theory of Relativity	
1921	Schönberg: twelve-tone rows	
1924	Beginnings of Surrealism	
1927	Schrödinger: electrons move like waves lapping the proton	
1929	Hubble: our Galaxy is one among many	
1930	Calder's first mobiles	
1931	Virginia Woolf's *The Waves*	
1939	Saint-Exupéry's *Terre des Hommes*	12
1945	First atomic bomb	11
1950	Many reported sightings of UFOs	13
1950–60	Spread of television	
1952	First hydrogen bomb	11
1961	Yuri Gagarin orbits the Earth	12
1963	Surtsey rises from the Atlantic	14
1963	First quasar discovered	
1968	*2001: A Space Odyssey*	13
1969	Man lands on the moon	12
1971	Discovery of a black hole (Cygnus X-1)	14
1974	O'Neill proposes habitable cylinders in space	13
1976	Viking spacecraft land on Mars	14
1979	Voyager photographs Jupiter and its moons	

Appendix

THE PROBABILITY OF LIFE: AN ASTRONOMER'S VIEW

BY

NIGEL HENBEST

Life exists on Earth; and that is all the astronomer *knows* about life in the Universe. Eventually, when man – or his automatic spacecraft – investigates other stars and their planets, we shall discover if other planetary systems are common or rare; whether planets like the Earth are rule or exception; whether the evolution of life on an Earthlike world is inevitable. Then we shall know our own place in the cosmic scale of things.

But it *is* possible to make some calculations even on the basis of our present knowledge. To define the target, I shall take as my starting-point the question: what is the *a priori* probability of a life-bearing planet orbiting the Sun? (Other authors have preferred to ask different, but related, questions, along the lines of 'How many inhabited planets are there in our Galaxy?' Many of these calculations have erred, when possible, on the side of optimism, and I shall attempt to redress the balance by taking a more conservative view.)

The question can conveniently be broken down further. What is the chance that (1) the Sun should have planets; (2) one of these planets should be a suitable place for life to form; (3) life *will* form there; (4) life will subsequently survive the various catastrophes that an unkind cosmos may inflict on it. And there is the key question for the philosopher: How is it that the laws of nature, the rules that govern the Universe, are just right to allow 'life' to exist? I'll conclude with that one.

No one knows how commonplace planetary systems are. With current technology, astronomers cannot photograph the planets of other stars directly, and claims for planets revealed indirectly have recently been severely criticized.

The pioneering worker in the search for other planetary systems has long been Peter van de Kamp of the Sproul Observatory, Pennsylvania. He has measured carefully the positions of nearby stars, and looked for small, slow movements as they are perturbed by 'unseen companions'. Some of these companions are undoubtedly lightweight dim stars, but the lowest weight companions may not have enough mass to shine at all: any companion less than 1/20 the Sun's mass (50 times Jupiter's mass) is not a star but a planet. And if a star has one such massive planet, it is reasonable to assume it has a family of planets. By 1975, van de Kamp and his colleagues claimed they had evidence for planets orbiting four nearby stars. Their strongest case was Barnard's Star. Van de Kamp claimed his data on the motions of this star showed it was being perturbed by two orbiting planets, one similar to Jupiter in mass, the other about half this weight (*Annual Reviews of Astronomy and Astrophysics*, vol. 13, p.295 (1975)).

Other astronomers have more recently begun to observe these stars, and to look at van de Kamp's data more critically. Since the changes in a star's posititition are minute – less than 1/1000 of a millimetre, which is 1/100 the size of the star's image in the photograph – even tiny faults in the telescope or measuring procedure could produce false evidence. Statistical studies have in fact revealed systematic trends in the positions of stars measured at the Sproul Observatory, and these indicate that the data are not reliable enough to determine planetary orbits: errors from various causes seem to be larger than the tiny displacements that van de Kamp claims as evidence for planets (for references, see G. Gatewood, *Icarus*, vol. 41, p.205 (1980)). Particularly damning is a study made by Wulff Heintz in 1976 (*Monthly Notices of the Royal Astronomical Society*, vol. 175, p.533 (1976)). He points out that the Sproul observations have turned up 11 cases of unseen, low-mass companions to nearby stars – either planets or dwarf stars. In all 11 of these, we apparently see the orbit almost 'edge-on', and in 9 out of the 11 the companion's orbit seems to be orientated in approximately the same direction in space. The odds against this happening by chance are higher than 10,000:1. Heintz concludes that the tiny changes in these star's positions as measured at Sproul are not real, but errors which affect all the star images similarly. He backs this up by showing that – if we believe the Sproul data – all these stars appear to swing back and forth in unison with one another as the years go by!

Meanwhile other astronomers were looking for the claimed position changes of Barnard's Star and the others – without any success. Barnard's Star has naturally attracted the most attention. George Gatewood and H. Eichhorn used data from the Allegheny and Van Vleck Observatories, which have long refracting telescopes similar to the Sproul instrument. In the combined results, the star showed no sign of the 'wobble' that van de Kamp had claimed to find (*Astronomical Journal*, vol. 78, p.769 (1973)). Gatewood later repeated the analysis, including van de Kamp's data from the Sproul telescope, and again found no significant evidence for a planet (*Icarus*, vol. 27, p.1 (1976)). After a further three years' observations of Barnard's Star at Allegheny with much more sensitive techniques than van de Kamp's, Gatewood is even more certain that the already dubious 'wobbles' do not exist (*Icarus*, vol. 41, p.205 (1980)). His refined measurements also contradict van de Kamp's 'evidence' for planets to the stars Lalande 21185 (Gatewood, *Astronomical Journal*, vol. 79, p.52 (1974)) and epsilon Eridani (as yet unpublished).

Gatewood's techniques are more modern and sophisticated than van de Kamp's, and his results are undoubtedly more accurate – a conclusion reinforced by the statistical inconsistencies in van de Kamp's data. By the time the Allegheny observations have run for 12 years, they should unambiguously show whether or not 20 of the nearest stars have planets like Jupiter; and even more sensitive searches are being planned (Gatewood (1980) *op. cit.*; Brian O'Leary, *Sky & Telescope*, vol. 60, p.111 (1980)).

At present, the general verdict is that van de Kamp's claimed planetary systems do not exist: there is no good evidence for planets to nearby stars. Current techniques are, however, near the limit for detecting planets even as heavy as Jupiter. The present weight of negative evidence does not necessarily mean that planetary systems are *rare* – particularly if they happen to consist of low-mass planets like the Earth. The observational results leave us in limbo.

Current theories of planetary formation are probably the best guide to the likelihood of a star having planets. Unfortunately, it is still an area of some dispute. Stars form from denser pockets of gas in the gas clouds – *nebulae* – in space; these gas pockets are undoubtedly rotating, and the newly-born stars are probably surrounded by a rotating disc of gas and dust which has too much 'spin' (angular momentum) to fall into the central star. The gas-and-dust disc then breaks into fragments, each small portion condensing to a *planetesimal* a few kilometres across. These planetesimals accumulate into a few planetary-sized bodies.

This is the broad scenario currently accepted by astronomers. It suggests that stars are likely to be born with planetary systems; but the details of the fragmentation of the disc and aggregation of planetesimals are still controversial. Computer simulations have, however, provided some suggestive answers. Stephen Dole (S. Dole & I. Asimov, *Planets for Man* (1965)) has used a very naïve computer program to generate some 'planetary systems', and his results are remarkably similar to the solar system. Richard Isaacman and Carl Sagan later extended Dole's method (*Icarus*, vol. 31, p.510 (1977)) and found that solar-system-type planetary systems form around a Sun-like star when the mass of the disc was initially between 1/50 and 1/5 the mass of the central star. Such a disc is indeed likely to form, according to the most widely-accepted theories of star formation, although it is higher than the present mass of the planets because much of the original gas is lost into space). But they find that the formation of these planets does depend critically on how the density of the disc changes from centre to edge.

Richard Greenberg and his colleagues have made more realistic simulations of planet formation in the solar system (*Icarus*, vol. 35, p.1 (1978)). The results of their rigorous analysis turn out to be surprisingly similar to the simpler models of the Dole school.

If current theories are correct, there is thus a high probability of a Sun-like star having a planetary system. On the other hand, rival theories which make planetary systems a rarer event are not yet ruled out (I. P. Williams, *The Origin of the Planets*, Bristol (1975)). Michael Woolfson, for example, in one of the most 'pessimistic' of theories, calculates the probability of planets around stars as being about 1 in 100 (*The Origin of the Solar System* (ed. S. F. Dermott), p.178, Chichester (1978)). I consider it reasonable to take 1 in 10 chance that a Sun-like star will have planets, without being wrong by an order of magnitude.

Slow rotation of a star was once taken as evidence that that star has a planetary family which holds the 'spin' missing from the star. Most astronomers now believe that stars lose spin by a variety of means, and discount this line of evidence. Myron Smith of the University of Texas has recently found that the Sun is the second-slowest rotator of 25 Sun-like stars he has investigated (reported in *New Scientist*, vol. 86, p.312 (1980)): *if* the slow rotation argument *is* correct, this indicates that the figure of 1 in 10 is reasonable.

Is such a planetary system likely to contain an Earth-like, habitable planet? To be habitable, a planet must lie at the right distance from the Sun, and be roughly Earth's size. A planet too close to the Sun suffers a 'runaway greenhouse effect': the stronger sunlight is trapped by carbon dioxide in the atmosphere; water evaporates and traps more heat; the temperature rises above the boiling point of water and the planet becomes arid and baked. Venus has suffered this fate, with a surface temperature of $475°$ C. If Earth were only 5 per cent closer to the Sun, it too would have gone this way (S. I. Rasool & C. de Bergh, *Nature*, vol. 226, p.1037 (1970)).

Any further from the Sun, and runaway Ice Ages would occur, freezing the planet solid. This has happened to Mars. Earth in fact suffers periodic Ice Ages triggered by relatively minor events such as continental drift and minute changes in its orbit. Michael Hart argues that Earth would have to be only 0·4 per cent further from the Sun for it to be permanently covered by ice (*Icarus*, vol. 33, p.23 (1978)).

The 'habitable zone' (or *ecosphere*) thus extends from 95 per cent to 100·4 per cent of Earth's distance from the Sun. Before calculating the probability of a planet lying within the ecosphere, let us look at the size requirement.

The essential point is a planet's *mass*. Too light, and it will not be able to retain sufficient atmosphere for life; too heavy and it accumulates too much.

It is difficult to set definite limits here, but a conservative judgment would put them around half and twice the Earth's mass respectively (Dole & Asimov, *op. cit.* pp.88–98).

To use the solar system as a basis for calculating the probability of a potential life-bearing planet is clearly not very satisfactory. Since Dole's method seems to mimic the process of planetary formation reasonably well, the ensemble of planetary systems he can generate appears to be a reasonable basis for estimating this probability. Alan Bond and Anthony Martin have created a larger ensemble of systems from Dole's formula (*Journal of the British Interplanetary Society*, vol. 31, p.411 (1978)). From their tabulated results, we can calculate that a planet within the above mass limits will lie within the ecosphere in 1 in 20 planetary systems. This is likely to be an over-estimate, because Dole's programme uses a particular form for the density distribution in the pre-planetary disc, and other distributions tend to produce 'pathological' planetary systems (Isaacman & Sagan, *op. cit.*), systems where the masses and arrangements of the planets are very different from the solar system. A realistic estimate would be more like 1 in 50.

Thus the chance that a Sun-like star would have an Earth-like planet must be around 1 in 500. To compare this with other estimates of 'habitable planets in the Galaxy', we must allow for the fact that about two-thirds of all stars lie in double star systems, where stable planetary orbits within the ecosphere are ruled out. This reduces the probability to 1 in 1500 for stars similar in constitution and mass to the Sun, but not necessarily single.

Habitable planets can only occur for stars very similar to the Sun in mass. A star lighter than 0·85 Suns has no ecosphere; a star heavier than 1·2 Suns changes its light output so rapidly that life would not have time to evolve (M. H. Hart, *Icarus*, vol. 37, p.351 (1978)). Suitable stars constitute about 1 per cent of the stars in the Galaxy (from data on pp.545, 567, 569 of K. R. Lang, *Astrophysical Formulae*, Berlin (1974)). All in all, the number of habitable planets in our Galaxy of 150,000 million stars (Lang, *op. cit.* p.547) is thus around *one million*. We may note that this is considerably smaller than other published estimates described as 'conservative': Ron Bracewell finds 1000 million (*The Galactic Club*, p.8 (1978)); Dole & Asimov 600 million (*op. cit.* p.171); Bond & Martin 10 million (*op. cit.* p.415).

Whether life inevitably evolves on a suitable planet is a question that cannot be answered at our present state of knowledge. Many authors have assumed that it does, but their grounds for optimism are diffuse. It is premature to try to estimate statistics on this problem until either we have evidence for life elsewhere, or until biochemists can establish just how the first primitive chemicals got together and formed living cells.

Certain factors were, however, essential for life, and probabilities can be 'guesstimated' for some of them. Water was essential, and not too much of it. Life probably began in the shallow continental margins of the oceans, and if Earth had had too much water these would not exist. The same problem would have arisen if the planet were too flat, for the Earth has an average ocean depth of 3 km. and if our planet's relief were not so extreme all its surface would be flooded. Some use can be made here of a comparison with the other similar planets in the solar system, Venus and Mars. Both of these have sufficient relief that they too would have dry land 'continents' if they were flooded to an average depth of 3 km. All three planets would still have small continents – and fairly extensive continental shelves – if they were flooded to twice this average depth. It seems reasonable to take a 1 in 2 chance that a planet of suitable size and location in the ecosphere has a suitable amount of water.

Continental drift may play a role in life's origin, and certainly in its evolution. Judging by the preliminary evidence from radar observations of Venus, that planet too suffers a mild form of continental drift (*Scientific American*, vol. 243, p.46 (1980)), confirming theoretical expectations that any approximately Earth-mass planet has drifting continents.

The role of the Moon is obscure. Tides, covering and uncovering rock pools on the early Earth, may have been important in instigating life; and the Moon was at its closest to the Earth at the time of the oldest-known living cells, 3500 million years ago (reported in *New Scientist*, vol. 86, p.376 (1980)). If – and it is a big 'if' – the Moon's presence was essential, then the formation of life must be a comparatively rare event, for it is unusual for a planet to have a satellite as relatively large ($\frac{1}{4}$ Earth's size) as our Moon. The only comparable case in the solar system is Pluto and its $\frac{1}{3}$-size moon Charon.

During life's evolution on to land, it has been protected by the ozone layer. Plant cells long ago converted Earth's original carbon dioxide atmosphere to oxygen, and 50 km. high in the atmosphere sunlight changes some oxygen molecules to the three-atom ozone form. This protects us from the Sun's lethal ultra-violet radiation. Despite exaggerated reports that aerosol sprays are destroying the ozone layer, recent research shows that it is remarkably resilient – in fact the amount of ozone is currently *increasing* slightly (M. Allaby & J. Lovelock, *New Scientist*, vol.87, p.212 (1980)).

What *may* have endangered the ozone layer in the past is the explosion of a nearby supernova. Earth's life may also be harmed by the infall of large meteorites. Such 'cosmic catastrophes' have been widely discussed recently (see for example D. Clark, G. Hunt & W. McCrea, *New Scientist*, vol.80, p.861 (1978)). There is good evidence that the extinction of the dinosaurs – and many other species simultaneously – was caused by a 10 km. diameter asteroid hitting the Earth 65 million years ago (reported in *New Scientist*, vol.85, p.59 (1980)). Evolution quickly ensured that life continued, however, as it undoubtedly has from earlier meteorite impacts, and from nearby supernova explosions that have temporarily destroyed the ozone layer. To eliminate life on Earth entirely would require a catastrophe orders of magnitude worse. Asteroids ten times larger than the one thought to have destroyed the dinosaurs should hit the Earth on average once every 6000 million years (from data in D. W. Hughes, *Nature*, vol. 281, p.11 (1979)), rather longer than the Earth's present age of 4600 million years. A supernova will explode within 10 light years of the Sun every 5000 million years (from data in W. McCrea, D. H. Clark & F. R. Stephenson, *Nature*, vol. 265, p.318 (1977)) – although it is not clear whether this would necessarily destroy life in the oceans. Since the frequency of this kind of catastrophe is comparable to the age of the Earth, there's a roughly 1 in 2 chance that life should have survived until today.

Multiplying all these numbers together, we find that there is a 1 in 2000 chance of a continuously-habitable planet circling a Sun-like single star which is about 5000 million years old. If we assume that every planet capable of having life on it does have life, and that there are no other factors to take into account, then the same odds apply to the existence of a life-bearing planet.

But from a philosphical point of view, the most unlikely fact about life is that the Universe is built in just such a way as to allow its fantastically complex reactions to operate. So far we have taken it for granted that we can guarantee that stars will shine steadily for 5000 million years and more; that there are atoms – namely carbon – that can react together to function as living cells. It is actually very odd that the Universe is made this way: only very slight changes in many of the fundamental constants of nature would alter the Universe in such a way that life would be impossible.

Freeman Dyson, of Princeton's Institute of Advanced Studies, has approached this problem by looking at energy flows in the Universe (*Scientific American*, vol. 225, p.50). He points out that the Universe is basically unstable, because every particle of matter is attracting every other by gravitation. We should expect all matter to just collapse together into 'dead' compact objects – what we would now call black holes.

Dyson replies that the Universe suffers 'hang-ups' which have prevented its matter from collapsing, even 10,000 million years after the Big Bang. First is the fact that its matter is so spread out: although the Universe seems extravagantly large, its very size means that the time it would take to collapse is extremely long – some 100,000 million years (even if it were not expanding with the impetus from the Big Bang). Second is the hang-up of spin. Any rotating or revolving body produces a centrifugal force which opposes gravitation. This is why the Sun has not fallen into the centre of our rotating Galaxy; and why the planets have not dropped into the Sun.

The third hang-up is that the Universe's original matter – predominantly hydrogen – can produce radiant energy by nuclear reactions. In stars, the inpull of gravitation is balanced by the outflow of energy from its central nuclear reactions. For this reason stars are stable for the 10,000 million years or so that life needs to evolve on an accompanying planet.

The actual forces involved in the nuclear reactions are in fact critically balanced to make this possible. The Sun's reactions are similar to a hydrogen bomb's. The essential difference is, however, that a bomb is made of heavy hydrogen (deuterium), which reacts by the *strong* nuclear force extremely rapidly, while the Sun's 'fuel' is ordinary hydrogen, and the crucial first step of converting this to heavy hydrogen takes place at a rate controlled by the *weak* nuclear force. Since the weak force operates a million million million times more slowly than the strong variety, the Sun's nuclear reactions are correspondingly slowed. On the simplest level, physicists see the need only for one nuclear force, the strong force that binds protons and neutrons into the nuclei of atoms. Yet without the existence of a second type of nuclear force, much weaker, there could be no slow-burning stars, and hence no life.

The actual strength of the strong force is also critical. If it were only 3 per cent stronger, then ordinary hydrogen atoms could react by the strong force – whether or not the weak force existed – and stars would spontaneously explode as soon as they were born.

Dyson's analysis was not the first to consider the strange attributes of the Universe which make it uncannily suitable for life, and recent work has revealed even more. Fifty years ago, Sir Arthur Eddington noticed that the biggest ratios he could make from natural quantities were almost identical, and equal to 10^{40} (1 followed by 40 zeros). One was the ratio of the *electrical* attraction of a proton and electron to their *gravitational* attraction; the other ratio the age of the Universe compared to the timescale of nuclear reactions. Yet the first ratio is fixed, while the second must increase as the Universe gets older.

In 1961 Robert Dicke explained this odd coincidence by the unusual argument that consciousness is involved (*Nature*, vol. 192, p.440). Life cannot form from the hydrogen and helium of the early Universe, but only after stars have converted some into heavier elements – particularly carbon and oxygen – and spewed them back into space to make up a new generation of stars with planets. The time between the Big Bang and when a conscious being evolves to view the Universe is thus very roughly equal to the age of an average star. The latter is in turn governed by the interplay of gravitational, electro-magnetic and nuclear forces. So although Eddington's two ratios do not stay equal to one another, it is not 'pure coincidence' that *we* measure them to be equal. The laws of nature dictate that they are equal at about the time that intelligence can evolve to measure them.

Since then, astrophysicists have found other apparent coincidences even more essential for life, along the lines of Dyson's hang-ups. In the early 1970's, Brandon Carter was a leading figure in this field. He pointed out that the force of gravitation has just the right strength: it is balanced on the thin knife-edge where it must lie if life is to exist in the Universe (in *Confrontation of Cosmological Theories with Observation Data* (ed. M. S. Longair) p.291, Dordrecht (1974)). If the force of gravitation were only a few per cent stronger, all stars would be like blue giants; one or two per cent weaker, and all would have the internal structure of red dwarfs. Carter argued that in the first case, stars could not have planets; in the second, the stars could not explode as supernovae, and so the heavier elements necessary for life would not have been spread through space from their formation sites at stellar cores.

Bernard Carr and Martin Rees took these ideas further in a seminal article in 1979 (*Nature*, vol. 278, p.605 (1979)), where they reviewed and extended the list of 'coincidences' (for original references, see their article). Concerning the forces of nature, they cite Carter's argument that the gravitational force is just the right strength. The weak nuclear force must exist, for the reasons that Dyson has put forward, but Carr and Rees point out that it too has just the necessary strength. Supernova explosions are necessary for scattering the elements which will make up life, and the power behind the explosion comes from a sudden burst of neutrinos liberated in the star's core. These tiny particles lift off the outer layers of the star like steam lifting off a kettle lid, but more explosively. Neutrinos interact with the star's gases through the weak nuclear force: if this force were ten times weaker, the neutrinos would shoot straight out of the star without blowing off the gases; if it were ten times stronger, the neutrinos would be trapped in the star's core, and again there would be no supernova explosion.

The other, strong, nuclear force is even more constrained. As Dyson noted, a 3 per cent increase would mean that all stars were unstable; in fact, according to modern cosmological theories, such a modest increase would have led to these runaway reactions occurring in the original Big Bang, and there would be no hydrogen around now to make up slow-burning stars, or to be incorporated into water and other compounds essential to life.

If the strong force were any weaker, atomic nuclei could not exist because its attractive force would not be sufficient to hold the nucleus together against the repulsion of the electrically charged protons in it. The criterion is not so stringent here, but if the strong force were less than half its actual strength most nuclei could not exist.

On more purely astronomical matters, the expansion of the Universe involves two quantities hitherto unexplained. Einstein's equations for the expansion predict a kind of universal force acting on everything in the universe, called the cosmological constant. Astronomers, however, find no evidence for its effects in the real Universe. Carr and Rees show that such a universal force would not allow life to exist: if its effects were repulsive, gas clouds could not have condensed as galaxies; while an attractive type of cosmological constant would have made the expanding Universe collapse again before life had a chance to evolve. The other oddity is the amount of universal background radiation left over from the Big Bang. If it were a thousand times less, galaxies would be appreciably smaller. Although it is not clear whether this would have precluded life, a larger amount of radiation, by a factor of a thousand or more, would have meant that galaxies could not have formed at all. According to many theories of the Big Bang itself, the amount of radiation produced was indeed surprisingly small (J. D. Barrow & J. Silk, *Scientific American*, vol. 242, p.98).

Carter, Carr and Rees use these arguments to support the anthropic view of the Universe – the idea that we observe a Universe around us that is fit for life *because* we are here to observe it. In other words, one could imagine an infinite number of possible universes, each with different physical laws. But in most of them conditions would not be suitable for life, and so no life could evolve. Only in a universe (like ours) where physical laws were suitable, could life actually evolve; and therefore living creatures would perforce find themselves in a 'Universe-fit-for-life'.

From the philosophical point of view, though, the weight of evidence produced by Dyson, Carter, Carr and Rees and others is open to more than one interpretation. The anthropic interpretation finds favour amongst present-day astrophysicists. Philosophers of another school could argue that introducing 'other universes', in principle unknowable, is a logically indefensible step: what the evidence clearly implies is that the Universe has been designed to be suitable for life. The reader may choose his own interpretation for the remote chance represented by the remarkable fact of his own existence.

Index